ANALYTICAL CHEMISTRY IN A GMP ENVIRONMENT

ANALYTICAL CHEMISTRY IN A GMP ENVIRONMENT
A Practical Guide

EDITED BY

James M. Miller   Jonathan B. Crowther

A WILEY-INTERSCIENCE PUBLICATION
JOHN WILEY & SONS, INC.
New York / Chichester / Weinheim / Brisbane / Singapore / Toronto

DISCLAIMER

SAFETY

The laboratory procedures described in this text are designed to be carried out in a suitably equipped laboratory. In common with many such procedures, they may involve hazardous materials. For the correct and safe execution of these procedures, it is essential that laboratory personnel follow standard safety precautions.

Although the greatest care has been exercised in the preparation of this information, the authors, speaking for themselves, and for the classroom and laboratory instructors, expressly disclaim any liability to users of these procedures for consequential damages of any kind arising out of or connected with their use.

The analytical procedures detailed herein, unless indicated as such, are also not to be regarded as official, but are procedures that have been found to be accurate and reproducible in a variety of laboratories.

APPARATUS

The items of apparatus described in this manual are intended to illustrate proper techniques to obtain a quality analysis and are not to be considered as official and/or required. Any equivalent apparatus obtained from other manufacturers may be substituted.

This book is printed on acid-free paper.[⊗]

For ordering and customer service, call 1-800-CALL-WILEY.

Library of Congress Cataloging-in-Publication Data:

Analytical chemistry in a GMP environment: A practical guide / edited by
    James M. Miller, Jonathan B. Crowther
    p. cm.
    "A Wiley-Interscience publication."
    ISBN 0-471-31431-5

Printed in the United States of America.
10 9 8 7 6

CONTENTS

v

2 Laboratory Controls and Compliance

Henry Avallone

Perlette Abuaf and Alvin J. Melveger

7   Chromatographic Principles   **185**

James M. Miller

8 Gas Chromatography   **217**

James M. Miller and Harold M. McNair

12 Analytical Method Development for Assay and Impurity Determination in Drug Substances and Drug Products 331

Jonathan B. Crowther, Paul Salomons, and Cindi Callaghan

APPENDIXES

CONTRIBUTORS

Perlette Abuaf, IRI*Trials Management Center, Annandale, NJ, 08801

Sagar Adusmalli, Janssen Pharmaceutica, P.O. Box 200, Titusville, NJ 08560-0200

Henry Avallone, Janssen Pharmaceutica, P.O. Box 200, Titusville, NJ 08560-0200

Cindi Callaghan, Janssen Pharmaceutica, P.O. Box 200, Titusville, NJ 08560-0200

Jonathan B. Crowther, Janssen Research Foundation, Titusville, NJ 08560-0200

Jenny G. Feldman, Cilag A. G., Hochstrasse 201, 8205 Schaffhausen, Switzerland

Richard Hartwick, PharmAssist Analytical Laboratory Inc., Box 248A, South New Berlin, NY 13843

M. Ilias Jimidar, Janssen Research Foundation, Turnhoutseweg 30, B-2340 Beerse, Belgium

Ross Kirchhoefer, Gateway Analytical Laboratories, 5703 Hidden Stone Drive, Saint Louis, MO 63129

William Lauwers, Janssen Research Foundation, Turnhoutseweg 30, B-2340 Beerse, Belgium

Thomas Layloff, Division of Drug Analysis-FDA, 1114 Market Street, Room 1002, St. Louis, MO 63101

R. D. McDowall, McDowall Consulting, 73 Murray Avenue, Bromley, Kent, BR1 3DJ, UK

Harold McNair, Department of Chemistry, Virginia Tech, Blacksburg, VA 24061

Alvin J. Melveger, AJM Technical Consulting, 9 Patrick Court, Flanders, NJ 07836

James M. Miller, Department of Chemistry, Drew University, Madison, NJ 07940

W. Rorer Murphy Jr., Department of Chemistry, Seton Hall University, South Orange, NJ 07079-2694

Nico Niemeijer, Janssen Research Foundation, Turnhoutseweg 30, B-2340 Beerse, Belgium

Rudy Peeters, Janssen Research Foundation, Analytical Development, Turnhoutseweg 30, B 2340 Beerse, Belgium

Lee N. Polite, Axion Analytical Laboratories, Inc., 2122 North Bissell, Suite #3, Chicago, IL 60614

Paul Salomons, Janssen Pharmaceutica, P.O. Box 200, Titusville, NJ 08560-0200

Ponniah Shanbagamurthi, Janssen Pharmaceutica, P.O. Box 200, Titusville, NJ 08560-0200

Nicholas H. Snow, Department of Chemistry, Seton Hall University, South Orange, NJ 07079

FOREWORD

The laboratory is an extension of our senses, enabling us to obtain data on substances beyond what we can see with a naked eye and in amounts that our hands could never achieve. These data are compiled into reports and are ultimately used for making decisions, decisions that cannot be confirmed with our unaided senses. The quality of any decision is absolutely dependent on the quality of the data; junk data lead to junk decisions.

The process of acquiring valid data requires properly trained personnel using appropriately calibrated tools. In order to ensure the acquisition of high-quality data, one must be certain that all laboratory tools are suitable for their intended use [i.e., meet their standard operating procedure (SOP) requirements] within their validated limits. In addition, all involved personnel in the data gathering and information generation efforts must have the required knowledge, skills, and abilities (KSAs) to satisfactorily perform their tasks. As has been noted,* this is good business practice and, secondarily, necessary regulatory compliance.

In addition, however, our technological industry continues to churn out an ever-expanding array of almost magical analytical technologies, thereby creating a new group of incompetent laboratory personnel who are not familiar with or trained to use them.

Not surprisingly, these expanding technologies have posed a great and insurmountable challenge to our already much maligned educational system. The college/university curriculum continues with the traditional four-year program where the faculties are supposed to inculcate into the students the usual very strong foundation in the basic knowledge and skills of the science, packaged as a palatable academic program. Because all of this knowledge cannot be rationally delivered in a four-year curriculum, the assurance that those who generate data have the basic KSAs falls to the employers. Management must have confidence that all of the employees in the organization possess the required KSAs to perform their assigned tasks. As competent analysts performing in the laboratory reflect on the adequacy of the first-line management team, incompetent analysts in the laboratory reflect the inadequacy of that team.

Because of severe infractions in the practice of good science and science

---

* Alan Dinner, personal conversation.

management by several firms, the U.S. Food and Drug Administration found it necessary to issue regulations defining minimum appropriate standards for the performance of nonclinical studies submitted to the agency. This issuance of the "Good Laboratory Practices" regulations made the acronyms "GLPs" and "SOPs" "household" words in laboratories throughout the world. Subsequently, the agency issued the related regulations to provide administrative law guidance for pharmaceutical manufacture in the current good manufacturing practices (CGMPs).

In both cases the regulations were intended to provide broad guidance on appropriate scientific practices in the pharmaceutical industry while not stifling innovation and the evolution to superior practices that still satisfy the requirement. These regulations address many aspects of laboratory operations but only broadly address the skills and abilities of the primary practitioners: the management and bench scientists. This deficit was pointed out in "Analysts: The Unknowns in the Quality Assurance Equation".[†] That presentation and many subsequent ones focused on the fact that college science graduates do not in general have all the skills required to competently function in an FDA-regulated environment. This poses a crisis for first-line managers who must have absolute confidence that their staff members possess the required KSAs to competently perform the tasks that they are assigned.

In order to ensure that the scientists have acquired the required competencies to adequately perform their assigned tasks, management must establish quality systems structured to provide necessary training and education. It appears that one company, Johnson & Johnson, has taken a direct approach to meeting this challenge by establishing a laboratory analysts training and certification program for its employees.

This text has emerged from that program. It is designed to establish a basic knowledge and skill base in the technologies that are most prevalent in "product control laboratories" of the pharmaceutical industry. The laboratory supervisors who employ the individuals who successfully complete this course can have confidence that they have this well-defined starting point from which they may begin to evolve the individual employee's skills to journeyman performance levels in their specific organization.

THOMAS P. LAYLOFF

June 1999

[†] T. P. Layloff, AOACI Referee, December 1990, p. 6.

PREFACE

In his Foreword and elsewhere,* Layloff has described the need for more and better training of pharmaceutical laboratory analysts, as perceived by the Food and Drug Administration (FDA). To meet their own needs, the FDA produced a series of self-training aids that could be used in their testing laboratories. Many others are equally aware of the need for training because of the constant introduction of new methods, the increasing demands for better analyses, and the fact that little or no discussion of government regulations is presented in the traditional undergraduate educational program of chemists. Johnson & Johnson recognized this need in the spring of 1996 and began the development of an in-house training course. With the help of academic and industrial consultants, the course was first offered in October 1996 and became the basis for this text.

From the onset, the Johnson & Johnson Laboratory Analyst Training and Certification Program's (LATCP) objective has been to provide lecture and laboratory work in analytical chemical methods and in government regulations (CGMPs) and procedures. The two-week-long course has been presented over 20 times to over 300 analysts, selected from J&J sites around the world. A special facility was constructed for this purpose at the IRI Trials Management Center in New Jersey; more details on the program can be found in a recent trade publication.†

This book is a natural outgrowth of the LATCP and is being published to make the contents of the program available more widely. The level of the material is that which has been found suitable for the participants in the course, who, on average, hold bachelor's degrees in chemistry and already have some experience in the pharmaceutical laboratory; these are typical recruitment criteria for J&J analysts.

The introductory chapter provides an orientation to the drug development process that might not be familiar to new employees in the pharmaceutical industry. Two chapters follow on regulations and compendia. Together these chapters should serve as a basis for understanding the issues in this regulated industry.

---

* A. S. Kenyon, R. D. Kirchhoefer, and T. P. Layloff, *JAOAC Int.* **1992**, *75*, 742–746.

† N. Corkum, H. Avallone, and J. Miller, *Inside Lab Management, AOAC*, **2000**, *4*, 26–29.

The middle chapters cover some basics of analytical chemistry of relevance to this audience, beginning with statistics and a quick review of equilibrium and solution chemistry. While this material may be too elementary for some, we have discovered that many students in our course are deficient in basic concepts such as significant figures, so such topics are included. The major quantitative techniques covered next are spectroscopy (UV and IR), chromatography (GC, LC, HPLC, and TLC), and dissolution. Of these, HPLC is unquestionably the most important and is the focus of much of the material throughout the book.

The final chapters cover detectors (mainly chromatographic), quantitative analysis, and data systems, plus the special topics of method development (based mainly on HPLC), qualification of instruments, and validation and method transfer.

A multiauthor work such as this one runs the risk of being fragmented and uneven. We have tried valiantly to make it as unified as possible, drawing on our shared teaching experience with the LATCP course. It is, of course, impossible to define a single set of symbols when the topics are so diverse. Appendix I lists the terms and symbols used, noting overlaps in an attempt to keep confusion to a minimum. In chromatography, the IUPAC symbols are used, not those of the USP. This anticipates that USP will eventually adopt the IUPAC system in the spirit of unity and international cooperation. Other appendixes include the terms and some procedures used by another international group, ICH.

Being written to accompany the LATCP course, this book is intended for individual use by laboratory analysts. We have attempted to keep it as succinct as possible and provide sufficient detail, given the wide range of subject matter. The editor and the publisher welcome suggestions and comments for future editions.

We want to acknowledge the three persons who are most responsible for initiating and guiding this project: Hank Avallone, Juanita Hawkins, and Nancy Corkum. Their vision, commitment, and support were crucial. In addition, we want to acknowledge the efforts of the LATCP Managers, Pat Magliozzi and Tom Caglioti.

Each of the authors is lauded for her/his efforts to produce a concise chapter within the limitations of time and page length. We also wish to thank the many content reviewers for their valuable time and expertise. Of course, none of this would have been possible without the tedious clerical support by IRI, especially Diane Kelly, Katherine Miles, and Patty Raymondi.

JAMES M. MILLER
JONATHAN B. CROWTHER

Madison, New Jersey
Titusville, New Jersey
March, 2000

1

THE LABORATORY ANALYST'S ROLE IN THE DRUG DEVELOPMENT PROCESS

Jonathan B. Crowther, William Lauwers, Sagar Adusumalli, and
Ponniah Shenbagamurthi

1.1. INTRODUCTION

1.1.1. The Importance of Analytical Methodology in the Drug Development Process

While many chapters in this text will focus on analytical methods used in release and stability of marketed drug substance and drug products, this initial chapter attempts to review analytical methodology used during the drug development process. One additional objective is to review some basic components of the overall development process from an analytical testing perspective because final regulatory methods, specifications, dosage forms, and so on, are based upon information gained in these earlier development stages.

During the drug development process, key decisions are based upon data obtained from analytical test methods. Results generated from these test methods are expected to be both accurate and reliable. In later stages of development, the methodology is expected to be robust because the methods will be

Analytical Chemistry in a GMP Environment. Edited by J. M. Miller and J. B. Crowther
ISBN 0-471-31431-5   © 2000 John Wiley & Sons, Inc.

ultimately transferred to a control laboratory. Throughout the development process, individual test methods are used in a variety of laboratory settings in the evaluation of:

1. Product safety
2. The dosage form's bioavailability
3. The setting of specifications for the drug substance, intermediates, and drug product
4. The shelf-life of the product (stability)
5. Determining optimum formulation.
6. Identification, quantitation, and "qualification*" of impurities and degradants
7. Determining optimum crystalline and salt form of the drug substance, testing support to process development and cleaning validation
8. Support to preclinical and clinical studies (product safety and efficacy)

Additionally, data from analytical test methods are required to support the regulatory filings. In brief, the chemistry, manufacturing, and controls section of a U.S. regulatory filing is required to contain the following directly related to analytical testing [1]:

1. Methods validation
2. Physical description and characterization of the drug substance and drug product—proof of the chemical structure
3. In-process controls
4. Characterization of reference standards
5. Specifications—description of analytical methods
6. Evaluation of container/closure system for storage
7. Drug product and drug substance stability
8. Justification of drug product development
9. References to compendial test methods for inactive ingredients
10. Bioequivalence
11. Pharmaceutical development report†

1.1.2. Interdiscipline Use of Analytical Methodology

Fundamentally, the role of the laboratory analyst in the development process is to provide data to establish the identity, potency, purity, and overall quality of

---

* The process of acquiring and evaluation of data, which establishes the biological safety of an impurity/degradant or a given impurity profile at the level(s) specified.

† May be submitted in a regulatory filing but not required.

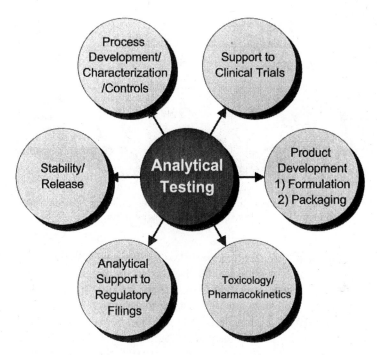

Figure 1.1. "Stakeholders" in quality analytical testing during drug development.

the drug substance and the drug product. As an example, suitable analytical methods are required to establish "equivalency" of different formulations and batch sizes during the development process. Similarly, analytical methodology also serves to support process development, characterization, scale-up, and process validation activities. Analytical methodology is also used extensively in the comparison of the bioavailability of drug in various dosage forms and formulations. Consequently, methods used to support development activities should have sufficient controls and documentation to ensure the accuracy and reliability of the data.

Analytical methods used in development activities are typically used in specialized laboratories beyond the central analytical development group. Figure 1.1 displays typical recipients or end-users of the validated methods.

1.1.3. Phases of Drug Development

The phases of drug development are often explained in clinical terms. They will be explained briefly in this section and will be detailed in other sections of this chapter. Upon discovery and the decision to develop the active pharmaceutical ingredient (API), the molecule is fully characterized with initial pharmacology, and toxicology studies in Prephase I. In the United States, after the firm files an investigational new drug (IND) application, the API in its simplest formulation

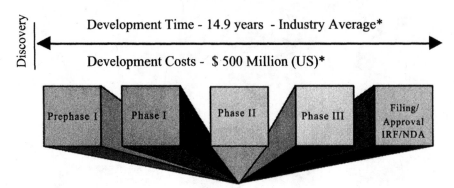

Figure 1.2. Phases of drug development, average duration, and costs. *Source:* PhRMA Industry Profile 1998 http://phrma.org/publications/industry/profile99/index.htm

is tested in healthy volunteers to determine its safety during Phase I. In Phase II, the drug is evaluated for clinical effectiveness in the target patient population; the proper and safe dose range is determined at this stage. If the drug candidate is safe and clinically effective, the API is formulated into the proposed-marketed formulation and tested in an extensive double-blind (placebo-controlled) study with typically hundreds to thousands of patients over a period of 2 or more years. Upon completion of the final Phase III trial (there may be two or more clinical trials at this stage), the "product" is ready to be submitted to regulatory agencies. This assumes that all components from clinical, pharmacokinetic, toxicology, regulatory, chemical, and pharmaceutical development groups are completed, successful, and properly documented. Figure 1.2 reviews the typical timeline and phases of the drug development effort required to bring a typical pharmaceutical to market. In discussions that follow, it becomes quite clear that the laboratory analyst participates throughout this process in a variety of roles and functional areas.

1.1.4. Introductory Summary

This initial chapter is designed to give a broad overview of the uses and requirements of analytical methodology during the drug development process.

The following sections of this chapter will discuss the various testing method types and regulatory requirements during various stages of the drug development process. In addition, the chapter also surveys the role of analytical methodology in the following:

1. Formulation development
2. Pharmacokinetic and toxicokinetic studies

Finally, the chapter also reviews both the regulatory and analytical requirements for stability testing of drug substance and drug products.

To simplify the discussion, the regulatory discussions are loosely based on requirements for pharmaceuticals developed for the US market with some references to European filing requirements. In all situations, current regulatory guidelines and international guidelines [International Conference on Harmonization (ICH)] should be referenced before proceeding on any development activity.

1.2. REQUIREMENTS OF AN ANALYTICAL METHODOLOGY DURING THE DRUG DEVELOPMENT PROCESS

1.2.1. Introduction

This section provides information on the drug development process from a nonclinical point of view, more specifically on the development of analytical methods for release and stability in support of a (United States) new drug application (NDA) and/or international (European) registration file (IRF) submission.

At various stages of the drug development process, release and stability testing is required to guarantee the quality and safe use of a new-marketed medicinal product. The quality parameters and stability requirements imposed to the API and drug product (DP) must be carefully followed-up by means of appropriate and validated analytical methods. Depending on the stage of development, these analytical methods are standard screening methods* at the start of the development process, which over time are gradually upgraded to thoroughly validated methods for NDA/IRF application. The filed methods[†] must be simple, robust, and reliable—that is, easy to use and perform without deviations when appropriately applied in a qualified laboratory. The method validation should be in accordance with the requirements imposed at a particular phase of the development.

Although many quality attributes of the API and DP need to be controlled by a variety of analytical methods, this section will focus on the development of an HPLC method for the assay of the API and related impurities and degradants—that is, impurities that resulted from the manufacturing process and degradation products formed upon storage. The method should be applicable for both release and stability purposes.

As previously discussed, the drug development process can be divided into distinguishable phases. At the end of each phase appropriate registration documents are delivered allowing for a gradual extension of the clinical trials as

---

* A "screening method" is typically a very capable gradient high-performance liquid chromatography (HPLC) method that is suitable for separation of a variety of compound types. Screening methods are typically not optimized for speed or robustness.

[†] Referred to as VTR^{2}AP methods (*v*alidated, *t*ransferable, *r*obust, *r*apid, *a*ccurate, and *p*recise analytical methodology) in Chapter 12.

part of the development process. Each of these stages is briefly described below with respect to their particular analytical method development requirements.

1.2.2. Discovery Phase

Analytical methods are at first applied in the discovery group for the analysis of new chemicals obtained from the Discovery Synthesis Department. The analytical discovery group may apply standard HPLC screening methods, often using an HPLC–mass spectrometry (MS) system. The main purpose of these analyses at this stage is restricted to the following:

1. Confirmation of the expected structure
2. A tentative content (assay) estimation

Current good manufacturing practices (CGMP) documentation of the analytical results is typically not required in discovery research.

1.2.3. Early Development

1.2.3.1. Prephase I (Pre-IND). New chemical entities displaying interesting pharmacological activities are selected in the late lead optimization process in discovery in collaboration with Nonclinical Development. This review team evaluates the development and market potential of a drug candidate using data presented in the discovery phase, prior to the launch of the development project. The drug candidate, or new molecular entity (NME), enters Prephase I for the preparation of a nonclinical information file for the Ethical Committee and ultimately for review and approval by the regulatory authorities prior to initial clinical trial approval in healthy volunteers. This file contains a toxicological and general pharmacological profile, the preclinical pharmacokinetic profile, and a brief summary of some chemical and pharmaceutical properties of the NME. These studies are performed on small fractions from discovery synthesis and on a pilot laboratory batch (PLB). Data from Prephase I is reported to the regulatory authorities (the United States—IND application); once approved, clinical trials in healthy volunteers can begin.

A PLB is manufactured in a pilot laboratory to provide sufficient NME material (API) for characterization and evaluation. The PLB is synthesized under Phase I GMP for Prephase I toxicity and human volunteers studies. This PLB of a typical amount of 100 g up to 300 g is analyzed by HPLC or HPLC/MS screening methods to facilitate the necessary identification work for impurities and degradants. These methods may be further applied for other purposes— that is, investigation of the stability of the NME in the solid form and in aqueous solutions, for the selection of the appropriate salt (base), and for the preformulation work performed to develop a stable Phase I formulation.

At this stage, it is prudent to evaluate solution stability to investigate the influence of pH, metallic ions, light, and an oxidative medium on the stability of the NME. The assay of API and degradants is determined. A tentative stability profile can be established and the significant degradants, observed in the various conditions, can be determined case by case depending on the stability profile. A rationale for the selection of these significant degradants needs to be provided because they influence the selectivity criteria of the filed purity- and stability-indicating method.

The stability study on the solid form allows investigating the influence of enhanced temperature and relative humidity on the stability of the NME. Besides the stress stability studies, a study is started up to determine the stability of the NME in the controlled condition 25°C/60% RH (relative humidity) in order to assign a provisional shelf-life. This is a regulatory requirement in support of the various pharmacological, pharmacokinetic, and toxicological studies carried out under good laboratory practices (GLP).

In the salt selection program the most suitable salt for formulation development is searched for in a comparative stability study. Sometimes different hydrates of the API are also investigated. In this stage, special attention is also paid to the occurrence of polymorphism for the NME by applying infrared, differential scanning calorimetric, and X-ray diffraction techniques.

Preformulation work with the PLB requires appropriate synthesis documentation and GMP-compliant analytical methods. The release of the PLB is the responsibility of the Quality Assurance (QA) Department.

Even at this early stage, the NME is placed in stability chambers along with potential formulation ingredients to evaluate excipient compatibility. The stability testing requirements will be discussed in a later section.

The various analyses accomplished in Prephase I preferably make use of the same screening HPLC method for the benefit of mutual comparison of analytical results. This HPLC method enables us to acquire a broad knowledge on the quality and the stability of the NME. During the next phases of development the method is modified to provide a simple, robust, and reliable method for use in quality control and stability studies.

1.2.3.2. Phases I and II.

The Phase I formulations obtained from the Pharmaceutical Product Development Group are put on stability to select the most stable formulation that will be used to study the safety, tolerability, and kinetics in healthy volunteers (single and multiple dose).

The assay determination is a principal quality parameter of the Phase I formulation which may be determined by a nonselective ultraviolet method. At this stage the formulation* is released by QA applying broad specification limits (assay value >90%).

As shown in Figure 1.3, in Phase II the early safety and efficacy trials in target patients are performed. The Phase II formulations are prepared with pilot

---

* At this stage in development, the formulation is simplified; for example, if the NME is targeted to be administered orally, the formulation in early clinical studies may be as simple as a suspension of the NME in water.

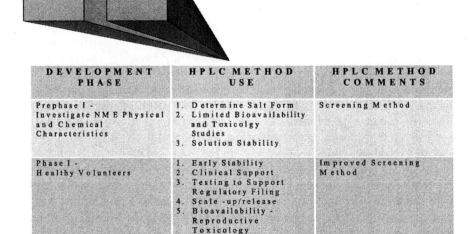

DEVELOPMENT PHASE	HPLC METHOD USE	HPLC METHOD COMMENTS
Prephase I - Investigate NME Physical and Chemical Characteristics	1. Determine Salt Form 2. Limited Bioavailability and Toxicolgy Studies 3. Solution Stability	Screening Method
Phase I - Healthy Volunteers	1. Early Stability 2. Clinical Support 3. Testing to Support Regulatory Filing 4. Scale-up/release 5. Bioavailability - Reproductive Toxicology 6. Formulation Development	Improved Screening Method

Figure 1.3. Methodology used in Prephase I and Phase I of development.

batches manufactured in the pilot plant. Although the synthesis method is still in full development, the production proceeds under good manufacturing practice (GMP) conditions. Qualification of the batches is obtained by additional toxicological studies if aberrant impurity profiles are observed in comparison with the PLB profiles.

Phase II concludes with the submission of a clinical trial approval document (CTA)/IND to the registration authorities (Figure 1.4). This document refers to the analytical methods and specifications applied in the release and stability studies of API and DP that will be used in Phase III clinical trials.

The CTA stability- and purity-indicating method should be capable of the quantitation of the major impurities and degradants. These are preferentially identified, but need not necessarily be synthesized and qualified as a reference standard because impurity and degradant content calculations are performed against the API (area %).

The validation package of the analytical methods during the various development stages toward the NDA/IRF submission gradually increases. A CTA/IND can be filed with a restricted set of validation data that, at a minimum, includes the specificity and sensitivity of the method toward the impurities and degradants, accuracy, linearity, repeatability and intermediate precision. Appropriate system suitability tests are included in the method description to check for proper performance of the method over time and on different analytical chromatography systems.

DEVELOPMENT PHASE	HPLC METHOD USE	HPLC METHOD COMMENTS
Phase II - Early Clinical Trials	1. Begin Formal Stability on the Drug Substance 2. I.D. Major Impurities 3. Formulation Support - Refine Formulation	1. Begin Development of Specific Methods. 2. Moderate Levels of Validation.

Figure 1.4. Methodology used in Phase II.

1.2.4. Final Development (Phase III)

Late Phase II and Phase III includes the international safety and efficacy (dose ranging) trials in a large number of patients. These clinical studies are conducted using DP batches manufactured according to the processes described in the NDA/IRF file for the marketed formulation. In turn, the DP contains API material manufactured according to the filed synthesis process.

The knowledge gained in the previous stages of the development process with regard to impurities and degradants must be consolidated into the NDA/IRF stability- and purity-indicating method. Based upon the acquired stability data, profiles of the API and DP are established; selection criteria are used to determine specified impurities and degradation compounds.

The specified degradants occur in the normal and accelerated stability conditions at concentrations equal to or greater than 0.1% with regard to the API [2, 3]. These degradants are typically identified and synthesized for the elaboration of the method particularly for purposes of further method development and validation. A similar approach is applied when the selection of the specified impurities is based on the various batch results and impurity batch surveys [4, 5]. The specified impurities and degradants are taken up in the set of the API and DP quality specifications with their corresponding upper limits. These limits are determined from the actual impurity levels observed in batches and stability samples (manufacturing process capability) and also from their toxicological properties.

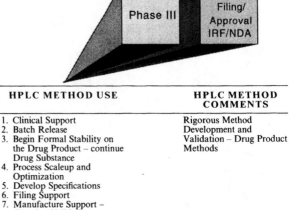

DEVELOPMENT PHASE	HPLC METHOD USE	HPLC METHOD COMMENTS
Phase III – Large Double Blind Clinical Studies	1. Clinical Support 2. Batch Release 3. Begin Formal Stability on the Drug Product – continue Drug Substance 4. Process Scaleup and Optimization 5. Develop Specifications 6. Filing Support 7. Manufacture Support – Cleaning Validation 8. Finalization of Impurities and Degradants	Rigorous Method Development and Validation – Drug Product Methods
Filing (IRF/NDA) Preapproval Inspections – Post Approval Activities	1. Support to Stability and Marketed Batches 2. Process Validation 3. Ongoing Stability/Release 4. Possible Method Optimization	Method Transfer Simplify and Optimize Methods (Rebustness and Ease)

Figure 1.5. HPLC methodology in Phase III onward.

The "selectivity" impurities and degradants occur at concentrations below 0.1% and may be potential degradants observed in the conducted stress degradation studies. The analytical method must be capable to separate the selectivity impurities and degradants as well.

The NDA/IRF purity- and stability-indicating method must be available before the start of the DP NDA/IRF stability development program. This implies that the validated method is transferred to the Development Stability Group approximately 18 months before the NDA/IRF submission. The availability of one year of DP stability data of the marketed formulation is typically a regulatory prerequisite for the NDA/IRF application. These analytical methods used in late Phase III are expected to be robust and validated (Chapter 12) as shown in Figure 1.5.

The analysis of the API and DP verification batches—that is, batches manufactured to the final manufacturing specifications for the marketed product—completes the activities of the analytical method development group unless new unexpected impurities are observed that require further investigation and method adaptation. Hence, the specifications for the API and DP which include the parameters, methods, limits and frequency of testing can be finalized [6, 7], and the analytical methods can be transferred to the quality control sites.

The specified impurities and degradants of the DP need to be qualified according to the guidelines on qualification and identification of impurities [2–5]. Qualification can be achieved using the analytical results of a comparative batch survey, performed with the NDA/IRF method on all the batches previously released, including those used in the various toxicological studies (starting from the PLB). Alternative qualifications involve (a) evaluations making use of toxicological data of structurally related compounds and (b) additional toxicological trials with the synthesized impurity or API spiked with the impurity(ies). In case the DP shows a degradant not previously detected, the DP might be examined in toxicological trials.

NDA/IRF method is validated according to the guidelines [8–16]. The requirements for validation are reviewed in detail in Chapter 15. For HPLC methods used for assay and purity, the validation requirements specify the following validation parameters:

- Method specificity toward related compounds, placebo, and stressed placebo
- Stability of the analytical solutions under the described analytical conditions
- Accuracy
- Linearity and the related range
- Precision (injection repeatability and intermediate precision)
- Limit of detection and quantification of the specified impurity or degradation compound

1.2.4.1. Method Robustness. Special attention must be given to the robustness of the analytical methods, especially for methods that will ultimately be transferred from a development laboratory to quality control sites and stability laboratories. In this respect, evaluation of the method robustness is a part of the early NDA/IRF method development.

An additional tool in the development of a robust analytical method is a swift check on the performance of a method at various application laboratories. In this evaluation test, prior to formal validation and method transfer, some qualitative aspects of the method—for example, a visual check of the baseline, selectivity of the methods toward the specified impurities and degradants, estimation of the limit of quantitation, and so on—are examined in the control laboratory. This preliminary evaluation can be used to improve the quality of the method through understanding possible method performance deficiencies and enable sufficient time for the concerns to be addressed prior to method validation and transfer.

There are other factors required for successful application of an analytical method in quality control and stability groups. The major factors that require attention are as follows:

- The involvement of skilled technicians who are adequately trained
- The use of qualified analytical instrumentation that is subject to periodic calibration and maintenance
- The lab and lab management are CGMP compliant

These factors warrant consideration throughout the development process to ensure successful application of an analytical method worldwide as required.

1.3. THE ANALYST ROLE IN FORMULATIONS DEVELOPMENT

1.3.1. Overview

This section is designed to be a very brief review of analytical participation in the formulation development process. Pharmaceuticals are rarely administered into the body as the API alone; instead, they are formulated with inactive ingredients or excipients, into an effective and convenient dosage form. In simplest terms, the final dosage form is based upon the following:

1. The proposed indication of the drug
2. The bioavailability profile
3. The potency
4. The stability of the API—shelf-life
5. Patient convenience (marketing expectations)
6. Manufacturing considerations
7. Release characteristics (delayed dosing)

The formulation development activities are based not only upon the considerations listed above, but also upon the proposed and sometimes-required dosage form. Typical routes of administration of a drug include:

- Oral (tablets, suspensions)
- Rectal (suppositories)
- Inhalation—intranasal
- Parenteral (I.V., subcutaneous, I.M.)
- Topical (creams and ointments)
- Ophthalmic

The laboratory analyst needs to be aware of (a) the rationale behind formulation development efforts and (b) the effect that the formulation may have on testing of the final product. As safer and more potent drugs are developed, the concentration of the pharmaceutical "active" becomes less compared to that of bulk ingredients. Consequently, the quality and composition of the inactive materials begin to have more importance to analytical testing. Furthermore, new advances in formulation design enable very unique formulations to be developed to improve the API's apparent solubility, stability, or bioavailability. These are often complicated formulation "systems" (extended release, polymer extruded or coated, microspheres, complexes) that present further analytical challenges during formulation development and characterization.

1.3.2. Analytical Testing in Formulations Development

Analytical testing is a fundamental component of formulation development effort. Even before development activities can proceed, the API must be thoroughly characterized. Depending upon the proposed administration route, the following properties of the API must be determined:

- Solubility in a number of solvents and pH range
- Crystal structure/size
- Particle size
- Stability (heat, light, oxidation, pH)
- Compatibility with the expected formulation ingredients

Once the drug substance is characterized and understood, formulation efforts can begin to improve on characteristics intrinsic to the compound. For example, in an oral suspension, one may want to enhance the drug substance insolubility in the formulation. This may be accomplished through pH control, proper selection of counterion, choice of solvent or suspension vehicle, and so on. On the other hand, solubility may need to be enhanced for a poorly soluble drug that is to be administered parenterally. In another example, drug substances determined to be susceptible to light degradation may require an ultraviolet (UV)-absorbing tablet coating or amber bottle to offer some protection. The susceptibility of the API and the degree of improvement rendered in the formulation are measured through analytical (typically HPLC) testing.

1.3.3. Pharmaceutical Excipients

As stated above, the role of formulation excipients becomes increasingly important as more potent drugs are developed and more elaborate formulations are developed. Table 1.1 is a generalized list of pharmaceutical ingredients and their function in formulation development. Many of these complex materials (flavors, colorants, surfactants) provide a challenge to the method development effort because potential interferences should not hinder the assay of the API or related substances. Other ingredients, such as antioxidants and preservatives, may require assay upon release and during stability. Clever method development enables simultaneous assay of the API, related substances, antioxidants, and formulation preservatives in a single HPLC analysis.

1.3.4. Pharmaceutical Development Summary

Formulation development is a balance of improving the inherent properties of the API by enhancing stability, improving solubility, simplifying manufacture, and improving bioavailability without imparting negative characteristics or providing too complex a formulation. This difficult balance of properties must

Table 1.1. Example Pharmaceutical Ingredients[a] and Their Function in Formulation

Pharmaceutical Ingredient—Category	Example
Acidifying agent	Acetic acid, citric acid, HCl, phosphoric acid
Aerosol propellant	Butane, propane, dichlorodifluoromethane
Alkalizing agent	Ammonia, ammonium carbonate, sodium borate
Antifoaming agent	Dimethicone
Antimicrobial preservatives	Benzalkonium chloride, benzyl alcohol, cresol, parabens, phenol, sodium/potassium benzoate, sodium sorbate, sorbic acid
Antioxidants	Ascorbic acid, butylated hydroxyanisole (BHA), butylated hydroxytoluene (BHT), potasium/sodium metabisulfite
Buffering agents	Acetic acid, ammonium carbonate, lactic acid, sodium citrate, sodium phosphate
Chelating agent	EDTA
Coating agent	Carboxymethylcellulose, gelatin, polyethylene glycol, shellac wax
Color	Caramel, ferric oxide
Desiccant	Calcium chloride, calcium sulfate, silicon dioxide
Emulsifying and/or solubilizing agent	Lecithin, oleic acid, polysorbates, polyoxyethylene 50 stearate, sodium lauryl sulfate, sorbitan monolaurate, stearic acid
Flavors	Peppermint oil, thymol, rose oil
Glidant	Magnesium silicate
Ointment base	Lanolin, white petrolatum
Solvents	Methanol, corn oil, polyethylene glycol
Suspending–viscosity-increasing agent	Carbomers, methylcellulose Povidone
Sweetening agents	Saccharin, aspartame, sucrose
Tablet binder	Gelatin, polyethylene oxide, cellulose—microcrystalline
Tablet and/or capsule dilutent	Calcium carbonate, cellulose, starch
Tablet disintegrant	Crosprovidone, starch
Tablet and/or capsule lubricant	Magnesium stearate, polyethylene glycol
Water repelling	Dimethicone
Wetting/solubilizing agent	Benzalkonium chloride, Nonoxyl 9, polysorbates, sodium lauryl sulfate

[a] *United States Pharmacopeia*, USP 24, Excipients, pp. 2404–2406.

be accurately measured with performance quantitated by suitable methodology on qualified instrumentation by a capable analyst.

1.4. REVIEW OF THE ANALYST ROLE IN PHARMACOKINETICS, TOXICOLOGY, AND CLINICAL SUPPORT

1.4.1. Introduction

Bioanalytics is the application of analytical techniques to determine drug concentrations in biological samples, mostly plasma, serum, or urine samples. It plays a key role throughout drug development from discovery to drug approval.

> Pharmacokinetics is the study of the time course of drug absorption, distribution, metabolism and excretion.

Application of bioanalytical methods and pharmacokinetics has led to the shorter drug development timelines and rational drug therapies. These disciplines have also contributed immensely to the field of routine therapeutic drug monitoring by providing rationale and tools, especially in the management of anticonvulsant, antiasthmatic, and cardiovascular drug therapies. Monitoring of blood drug concentrations during these therapies have significantly reduced drug toxicity and improved treatment outcomes and, hence, patients' quality of life. The growing role of bioanalytics and pharmacokinetics in the pharmacokinetic/ pharmacodynamic evaluations is likely to continue to drive the industry and regulatory efforts to reduce developmental and review times for novel therapeutic agents.

1.4.2. Bioanalytical Considerations

The use of bioanalytical methods in drug development is not a new one. However, recent advances in bioanalytical technologies have allowed much greater sensitivity and specificity, which enabled pharmacokinetic discipline to be a guiding force in nonclinical and clinical development of drugs and biologics. These disciplines also play major roles in development and approval of bioequivalent generic drug products. For scientists engaged in these disciplines, it is very gratifying to see their contributions leading to significant progress in drug development, therapeutics, and generic drugs.

The analytical methods that generate bioanalytical data are an essential component of the drug development process. A very large amount of bioanalytical data is required to support toxicology and clinical studies by confirming the integrity of the various tests and trials, thereby aiding interpretation of the results. Active participation of bioanalytical scientists in development teams is essential to avoid pitfalls and to allow for expeditious adjustments in analytical and development strategy when needed.

Representative examples of analytical methods employed include liquid chromatography, mass spectroscopy, capillary electrophoresis, and immunoassays. Mass spectrometry (MS) is a powerful analytical technique that is often coupled to gas chromatography (GC) or liquid chromatography (LC) for identification of metabolites and for routine quantitation of drug and its metabolites.

1.4.2.1. Bioanalytics in Development of Classical Pharmaceuticals.

During drug development, among the highest priority items is to establish sensitive and robust analytical procedures. It is very likely that sample analyses may potentially become the rate-limiting step throughout the development of the drug, thereby complicating the conduct of safety and metabolism, pharmacokinetic, pharmacodynamic, and efficacy studies. It is important to realize that the results of these studies depend on the sensitivity of the analytical method employed; and if the analytical methods do not provide accurate quantitation of the drug and metabolites, all results and derived conclusions will be questionable. With increasing demands for information being placed on the pharmacokinetics and toxicokinetics of both drugs and metabolites, the strain on resources for bioanalytical development has become severe.

To avoid pitfalls in drug development, the following bioanalytical factors need careful consideration:

1. Methodologies for estimating the drug and its metabolites
2. Methods validation that will assure the accuracy of the assay
3. Suitable method transfer evaluation prior to methods utilization at alternative test sites
4. Feedback and adjustments when needed on the assay's sensitivity requirements during different stages of drug development
5. Fine tuning and modification due to interfering endogenous and exogenous compounds in different populations, drug interaction studies, and disease states, assuring adequate sensitivity and/or the specificity of the assay
6. Analysis turnaround time, which may determine the pace of drug development
7. Quality assurance

1.4.2.2. Bioanalytics for Biotechnology Products.

The analytical task changes markedly in comparison with conventional drugs, because therapeutic concentrations are in a very low range. Clearly, no single analytical technique provides comprehensive characterization of protein and peptide therapeutics. In fact, several methods are used in protein pharmacokinetic studies, namely, immunoassays, bioassays, radiolabeled proteins, HPLC, electrophoresis, and/or mass spectrometry. Mass spectrometry has recently been used to characterize pharmacokinetics and metabolism of several proteins and peptides. The high potency of many protein therapeutic agents, resulting in very low therapeutic concentrations (picogram/mL or less), offer real challenges to bioanalytical scientists in developing routine assay methods for pharmacokinetic studies.

1.4.2.3. Bioanalytical Methods and Regulatory Compliance. Bioanalytical methods used to quantitate drugs and drug metabolites for human bioavailability, bioequivalence, and pharmacokinetic studies and animal toxicokinetic studies have recently become a focus of increased concern within the regulatory community. This led to organization of the December 1990 conference on "Analytical Methods Validation: Bioavailability, Bioequivalence, and Pharmacokinetic Studies" by a number of regulatory agencies. The conference report formed the basis for regulatory expectations. As a result, companies need to develop standard operating procedures (SOPs) for analytical methods validation, for cross-validation of collaborating laboratories, and for routine sample analysis.

Bioanalytical method validation requirements include documented data to support assay accuracy, precision, sensitivity, and specificity. Intra- and inter-day accuracy and precision determined from five or more replicates at each concentration should have bias no more than 15% (20% at limit of quantitation) and coefficient of variation no more than 15% (20% at limit of quantitation). Sensitivity was based on limit of quantitation (LOQ) established and was included in the calibration curve. Specificity was determined from six independent sources of the same matrix for background interference (demonstrated by lack of interference from known metabolites) while demonstrating lack of interference from concomitant medications. Response function was determined from five to eight calibration standards for each analyte run and for each analyte; the simplest relationship should be used and goodness of fit should be tested. An analytical method should be validated for each different biological fluid being analyzed.

Stability of drug in biological specimen and stability of the drug in long-term storage, in freeze/thaw, and in process should be determined and documented. Quality control (QC) samples should be used with each analytical run for three main reasons: to establish the precision and accuracy of an analytical method during validation, to establish long-term stability of frozen samples, and to provide a basis for accepting or rejecting the analytical run.

The following considerations should be addressed during sample analysis:

1. Order of sample analysis/special blinding; test and reference run concurrently for bioequivalence studies
2. Standard curves for each analytical run
3. Quality control samples handled exactly as study samples and run with each analytical run
4. Source of quality control samples matrix
5. Source of analyte standard and Certificate of Analysis
6. Handling and reporting of all repeat determinations
7. Procedure for selecting reported value

For routine application of methods in sample analysis, calibration-curve quality control samples at three concentrations (near LOQ, mid-range and high end) in duplicate should be run along with the samples. Four out of the six quality control samples should be at less than 20% bias, at least one acceptable at each concentration and serving as the basis for run acceptance or rejection.

These method validation criteria should be applied to immuno- and micro-biological assays where possible. For these methods the following selectivity issues should be resolved: Compare with alternate method if available; evaluate matrix effects; evaluate parallelism of diluted clinical samples with calibrators; apply separation techniques to samples before assay; assess cross-reactivity of known metabolites or endogenous materials; and assess cross-reactivity to concomitant medications. The following quantitation issues should be resolved: Use replicate analysis to improve precision/accuracy; use more calibrators to better define standard curve; apply best curve-fitting model; back-calculate calibrator concentrations to validate model; and set acceptance criteria for lower LOQ and upper LOQ.

The analytical documentation to an NDA consists of assay methods, assay validation, and an analytical summary. The analytical summary is one way of presenting a consolidated picture of the analytical support of the pharmacokinetic section to an NDA. All of the analytical methods used in support of nonclinical and clinical pharmacokinetic and bioavailability data must be detailed. The validation of these assays must be provided. The key elements of assay validation include linearity, precision, accuracy, sensitivity, specificity, and sample stability.

1.4.3. Preclinical Pharmacokinetics/Pharmacodynamics

An effective pharmacokinetic program begins at the preclinical phase to enhance confidence in the extrapolation of animal toxicity data to humans by integration of pharmacokinetics with the safety studies. The objectives of preclinical pharmacokinetic and metabolism studies are to obtain information, which is useful for toxicity and safety evaluation studies in animals and initial safety and tolerance studies in humans. Informative preclinical information can be helpful in expediting the drug development process.

Pharmacokinetic evaluation of compounds at an early stage in discovery can provide valuable information regarding analytical strategy when the compound is studied in humans. In general, drug doses used in early animal pharmacokinetic studies are higher than those needed for pharmacological activity in humans. Knowledge of peak plasma concentration, time plasma concentration profile, bioanalytical assay accuracy, precision, and limits of quantitation from these early studies would permit extrapolation of assay sensitivity required for first-time (Phase I) studies in humans and, therefore, bioanalytical strategies to meet the anticipated challenges prior to the start of the Phase I clinical studies.

The availability of specific and sensitive analytical procedures is essential to start any pharmacokinetic (PK)/pharmacodynamic (PD) studies for a new drug. When a major metabolite is known, particularly if pharmacologically active, an appropriate method should be developed for its identification and quantitation in biological fluids. If the drug and/or its metabolite(s) exhibit chirality, the assay should be stereospecific. Moreover, determination of systemic drug concentration ranges that are associated with pharmacological action and toxicological effects of a drug or its metabolites may aid in development of human dosing regimens and may indicate the likely steepness of the dose response curve in humans.

1.4.4. Preclinical Safety Studies

Toxicokinetics has been defined as the set of safety studies aimed to describe the absorption, distribution, metabolism, and excretion of drugs at doses within the range of toxic doses (Figure 1.6). Toxicity studies, which may be supported by toxicokinetic (TK) information, include single and repeated dose-toxicity studies and reproductive, genotoxicity, and carcinogenicity studies. TK information may also be of value in assessing the implications of a proposed change in the clinical route of drug administration.

Bioanalytical support of toxicokinetic studies is vital to ensure that sensitive assay is used to collect sufficient time–concentration data to calculate the area under the time–plasma concentration curve (AUC) and maximum plasma concentration (C_{max}) at each dose level of the test drug in each animal species. Such data are required to demonstrate that concentrations of drug are present in all animal species studied, and that the relationships of drug concentrations to dose level are acceptable. Concentration monitoring during safety evaluation studies and determination of extent of exposure allow better interspecies comparisons. Knowledge of the drug concentrations (drug exposure) is essential for substantiating safety assessments and will assist in interpretation of unanticipated toxicity.

The principal goals of toxicokinetic studies should be to (a) establish quantitative relationships between toxic doses and time course of drug in body fluids and/or tissues for selection of dose levels in subsequent studies, (b) obtain information on the extent of absorption and potential for drug accumulation in relation to toxic manifestations, (c) select the most adequate or sensitive species for characterization of toxicological characteristics, (d) select the mode of administration, and (e) define the target organ profile where further toxicological investigations could be focused.

Effective integration of pharmacokinetics, metabolism, and toxicology evaluations constitute pivotal elements in drug development and are essential to avoid problems. The benefits of this integration have enhanced our understanding of the metabolic alterations that take place under conditions of exaggerated exposures. A toxicity study gathers important data on clinical reactions and on time-related changes in blood elements and tissues, and in many cases it iden-

20

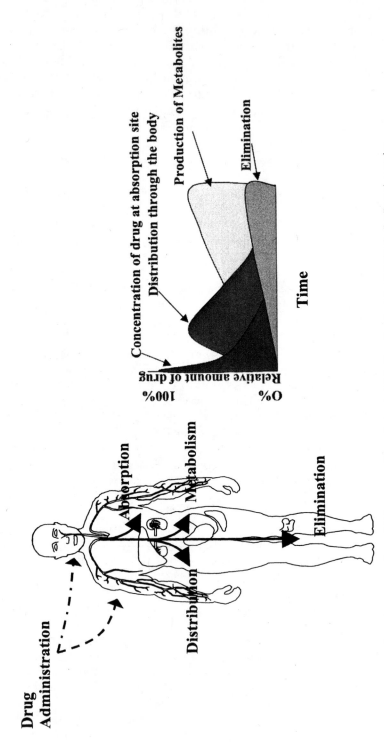

Figure 1.6. The ADME process: Upon drug administration, the drug is absorbed, distributed, metabolized, and then eliminated. The time dependances of the relative amounts of these processes are indicated in the graph.

tifies the effects of the compound on the target organ. In many cases the target organ of toxicity is not the therapeutic target. When PK data are generated at toxicologically relevant doses, excellent guidance in the interpretation of un-anticipated toxicity following saturation of metabolism is provided.

The initial projection of a therapeutic ratio to support the initiation of Phase I clinical safety and tolerance studies requires an understanding of the relationship of doses utilized in safety and efficacy studies and their projected relationship to the clinical dosing regimen. Experience with multiple recombinant protein products indicates that the PK behavior of many biomacromolecules is predictable across species. The predictable behavior of pharmacokinetics of many drugs and biologics across species permits extrapolation of preclinical safety and efficacy data to the clinical setting when doses are related on the basis of PK equivalence rather than on a body weight (mg/kg) basis.

1.4.5. Mass Balance and Metabolism

In order to fully understand and interpret toxicology studies, it is essential to determine the fate (ADME) of the drug in the species used in toxicology testing; it is important to discover interspecies differences, including differences from humans. Identification and pharmacological characterization of individual metabolites is key in comparing results of preclinical studies with those of human studies.

In order to understand the disposition of a drug, a detailed profile of the drug's metabolic fate is required. A variety of methods have been developed for use in metabolite identification and characterization analyses. These methods include HPLC coupled with UV, diode array, electrochemical, radioactive, and mass spectroscopic (MS) detectors, gas–liquid chromatography (GLC) coupled with MS detector, and high-performance thin-layer chromatography (HPTLC).

1.4.6. Clinical Support

The clinical pharmacokinetics program encompasses determining pharmacokinetics of a drug in early Phase I safety and tolerance studies in humans and in special populations such as women, geriatric, pediatric, and hepatic and renal compromised patients in late Phase I program. The late Phase I study also includes intravenous (I.V.) and oral bioavailability studies, food–drug and drug–drug interaction studies, and bioequivalance of dosage forms. Pharmacokinetics is also extensively applied to facilitate Phase II and III trials and to establish therapeutic dosing regimens in dosage form development, in drug targeting and delivery technology, in therapeutic drug monitoring, and in understanding variability in drug response due to genetics and circadian changes. Materials used in clinical studies must be "released" using test methods recorded in specifications.

1.4.6.1. Phase I Safety and Pharmacokinetic Studies. The objectives of Phase I clinical development is to define the initial parameters of toxicity, tolerance, and their relation to dosage and the relevant pharmacokinetics of the drug in healthy volunteers. Immediately following acceptance of an IND application, the Phase I safety and PK studies are initiated followed by the late Phase I studies. Doses for the early Phase I safety-tolerance study in humans is proposed based on drug pharmacology, pharmacokinetics, toxicokinetics, and the anticipated therapeutic dose range in humans. For the planning of these studies, it is extremely valuable if plasma concentration data are available from animals at doses that equate to efficacy, tolerance, and the onset of toxicity in humans.

These initial rising dose-tolerance and pharmacokinetic data (half-life and clearance), are often the key to deciding dosage regimens for the clinical program and are utilized to establish the appropriate dosing program to be incorporated in Phase II studies. Of equal importance are bioanalytical data generated between each dose to help decide appropriate dose escalations considering the unpredictability of metabolism and nonlinear pharmacokinetics of first-time studies in humans.

Further Phase I studies in volunteers and patients, usually carried out during and even after the Phase II and III clinical studies, are intended to characterize the drug's PK/PD relationships in special populations, to optimize drug delivery system, and to probe potential drug–drug, drug–food, and drug–disease interactions that might be expected to perturb the PK/PD relationship of the drug.

1.4.6.2. Phase II and Phase III Efficacy Studies. Phase II studies in patients are to assess the drug's therapeutic effectiveness and to develop a rational dosing strategy for Phase III studies. Phase III studies are designed to document the clinical safety and efficacy.

Throughout Phases II and III, plasma concentration data should be routinely obtained on a survey basis to help explain unusual responses, to suggest the possibility for drug–drug, drug–food, and drug–disease interactions, and to identify metabolic heterogeneity. In addition, plasma drug concentrations data are essential to determine dose-concentration and concentration-response information, an estimate of the lower useful concentration, the concentration beyond which greater response is not seen, the highest concentration that is tolerated, and understanding of the concentration-effectiveness and concentration-toxicity curves.

1.4.6.3. Therapeutic Drug Monitoring in Clinical Practice. The most practical application of the knowledge of plasma concentration versus pharmacological effect relationships learned throughout the clinical development (Phase I to III) for individualization of drug treatment has revolutionized the therapeutic drug monitoring discipline. The concept that the variability in drug plasma concentration is primarily due to individual differences in drug absorption, metabolism, and excretion is well accepted. These differences in plasma drug levels may mean therapeutic or toxic effect in patients. Therefore, an optimal therapeutic effect without toxicity could be achieved in a variety of patients by monitoring the drug plasma concentration versus dose data. The plasma concentrations could also be used to assess compliance and to adjust dose for concomitant use of interacting drugs.

Therapeutic drug monitoring has been made possible by the development of highly selective and sensitive analytical techniques for determination of plasma levels quickly. Since the early 1970s, therapeutic drug monitoring has been extensively used in the clinical setting, especially for low therapeutic index drugs such as theophylline, digoxin, procainamide, and antiepileptic drugs.

1.4.6.4. Bioequivalence Studies. Bioequivalence studies compare the systemic absorption of the proposed generic product to achieve comparable plasma levels of the active ingredient and, in some cases, their metabolites with those achieved by their brand name counterparts, when administered to the same individuals under as identical conditions as possible. A well-validated, specific, precise, and accurate analytical method is required for confidence in the bioequivalence data. The analytical method validation is, therefore, a critical component of a bioequivalence study. Inadequate validation of the method is one of the major causes of a deficient submission.

The above applications illustrate clearly how the bioanalytical discipline can play a key role in drug development and how effectively it can meet clinical needs. The effective use of the data undoubtedly helps realistic planning of drug development based on the potential limitations of the compound and the formulation from the outset. The critical nature of such data to guide drug development provides a challenge to the bioanalytical scientist. The promise of developing new drug and biologic therapeutic agents will continue to be a driving force for new bioanalytical technologies in the future. The future is very bright for the role of bioanalytics in developing new therapeutic agents.

1.5. STABILITY PROGRAM IN PHARMACEUTICAL INDUSTRY

1.5.1. Introduction

The purpose of stability testing in pharmaceutical industries is to provide evidence on how the quality of a drug substance or drug product varies with time under the influence of a variety of environmental factors such as temperature, humidity, and light and enables recommended storage conditions, retest periods, and shelf lives to be established.

Shelf life (expiration dating period) is the interval that a drug product is expected to remain within the approved specifications after manufacture. The shelf life is used to establish the expiration date of individual batches.

The performance of a drug when given as a tablet, capsule, syrup, or injection depends not only on the content of the drug substance but also on its pharmaceutical properties such as dissolution, disintegration, hardness, and so on. All of these aspects are therefore part of pharmaceutical stability programs [17, 18].

1.5.2. Goals of the Stability Program

The goal of a stability program depends on the stage of development of the drug product. At the very beginning of product development, it is necessary to understand the inherent stability of the drug substance and its interaction with the proposed excipients. At this stage the effect of pH, moisture, air (oxygen), and light on the stability of the drug substance is also studied. The accelerated stressing of the drug substance and drug product provides information to the intrinsic stability of the molecule/formulation and may establish likely degradation pathways. The formulation group also has the responsibility for recommending to the toxicology group about the stability of the drug substance in the vehicle used in the animal trials. On the analytical side, the analytical research group supports the preformulation stability program, which is ultimately responsible for developing and validating the stability-indicating assay that will be included in the NDA.

In the preclinical formulation stage, the selection of a stable drug product formula is the primary goal. This temporary preclinical formula is included in the IND.

The goal of the stability program in the clinical trial stage is to ascertain that the drug product batches tested in the clinical trials are stable, and these data will be subsequently included in the NDA. So far the analytical support was obtained from the analytical research group. At the NDA approval stage, the validated stability-indicating analytical method will be transferred to the quality control group, to ascertain that it works well in the hands of those who have to monitor the stability of the marketed product.

The marketed product stability program fulfills the commitment part of the NDA and also ensures that the marketed drug products are stable (potent) until the expiry date stamped on the product label. Usually the first three marketed batches and at least one batch per year are subjected to stability monitoring.

1.5.3. ICH Guidelines on Stability Testing of Drug Products

The Expert Working Group of the International Conference on Harmonization of the Technical Requirements for Registration of Pharmaceuticals for Human Use (ICH) developed a guideline on stability testing for a registration application within the European Union, Japan, and the United States. This guideline summarizes in a nutshell the stability study requirements for the drug products and drug substances. The requirements for a drug product application are listed below.

- *Selection of Batches:* Provide accelerated and long-term stability data on three batches of the same formulation and dosage form in the containers and closures proposed for marketing. Two of the three batches should be at least pilot scale. The third batch may be smaller—for example, 25,000 to 50,000 tablets or capsules for solid oral dosage forms. Long-term testing should cover at least 12 months' duration at the time of submission.

- *Test Procedures and Test Criteria:* Use validated stability-indicating analytical test procedures. Testing should cover chemical and biological stability, loss of preservative, physical properties and characteristics, organoleptic properties, and, where required, microbiological attributes.

- *Specifications:* Shelf-life specifications based on release limits and stability data. Include upper limits of degradation products justified based on material used in preclinical and clinical studies. Limits for tests such as dissolution and particle size require reference to results of bioavailability and clinical batches.

- *Storage Test Conditions:* Length of studies and storage conditions should cover storage, shipment, and subsequent use—for example, reconstitution or dilution as recommended in the labeling. Maximum data at time of submission: 12 months of long-term testing at $25°C \pm 2°C/60\%$ RH $\pm 5\%$ RH and 6 months accelerated testing at $40°C \pm 2°C/75\%$ RH $\pm 5\%$ RH. Significant change under accelerated conditions leads to testing at an intermediate condition—for example, $30°C \pm 2°C/60\%$ RH $\pm 5\%$ RH. Significant change under accelerated conditions is defined as: 5% potency loss from initial assay of batch; any specified degradant exceeding specification; out of pH limits; dissolution out of limits for 12 capsules or tablets; failure to meet specifications for appearance and physical properties. Long-term testing should continue to cover expected shelf-life. Other justified storage conditions are allowed. Store heat-sensitive drug products under alternative lower temperature for long term testing.

- *Testing Frequency:* Normally test every 3 months in first year, every 6 months in second year, and then annually. Apply matrixing and bracketing where justified.

- *Stability Data Evaluation:* Acceptable approach is to determine the time at which the 95% one-sided confidence limit for the mean degradation curve intersects the acceptable lower specification limit. Data can be combined where there is low batch-to-batch variability after applying appropriate statistical tests. An appropriate statistical test (e.g., p values for level of significance of rejection of more than 0.25) is applied to the slopes of the regression lines and zero time intercepts for individual batches. Where inappropriate to combine data, shelf-life may depend on minimum time batch expected to remain in specifications.

- *Labeling Statements:* Use storage range in accordance with national/ regional requirements. Storage range should be based on stability evaluation of drug product. Terms such as "ambient conditions" or "room temperature" are unacceptable. There should be a direct linkage between the label statement and the demonstrated stability characteristics of the drug product.

1.5.4. Stability Monitoring

The chemical, physical, and microbiological aspects monitored during the stability studies on drug products depend upon the formulation. The tests usually performed on the different formulations are listed in Table 1.2.

1.5.5. Stability-Indicating Methods

The development and validation of stability-indicating assay and purity methods are discussed in detail in Chapter 12. Suffice to say all methods used to monitor stability assay should be specific for the API.

1.5.6. Pharmaceutical Packaging and Stability

The close relationship between a pharmaceutical preparation and its package is of major concern to the industrial pharmacist. Faulty packaging of pharmaceutical dosage forms can invalidate the most stable formulation.

The following are the criteria for a satisfactory pharmaceutical package: mechanical protection, environmental protection, security, functional adequacy, inertness, and cost. The package development team assures that the above criteria are met at the onset of the drug development program. This is accomplished by testing the container's effectiveness in protecting the product during extended storage under varying environmental conditions of temperature, humidity, and light.

The materials most commonly employed as components of container closure for pharmaceutical preparations include glass, metal, plastic, and rubber.

Glass has been the container of choice for pharmaceutical dosage forms because of its resistance to decomposition by the various environmental conditions or by solid or liquid contents of different chemical composition. In some cases the high-potency and, consequently, low-dosage drugs can readily be affected by release of soluble alkali from glass containers. A remedial measure is to buffer the product whenever possible to offset pH changes due to release of alkali by glass. Photostability studies have shown that amber glass offers a product better protection against light than does flint (clear) glass.

Metal tubes are the traditional choice for topical preparations. Tubes constructed of a single material can be tested readily for stability with a product. In the case of tubes with coating, it must be established that the coating material is inert for the preparation. Aluminum tubes have demonstrated reactivity with fatty alcohol emulsions and with preparations beyond a pH range of 6.5 to 8.0. Nonreactive epoxy linings have been found to make aluminum tubes more resistant to attack.

Table 1.2. Test Specification for Various Formulations

Formulation Type	Typical Specification Test
Tablets	Appearance, friability, hardness, color, odor, moisture, assay (potency), purity, and dissolution.
Capsules	Appearance, color, moisture, assay, purity, moisture, brittleness, and dissolution. For soft gelatin capsules, the fill medium should be examined for precipitate, cloudiness, and pH.
Emulsions	Appearance (phase separation), color, odor, pH, viscosity, assay, and purity. Storage on the side or inverted is suggested for assessment of closure system. It is also recommended that a heating/cooling cycle (between 4°C and 45°C) be employed.
Oral solutions and suspensions	Appearance (precipitate, cloudiness), color, odor, assay, purity, pH, redispersibility, dissolution (suspensions), and clarity (solutions). Storage on the side or inverted is suggested for assessment of closure system.
Topical ointments and creams	Appearance, color, odor, homogeneity, clarity, pH, assay, and purity. Ointments and creams should be assayed by sampling at the surface, middle, and bottom of the container. Tubes should be sampled near the crimp.
Oral powders	Oral powders are reconstituted prior to administration. The following characteristics of the powders are examined: appearance, color, odor, moisture, assay, and purity. The reconstituted product is examined for the following: appearance, pH, redispersibility, assay, and purity.
Metered-dose inhalation aerosols	Color, clarity (solution), assay, purity, delivered dose per actuation, number of metered doses, particle size distribution (suspension), loss of propellant, pressure, valve corrosion, spray pattern, weight loss, and absence of pathogens.
Small-volume parenterals	Appearance, color, assay, purity, particulate matter, pH, sterility, and pyrogenicity. Parenterals (except ampoules) should be stored inverted or on their sides in order to determine whether contact of drug product with the closure system affects the stability.

Starting in the mid-1960s, most pharmaceutical companies set out on a program of replacing glass bottles with plastic bottles, due to their lighter weight and durability. However, plastic containers did present the following potential problems: loss of product components to the plastic (sorption), loss of plastic components to the products (desorption), transport of water vapor or gases through the plastic (permeation), photodegradation of drug product, and alteration of container properties due to contact with product. Eventually, plastics became perfected to the point that the above problems are reduced to a minimum and did not seriously affect the expiration date.

Rubber of various compositions and formulations can also exhibit sorption and desorption problems when used with pharmaceuticals and biological. The presence of rubber closure extractives in the vial solution could interfere with the assay of the active ingredient or contribute to particulate matter in solution. Stability monitoring should assess these potential problems.

1.5.7. Stability Summary

There are few industries that have more technical talent dedicated to them than the pharmaceutical industry. Because of constantly changing regulatory requirements and improvement in analytical techniques (e.g., capillary electrophoresis, LC-MS) the opportunities and challenges that exist in pharmaceutical stability research are endless.

1.6. CHAPTER SUMMARY

This chapter attempted to survey the contribution of analytical testing to the drug development process. In each drug development specialty listed in Figure 1.7, suitable test methods run by capable analysts are required to ensure accurate

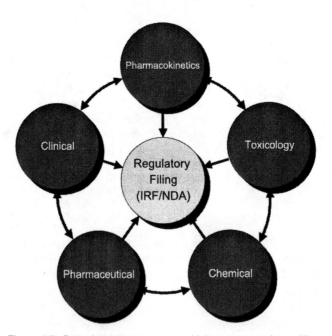

Figure 1.7. Drug development groups which support regulatory filing.

and reliable data at each stage of the development process. Subsequent chapters will provide useful information to ensure that data reported by the laboratory analyst is accurate, reliable, and in compliance with GMP requirements.

As reviewed throughout this chapter, the role of the laboratory analyst is to establish the identity, potency, overall quality, and purity of the product and drug substance throughout the development process. Certainly, the laboratory analyst contributes essential data during all phases of the drug development process. In most development stages, the testing effort is expected to be compliant, requiring proper documentation, qualification and calibration of instrumentation, validation, and documented training of laboratory personnel.

REFERENCES

1. *Federal Register, 21* CFR, 314.
2. *International Conference on Harmonization: Guidelines on Stability Testing of New Drug Substances and Products* (Q1A), December 1993.
3. *International Conference on Harmonization: Guidelines on Photostability Testing of New Drug Substance and Products* (Q1B), November 1996.
4. *International Conference on Harmonization: Guidelines on Impurities in New Drug Substances* (Q3A), January 1996.
5. *International Conference on Harmonization: Guidelines on Impurities in New Drug Products* (Q3B), November 1996.
6. *International Conference on Harmonization: Specifications: Test Procedures and Acceptance Criteria for New Drug Substances and New Drug Products: Chemical Substances* (Q6A), July 1997.
7. G. C. Davis, J. R. Murphy, D. A. Weisman, S. W. Anderson, and J. D. Hofer, *Rational Approaches to Specifications Setting: An Opportunity for Harmonization: Pharmaceutical Technology*, September **1996**, pp. 110–118.
8. International Conference on Harmonization; Guideline on the Validation of Analytical Procedures: Methodology; Availability, *Federal Register* **1997**, *62*(96), 27464–27467.
9. International Conference on Harmonization; Guideline on the Validation of Analytical Procedures: Definitions and Terminology; Availability, *Federal Register* **1995**, *60*(40), 11260–11262.
10. *United States Pharmacopeia*, 24th edition (USP 24), Validation of Compendial Methods, ⟨1225⟩, **2000**, pp. 2149–2152.
11. I. Krull and M. Swartz, *Validation Viewpoint, LC-GC* **1997**, *15*(6), 534–540.
12. I. Krull and M. Swartz, Method Validation Protocol—Required Data Elements, Method Transfer and Revalidation, Validation Viewpoint, *LC-GC* **1997**, *15*(9), 842–846.
13. J. M. Green, A Practical Guide to Analytical Method Validation, *Anal. Chem. News & Features* **1996**, 305A–309A.
14. J. D. Johnson and G. E. Van Buskirk, Analytical Method Validation, J. *Validation* **1996**, *2*(2), 88–103.

15. Food and Drug Administration, Center for Drug Evaluation and Research, *Validation on Chromatographic Methods*, November 1994.

16. G. C. Hokanson, A Life Cycle Approach to Validation, Pharm. Tech. **1994**, *18*, 118.

17. *Guideline for Submitting Documentation for the Stability of Human Drugs and Biologics*, Center for Drugs and Biologics, Food and Drug Administration, February 1987.

18. J. T. Carstensen (editor), *Drug Stability, Principles and Practices*, Marcel Dekker, New York, 1990.

2

LABORATORY CONTROLS
AND COMPLIANCE

Henry Avallone

2.1. INTRODUCTION

Throughout history, the U.S. Government has reacted to defective product and fraudulent practices by amending the law, issuing regulations, or issuing guidelines. Thus, in the late 1980s, when defective product and fraudulent activities reached an unacceptable level, the Food and Drug Administration (FDA) introduced the Pre-Approval Inspection (PAI) Program. This program formalized the manner in which FDA inspections of new or pending product applications were conducted.

One of the areas where considerable fraudulent activities were identified was the area of laboratory controls. This has resulted in increased FDA scrutiny of the laboratory, including more frequent utilization of FDA analysts in the conduct of these inspections. These analysts have brought increased expertise to FDA inspections because they are familiar with the emerging technology and data handling systems that are utilized in their own laboratories.

Coinciding with the increased scrutiny of laboratory controls by the FDA has been the advancement of laboratory technology and systems. The latter has resulted in the need for new standards, new methods, increased training, and increased capital expenditures to stay current.

The improvements in laboratory equipment and systems have resulted in increased sensitivity to detect and quantitate substances. This, coupled with the

Analytical Chemistry in a GMP Environment. Edited by J. M. Miller and J. B. Crowther
ISBN 0-471-31431-5 © 2000 John Wiley & Sons, Inc.

establishment and tightening of impurity specifications or standards, has led to the increase of compliance issues. The impurity profiles and methods detection limits have been particularly critical in obtaining new drug approvals. An even greater challenge has been to bring older products, processes, and methods that are nearing the end of their life cycle up to today's standards. Additionally, it has been necessary for laboratory records to fully document test procedures, demonstrating that the tests were performed correctly and in accordance with the method. Problems with fraudulent data and many legal cases have led the effort to require increased documentation in order to authenticate the data.

Failure on the part of industry to comply with GMP requirements has resulted in an FDA request for laboratory certification. In at least two legal cases [1, 2], laboratory certification by a third party was included in a court order (injunction); in other cases, companies have "voluntarily" submitted to third-party laboratory certification. These "voluntary" cases have been a direct result of FDA inspections and subsequent agreements with local FDA field (district) management.

This chapter will address some of the compliance issues identified above and provide guidance for maintaining a compliant laboratory.

In August of 1993 and July of 1994 FDA Injunctions against two drug manufacturers required certification of their laboratories [1, 2]. The court orders listed basic requirements and the certification of laboratory conformance to Current Good Manufacturing Practice (CGMP)* regulations by a third-party expert. At the time of the consent decree (injunction), the specifics of laboratory certification had not been delineated by FDA. Subsequently, an FDA memo in August 1994 [3] provided direction for the certification of a laboratory, and this document has been used by third-party "experts." Areas discussed in this document included the following:

- Management systems
- Operating procedures
- Personnel training
- Data accountability
- Method validation
- Equipment
- Facilities
- Certification documentation

All of these issues will generally be discussed in this chapter focusing on areas of laboratory management, laboratory controls, and laboratory compliance. In reflecting back on the FDA memo, it should be recognized that, while

---

* In the interest of simplicity the designation "GMP" will also be used in this book and is considered to be fully equivalent to "CGMP."

appropriate at the time it was written, there were issues addressed that may now be considered outdated by today's standards.

2.2. LABORATORY MANAGEMENT

2.2.1. Management Responsibility

The basic premise and foundation of the Federal Food, Drug and Cosmetic (FD&C) Act [4] pertains to management accountability. For example, Preamble Comment #432 to the 1978 current GMP regulations [5, 6] discussed the need for corporate officers and responsible managers to be aware of objectionable conditions. It states:

> Moreover, responsible officials already have very rigid legal duties to be aware of, and to take action upon, conditions contributing to drug adulteration [7].

While there is a movement on the part of some companies in the industry toward self-directed work teams, total quality management (TQM), and so on, this logic can be contradictory to FDA philosophy. The FD&C Act directing accountability and responsibility applies to the management of the laboratory. For example, at a recent training program an FDA manager was asked: "What are FDA thoughts on self-directed work teams?" He replied: "Firms are required to comply with GMPs and have individuals accountable and responsible. While we have no objection in some environments, in a GMP environment responsibility is a key issue. For example, in the review of analytical raw data, a manager or supervisor is held accountable for the work performed and review of the analysts' data. Certainly, a peer or even administrator could check calculations. However, a manager needs to review the data."

The area of resources, particularly regarding competent personnel, often presents a compliance issue. The typical laboratory has an analyst-to-supervisor ratio of a maximum of eight analysts to one supervisor. In an audit, problems can be identified through a review of out of specification (OOS) test results, deviations, and investigations. For example, incomplete investigations are often the result of inadequate laboratory resources. For the compliant laboratory, it is important to have data reviewed on a daily basis. If a supervisor is missing or not available, even on a limited daily basis, a qualified replacement should be available. As an example, replacement with an "acting" supervisor or a temporary supervisor not only covers the compliance issue, but serves as a developmental opportunity for personnel.

In the review of laboratory data, it is acceptable for the peer reviewer to ensure that data are labeled, calculations are verified, and data are complete. However, the supervisor must perform the general review of the complete data package. For example, the supervisor reviews raw data, such as chromatograms and their integration, and also evaluates the appropriateness of results. The supervisor also needs to conduct reviews and compare assays, content uniformity, and dissolution results. While data should be reviewed daily, the complete review of the data package should be performed within a reasonable time, typically five working days. Otherwise, it becomes extremely difficult to conduct any investigation of laboratory data.

In a laboratory training program, a senior FDA laboratory director was asked to comment on the responsibilities of the first line supervisor. He stated that responsibilities include the review of analysts' work, assurance of adequate procedures, assurance of safety requirements, planning of work, development of analysts, and certain administrative functions. By far, the most important of these is the assurance that analysts are trained and are performing the analytical work correctly.

From an auditing perspective, if a laboratory supervisor's office or work area is not in close proximity to the area where analysts are working, it is usually an indication that the supervisor is not involved with daily laboratory operations. If a supervisor is not aware of the samples that each analyst is working on, it could also be an indication that the laboratory is undermanaged.

2.2.2. Training

A major reason for laboratory noncompliance, as detailed in all of the FDA legal actions regarding laboratories, is the failure to have trained analysts. Training is necessary both from a technical basis and for GMP compliance. It is expected that a laboratory have a standard operating procedure (SOP), which provides some overall direction for laboratory training, including basic training on the content of the cGMP regulations. The sections of the regulations that detail laboratory requirements should be highlighted as well have as a discussion of basic SOPs, such as data handling and recording data in workbooks and/or on data sheets.

Specific training for each type of test to be performed should be reviewed during the technical training. Experienced laboratory directors have commented that the typical chemist graduating with a degree in chemistry has little practical laboratory experience. Therefore, for *general* tests, there must be some level of assurance as to the analysts' competency. It is also expected that each analyst who performs a relatively *complex* test, such as a high-performance liquid chromatography (HPLC) assay, have method-specific training. For example, in an audit it is not unusual to identify an analyst having difficulties conducting a specific HPLC assay, and then verify through a review of training records that the analyst has not been trained to conduct the product specific HPLC assay.

As new equipment is introduced, analysts should also be trained through a combined use of visual aids and hands-on use. The FDA St. Louis National Laboratory, for example, has a library of tapes for the training provided by the vendor each time a new piece of equipment is purchased.

A major part of the training program is the use of appropriate evaluation tools to certify that an analyst is trained. It is a relatively easy task to provide training. However, the real issue centers around the effectiveness of the training provided. During a review of laboratory deviations, management can identify analysts who have had problems conducting testing. It is the laboratory manager's responsibility to review analysts' training records and "certify" that the analyst has been found acceptable to perform the test in question. (By "certifying" the analyst, the manager acknowledges that the analyst has been evaluated or assessed and that the manager assumes accountability for the competency of the analyst in the area/method/procedure certified.)

There are basic documents that should be included in the training file of each analyst. These include:

- Analyst training file
- Resume or curriculum vitae (C.V.)
- Documentation of SOP training
- Documentation of GMP training
- Documentation of standard method training
- Documentation of product specific training
- General skill enhancement training

One of the ways in which management commitment can be demonstrated is in the resources allocated to training. Some companies have undertaken on their own a program for the certification of laboratory analysts. For example, Johnson & Johnson conducts a two-week hands-on laboratory analyst training and certification program. This program is conducted on a bimonthly basis for laboratory analysts during which they receive training in both compliance and technical skills. At the conclusion of the program each analyst is assessed and, if found satisfactory, is issued a certificate of satisfactory completion. As "certification" becomes more popular in the industry, it is expected that more companies will move toward some type of analyst/laboratory certification.

2.3. LABORATORY CONTROLS

2.3.1. Laboratory Records

Issues relating to data integrity and authenticity have raised the bar of laboratory records beyond what some consider "good science." Some analysts, and particularly research analysts, have a difficult time with regard to compliance with data handling requirements of the regulations. The CGMP Regulations, Section 211.194(a) [6], state:

> Laboratory records shall include complete data derived from all tests necessary to assure compliance with established specifications and standards, including examinations and assays, as follows:

1. Sample description
2. Method used
3. Sample weight
4. Complete records, including all graphs, charts, and spectra from laboratory instrumentation
5. Record of all calculations
6. Statement of results and comparison to specifications
7. The initials or signature of a second person showing that the original records have been reviewed for accuracy, completeness, and compliance with established standards

In order to assure compliance, it is important to have a SOP that specifically defines what raw data are, not what they can be. When dealing with computerized data handling systems and electronic data, it is important to define raw data up front. Generally, it is both the electronic signal and the hard copy that are generated and reviewed. As such, both are considered to be raw data. With movement toward paperless systems, there may be some consideration for paperless laboratory data. However, the volume of data and chromatograms generated in the typical laboratory is too great to affect review on a screen. For example, in an audit of a research laboratory that was considering the utilization of a paperless system, a manager replied that she was unaware of any supervisor or reviewer who would sit in front of a screen and review chromatograms and data. Furthermore, there have been some questions about the resolution of graphs displayed on a terminal or personal computer. Thus, at this time most laboratories continue to utilize both the electronic signal and hard copy. It is also recognized that some electronic data in the laboratory may be considered to be transient electronic data. For example, data from some pH meters could be considered transient data as included in the Electronic Signature Regulations (Part 11) [7]. While it is widely recognized that some clarification with regard to electronic data may be needed, it behooves the laboratory to have detailed procedures regarding the composition of raw data. (Further discussion of the definition of raw data can be found in Chapter 14).

Improvements in technology have also resulted in development of more complex HPLC gradient methods, particularly for quantifying impurities. Many of the newer, more specific methods have made it necessary to reprocess chromatograms. While the electronic signal is not changed, the parameters around integration may be changed to improve the clarity of the peak. In the assessment of data, it is important to recognize that specific results must be reproducible. In these situations, SOPs are needed including direction for the reprocessing of data.

SOPs should also define the data systems: notebooks, data sheets, electronic, and combinations. Current requirements for prenumbered pages and data sheets reflect the concern for data integrity. If results are derived via the treatment of raw data, this should also be defined. The processing of data by a computer using changeable software, such as a spreadsheet, requires SOPs as well as validation of the spreadsheet.

2.3.1.1. Computerized Data Acquisition Systems. The use and validation of computerized data acquisition systems and specifically computerized liquid chromatography (LC) systems is a major compliance issue. FDA policy on the validation of computerized LC systems was published in 1994 [8]. This document was written by FDA laboratory managers and circulated as an FDA policy document. The holistic approach presented in the document represents a practical application to validation and is the same philosophy expressed in the *United States Pharmacopeia* (USP). In the section on chromatography the USP states: "To ascertain the effectiveness of the final operating systems, it should be subjected to a suitability test prior to use. The essence of such a test is the concept that the electronics, the equipment, the specimens, and the analytical operations constitute a single analytical system, which is amenable to an overall test of system function" [9].

Years ago, with the advent of the Good Laboratory Practice (GLP) Regulations and increased FDA interest in nonclinical testing, the FDA provided guidance on computerized data acquisition systems that continues to represent basic policy. Dr. George James, FDA pharmacologist, FDA's Center for Drugs, commented on security and authenticity issues:

In reference to the growing trend towards computerized data acquisition systems, the Agency has provided the following guidance for computerized systems:

1. Provision must be made so that only authorized individuals can make data entries.
2. Data entries may not be deleted. Changes must be made in the form of amendments.
3. The database must be made as tamperproof as possible.
4. The Standard Operating Procedures must describe the procedures for ensuring the validity of the data.

Additionally, a basic aspect of validation of computerized data acquisition systems for a laboratory provides for a comparison of data from the specific instrument with that data electronically transmitted through the system and emanating on a printer. Once this functional testing has been accomplished over a period of time, periodic data comparisons would be sufficient.

The other basic aspect of validation is the qualification of equipment. In a laboratory, and particularly for an automated system, it is important that the test equipment, such as the HPLC unit and its components (pump, detector, injectors, etc.), be qualified (see Chapter 15). Generally, the laboratory equipment supplier will provide the aspects that need to be evaluated in the qualification exercise.

Improvements in LC technology increased interest in unknown peaks and sensitivity adjustments at the noise level and have made it difficult to resort to paperless LC systems. In a number of cases, regulators have asked for documentation of data integration and reprocessing. In some cases data are retained for each reprocessing, while in others only the final data are retained. Because it is a requirement that the laboratory supervisor review all data, it becomes a challenge for this supervisor to sit at a computer work station and review all chromatograms, along with the reprocessing data for each chromatogram. In any event, it would seem appropriate for some documentation each time data are reprocessed.

2.3.2. Out of Specification/Trend (OOS/OOT)

Perhaps the area that has received the most regulatory attention is that of retesting. FDA comments and observations concerning the practice of retesting without appropriate investigation and justification have been a major compliance issue for the last two decades. In the United States of America vs. Barr Laboratories court decision [10], Judge Wolin clarified the requirements.

The guidance that comes from the Wolin decision suggests that every OOS result must be investigated. There are two mechanisms for investigation: informal and formal. *Informal investigations* are appropriate for single-event OOS results. The informal investigation generally takes place at the departmental level between the analyst and the first-line supervisor. The investigation should be documented and follow a written procedure that directs the review of test procedures, calculations, equipment performance, and the data record. Typically, an informal investigation for an assay or impurity failure includes the following:

- A review of all lab data
- A review of calculations to ensure they are accurate
- A review of test procedure utilized
- A review of glassware utilized for the test; this includes pipettes
- A review of equipment, columns, charts and previous analyses of samples of the same product
- Reanalysis of dilutions
- A complete investigation and evaluation of initial results prior to a retest

If this investigation identifies a problem that resulted in the OOS result, the initial OOS result is invalidated and the test is repeated. In the context of this document, this repeat test is not considered a retest. It is documented in the laboratory deviation/nonconformance system (see below).

If the initial result is not invalidated, a laboratory manager or director usually becomes involved and a *formal investigation* is initiated. At this point, retests may be performed to either invalidate or confirm the initial results(s) utilizing preestablished procedures. Typically, two analysts perform two weighings and duplicate injections, thus generating eight results. It is then up to the responsible manager to review the laboratory and manufacturing data and history and either invalidate or confirm the initial results. If the analysis is of tablets, the investigation includes a retest of the same mix or grind. In some cases, tablets are not ground, but are placed in a flask, weighed, and diluted to volume where initial extraction is performed. For retest purposes, this equates to the same mix or grind. In the event that the initial tablet extraction flask is unstable, a new sample can be prepared from the initial composite sample of the batch. If the analysis is of a solution, a retest may be conducted of the same sample container. It should be pointed out that the retest procedure should be specific and include the number of sample weighings and number of duplicate tests.

When reviewing lab data associated with an OOS/OOT incident, there are some basic procedures that should be followed. This includes weighing every sample tested, even when content uniformity and dissolution testing has been conducted. Additionally, it would be difficult to conduct a meaningful investigation without duplicate injections of each sample preparation to rule out instrument/test system error. There should be acceptance criteria established to demonstrate the reproducibility of test results; for example, assay values should be within 2–3%.

At some point after the informal investigation is complete and the decision is made to embark on the formal retest mode, other issues have to be addressed and documented. These include the following:

- A review of production and control records.
- A review of the product history.
- A documentation of the investigation and the logic and recommendation of disposition of the batch by the manager/director.
- If the conclusion is that the value is an aberrant value, it should be reviewed by the laboratory director.

Retests for content uniformity and dissolution features are the second (S2) and third (S3) stages (for dissolution) testing, as provided for in the USP general test procedures. FDA policy has been that retesting is the second (or third) stage unless orginal results can be invalidated as described above. The second and third stage of testing follow the same formal investigation procedures. All S3 testing should include a review of production and control records and historical data.

2.3.3. Laboratory Deviations/Nonconformances

Another system associated with OOS/OOT investigations is that of laboratory deviations or nonconformances. They can be defined as laboratory situations, including deviations agreed upon prior to any analysis being performed, in which a departure has occurred from an SOP, method, specification, or protocol. These incidents include, but may not be limited to, equipment failures,

Table 2.1. List of Laboratory Deviations

System suitability failure	Failure to use bracketed standard when required
Reference standard not run	Wrong sample standard dilution
Degraded column	Changed injection volume
Incorrect wavelength	Described resolution not attained
Insufficient mobile phase	Described RSD (often <2.0%) not attained
Significant pressure drop	Described coefficient of correlation not attained
Pressure increase (shut down)	Change in programming time for wavelength change during HPLC run
Low theoretical plates	pH of buffers/Solutions used for HPLC analyses changed
Injector malfunction	Degradation compound not available or expired
Detector variability	Incorrect filter used
Change in flow rate	Incorrect sample preparation
Change in peak shape	
Retention time shift	

equipment calibration failures, documentation errors, and procedural deviations. Examples of typical deviations are included in Table 2.1. Examples that would not be classified as deviations are in Table 2.2.

The identification and tracking of deviations is an important management tool to determine overall laboratory compliance, to identify training needs, to track trends and highlight method problems and to specifically evaluate analyst performance.

Because both the laboratory deviation and laboratory OOS/OOT investigation procedures are significant laboratory management tools, it is important for management to account for and to periodically discuss and review these investigations.

In some cases a company may merge the laboratory deviation system into the OOS/OOT system. While from a compliance perspective it is acceptable, it involves considerably more work, particularly when the laboratory deficiency is obvious and documented and any result (if one is generated) can be easily invalidated.

All laboratories must have written procedures that describe how errors and failures are handled, investigated, and resolved. It is expected that failure in-

Table 2.2. Occurrences That Are Not Deviations

Power failure
Printer error/jam
Calculation error
Sample inadvertently tested twice
Information omitted on printouts

vestigations be specific enough to provide for clear direction and responsibilities for decision making with appropriate sign-off or documentation by management for investigation of failures.

2.3.4. Test Methods/Procedures/Specifications

Methods used should be written in company format. Copies of the compendia, which may allow for interpretations, are not to be used in lieu of a company written test procedure.

Procedures are appropriate for the operation, assembly, cleaning, regeneration, and inspection of some equipment. All laboratory procedures must be readily available to laboratory personnel.

Specifications for components, raw materials, reagents/solutions, standards, processing intermediates/subassemblies, and final product must be readily available to laboratory personnel. These specifications must be controlled documents, meaning that they should be uniquely identified, they should be subject to document change control policy, and not more than one version of a specification should be in effect at any time.

Another issue with regard to method specificity is the utilization of thin-layer chromatography (TLC) methods in lieu of more specific liquid chromatography/gas chromatography (LC/GC) methods. There should be a program for the review and evaluation of all semiquantitative TLC methods and replacement, if appropriate, by more specific quantitative methods.

2.3.5. Calibration and Maintenance

One area that is frequently cited on FDA 483 observation reports is the failure to calibrate and maintain laboratory equipment. Laboratory equipment must be calibrated. There should be a system to indicate calibration due date and to provide traceability to the calibration records. Calibration policy/procedures, calibration standards, and acceptance criteria must be written in controlled documents.

Equipment-specific logbooks detailing the use, cleaning, regeneration, calibration, repair, and so on, are required for complex equipment used for multiple product evaluations or used by multiple technicians.

Each laboratory should have a preventative maintenance program with responsibilities. Because maintenance records, particularly from outside vendors, are utilized in laboratory audits, there should be a system for senior laboratory management to be informed of any unacceptable calibration/maintenance.

As discussed in the next section, it is not unusual for the auditor to identify instances of maintenance work on equipment, and then review raw data generated during the maintenance period to ensure that there is correlation. From an auditing perspective, there can also be some review of correlation with laboratory deviations.

2.4. LABORATORY COMPLIANCE

2.4.1. General Notices

The USP [11] contains considerable information that clarifies compliance requirements. The General Notices section discusses tolerances, as well as the basis for their establishment, including overages for analytical error, manufacturing variability, and deterioration. Furthermore, the preface points out that the in-house quality release tests specified by a manufacturer should not be confused with official compendial standards. Basically, the trend among manufacturers has been to have tighter release specifications for all tests except for the sterility and identity tests. Tests that should have tighter in-house release limits include assay, impurities, content uniformity, dissolution, endotoxins, particulate matter, microbial limits, and pH.

For example, if the dissolution specification is set too tight by regulatory authorities, then there will be an excessive number of nonconformances and retests. Some regulatory managers believe that there should be a percentage of batches having S2 (second stage) level testing. Unfortunately, they fail to recognize that dissolution testing cannot be repeated if the OOS result cannot be clearly and directly attributed to analytical error. Having a low percentage of tablets below Q (the means of the dissolution values) increases the probability of having a dissolution failure that may indicate a product adulteration.

The USP also defines information regarding added substances. It points out that added substances cannot interfere with tests and assays. Because official (compendial and filed) tests are regulatory tests, they should be able to be conducted by a regulatory body on a stand-alone basis. This precludes the use of a placebo to blank out interfering substances.

The USP also provides information with regard to parametric release (release based on processing results, without testing). It points out that it is not a requirement to perform every compendial test to assure compliance. It states the following:

> Data derived from manufacturing process validation studies and from in-process controls may provide greater assurance that a batch meets a particular monograph requirement than analytical data derived from an examination of finished units drawn from that batch. On the basis of such assurances, the analytical procedures in the monograph may be omitted by the manufacturer in judging compliance of the batch with the pharmacopeial standards [12].

Depending upon the wording in an application, one could interpret the above to mean that filed tests do not have to be performed as long as compliance with the particular specification is assured. For example, it is not unusual for a company to delete performance of filed identity tests when there are other means to assure product identity. However, it would be difficult to make a case for deleting the very basic assay and impurity tests. These tests serve as a check on the process and operator performance and could identify foreign contaminants.

2.4.2. Method Development

The FDA PAI Program initiated in 1990 increased focus on the development of processes and the documentation of this development. The 1994 *FDA Guide to Inspections of Oral Solid Dosage Forms* [13] discussed product development reports. The *Guide* points out that while there are no statutes or regulations that specifically require a development report, companies are required to produce scientific data that justify processes. Clearly, the FDA could support an analogous interpretation for methods and the need for scientific data to justify methods.

This same *FDA Guide* addresses the laboratory and points out that the transfer of laboratory methods and technology from the Research and Development Department to the Quality Control Department should be reviewed during inspections. Any review of method validation and method development could include a review of the development trials and experiments that result in a laboratory method.

From a compliance perspective, there are some very basic issues that should be addressed in the development of a method. Ideally, the drug substance and drug product should have the same stability-indicating methods for assay and impurity testing. For example, it is not unusual to have a foreign contaminant appear in an assay or purity test. One of the sources of the foreign contaminant could be a foreign contaminant in the drug substance. Having the same method helps to identify the contaminant source. If the test methods are different, the dosage form manufacturer should look at the drug substance with the dosage form method.

Another general issue would include partnering with formulators to ensure that there will be no excipient interference. As discussed previously, it is unacceptable to utilize a placebo blank for a regulatory method.

With regard to method development, good practices would include a procedure for developing methods and a protocol detailing acceptance criteria for the method. The procedure should also include provisions for the referencing of raw data in the method development report.

In order to demonstrate robustness of a method, some basic areas that should be addressed in method development and included in the method development report are as follows:

- Stability of reagents, mobile phases, standards, sample solutions
- Temperature variations
- Mobile-phase composition variations
- Chromatographic gradient profiles and mixing systems
- Flow rate (pressure) variations

- Multiple lots of columns
- Alternate columns
- Instrument parameters (wavelength, bandwidth)
- Sources of reagents
- Limits of detection with different instruments

2.4.3. Method Validation

The GMP regulations require that test methods must meet proper standards of accuracy and reliability. Users of analytical methods described in the USP and in the *National Formulary* (NF) are not required to validate accuracy and reliability of these methods, but merely verify their suitability under actual conditions of use. However, for demonstrating and testing for stability, other noncompendial methods may be used in addition to, or in lieu of, compendial methods. Such methods do require validation.

The USP provides direction for parameters needed for the validation of test methods. Parameters include accuracy, precision, specificity, limit of detection, limit of quantitation, linearity, range, and ruggedness. The USP also defines and provides direction for the application of these parameters to the specific type of test procedure being validated [14].

In the validation of methods, the USP points out that method specificity is a requirement. It is defined as the ability to measure the analyte in the presence of components. Therefore, methods must be able to separate impurities and degradents from the active drug substances; components of the formulation must similarly be separated.

A major aspect of method validation, which has generated some discussion, is the establishment of the actual limit of detection (LOD) and the limit of quantitation (LOQ). In the development and validation of an analytical test method for impurities or related compounds it is important to establish the lowest limit of detection and quantitation. In a typical audit or FDA inspection, the review of chromatograms will include the determination of total impurities by adding together the area under all of the peaks. From a compliance perspective, it is important to quantitate *any* impurity peak above the noise level. However, from a practical aspect for dosage forms, it is very difficult and laborious to quantitate any impurity peaks at a concentration of less than 0.03%. For the most part, the International Conference on Harmonization (ICH) and previous references point out the need to quantitate impurities at levels of 0.1% or greater.

In order to establish 0.1% as a threshold, LOQ values of less than 0.1% are needed. As technology advances, the true LOQ for many substances is becoming less than 0.1%. In some cases, FDA reviewing chemists have been looking for impurities of less than 0.1% (0.03%). Thus, it is becoming increasingly important for analysts to quantitate impurities down to levels of 0.03% in purity methods that they use. It then becomes the responsibility of the supervisor to review the results, evaluate their significance, and report them as appropriate.

As stated earlier, one of the major issues in the industry is the compliance of older products under current standards and utilizing current technology. With improvements in analytical equipment, such as detectors, columns, and so on, previously unidentified impurities that are close to the level of 0.1% are being detected. For dosage forms where the daily dose is 10 mg to 2 g, impurities present at around 0.2% or more should be identified. In the review of stability data, this can present problems when the level of the unidentified impurity exceeds 0.2%. Thus, it is important to identify unknowns at levels of less than 0.2%. Industry needs to be proactive in the quantification and even the identification of impurities by having a plan for corrective action in the event that total impurities found in product and/or drug substance approach limits. The obvious corrective actions include improving the manufacturing process and/or changing the limit.

When addressing method validation, the titration method is a method that is frequently overlooked. The USP 24 states in Section ⟨1225⟩ [14] that general assays and tests (titrametric methods) should also be validated to verify their accuracy when used for a new product and raw material. From a compliance perspective, the use of titration methods, for the assay of drug substances, should be discouraged because the method is nonspecific and not stability-indicating. Furthermore, some impurities may not be discernible or may not absorb when chromatographic methods are used to detect impurities and degradants; the use of a nonspecific assay method may be of value.

Additionally, methods and specifications may not be married. While the USP states that the identity, strength, quality, and purity are determined by tests and assays relating to the article (USP 24, p. 7) [11], it also states that stability testing (Section ⟨1191⟩) [15] for both the drug substance and drug product should be performed by validated stability-indicating test methods. For drug products USP 24 states [16] that monograph assays may be used for stability testing if they are stability-indicating (capable of differentiating intact drug molecule from degradants).

There have been examples in which drug substances have been tested by the FDA utilizing specific LC methods, and not utilizing a nonspecific official titration method. In a few cases the FDA has utilized the assay limits in the monograph purity rubric statement as the specification for the substance, regardless of the method specified in the monograph. Such situations have resulted in improvements in the drug substance manufacturing process, methods, and specifications.

Other aspects of method validation, which have presented compliance problems, include robustness and ruggedness. In order to determine the ruggedness of a method, different laboratories would need to be used. One of the compliance activities associated with the introduction of a new product is the transfer of the method from the developing laboratory to the operational laboratory. A successful transfer demonstrates the ruggedness of a method. It is not unusual for a method-developing laboratory to issue a method validation report without demonstrating ruggedness (which includes testing at another laboratory). From a compliance perspective, it could be concluded that the method has not been validated because it hasn't been transferred or performed by another laboratory. (Refer to Chapter 15 for further discussion on method transfer.)

With regard to robustness, it is a measure of the capacity of an analytical procedure to remain unaffected by small but deliberate variations in method parameters. Issues that pertain to robustness are addressed in the development of a method. (Refer to Chapter 12 for further discussion on method development.)

2.4.4. Method Transfer

Prior to performing any method transfer for drug substance or drug product, there should be a protocol in place. Some compliance requirements to be addressed in each transfer protocol are as follows:

- The transfer should be performed on two batches, with three sample preparations for each batch. Ideally, the batches are not marketed and contain some level of impurities.
- Mean assay values between laboratories should be within 2% for drug substances and 3% for drug product.
- Precision should comply with intralaboratory requirements (system suitability).
- For known impurity determinations, the absolute difference between results should be no greater than 0.2%. Precision (system suitability) should be within 10–15% for the known impurities.
- For unknown impurities, there should be qualitative comparability.
- For oral solid dosage forms, mean dissolution values should be within 6% and mean content uniformity values should be within 5%.
- Linearity is performed during method validation and is not a transfer requirement.
- The LOD/LOQ provided for the method should be verified by utilizing standards at these levels.
- For compendial methods, the laboratory should verify that the method can be performed and that there are no interfering substances.

2.4.5. Auditing the Laboratory

As discussed in the introduction to this chapter, the major reason for conducting an audit of a laboratory is to ensure that data and results are not fraudulent. Also, audits are conducted to ensure that GMPs and commitment to regulatory filings are being followed.

There are some very general techniques that are utilized when auditing the laboratory. Basic to any audit is the observation of analysts. All of the areas of laboratory compliance can be evaluated by actual observation.

Other techniques, including having the responsible manager open drawers in the areas of analysts not present, can also be employed to identify data not reported or samples not tested.

A good audit will also include the review of maintenance reports and subsequent comparison to data in analyst logbooks or data sheets. If problems with instrumentation are identified during analyses, there should be some recognition of this in the analyst logbook or data sheets.

Generally, auditors look for missing electronic files, out-of-sequence data, and cut chromatograms as possible indicators of fraudulent data.

It is important to identify laboratory personnel and the laboratory management chain early in the audit. This will enable the auditor to identify responsible personnel and personnel who are outstanding performers. Typically, the best analyst will perform testing on the more difficult and/or problematic tests.

A good laboratory audit will generally include a review of a representative number of laboratory deviations and OOS/OOT investigations.

2.4.6. Use of Outside Testing Laboratories

A laboratory is responsible for the quality of the outside testing facilities that it uses. Laboratory management is responsible for assuring the quality of the work performed by these service laboratories; the use of audits and sending spiked/known samples, and so on, may be appropriate. Additionally, there should be some type of written agreement regarding the responsibilities, reporting of data, and supply of copies of raw data.

2.5. CONCLUSION

A number of issues central to laboratory compliance have been discussed in this chapter. These include laboratory certification and management, records, method validation, retesting, training, computerized data acquisition systems, calibration, and auditing. Certainly, there are other issues that need to be addressed, and this discussion was not intended to be all-inclusive. Although the establishment and monitoring of systems designed to support laboratory compliance may seem a daunting task, the company with a true commitment to quality will find a way to incorporate all issues in an effective quality plan.

REFERENCES

1. *United States of America vs. Warner Lambert Company*, Civil No. 93-3525, August 16, 1993.
2. *United States of America vs. Biocraft Laboratories, Inc.*, Civil No. 94-3499, July 28, 1994.
3. H. L. Avallone, *Laboratory Certification*, August 24, 1994.

4. Title 21, U.S. Code *Federal Food, Drug and Cosmetic Act.*

5. Current Good Manufacturing Practice Regulations, *Federal Register*, September 29, 1978, pp. 45067.

6. Current Good Manufacturing Practice Regulations, *Federal Register*, September 29, 1978, p. 45086.

7. 21 CRF, *Electronic Signature Records Part II; Electronic Signatures*, Part II, August 20, 1997.

8. T. Layloff and W. Furman, *AOAC Int.*, **1994**, *77*, 1514.

9. *United States Pharmacopeia*, 24th edition, ⟨621⟩ U.S. Pharmacopeial Convention, Inc., Rockville, MD, 2000, p. 1923.

10. *United States of America vs. Barr Laboratories*, Civil No. 92-1744, February 4, 1993.

11. *United States Pharmacopeia*, 24th edition, 2000, p. 7.

12. *Current Good Manufacturing Practice Regulation*, Federal Register, September 29, 1978, p. 45086.

13. Food and Drug Administration, *Guide to Inspections of Oral Solid Dosage Forms*, January 1994.

14. *United States Pharmacopeia*, 24th edition, ⟨1225⟩ 2000.

15. *United States Pharmacopeia*, 24th edition, ⟨1191⟩ 2000, p. 2128.

16. *United States Pharmacopeia*, 24th edition, ⟨1151⟩ 2000, p. 2106.

3

THE USP, ICH, AND OTHER COMPENDIAL METHODS

Jennifer G. Feldman

3.1. INTRODUCTION

In order to ensure the quality and uniformity of pharmaceutical substances throughout established geographical regions, several compendia were created. These lengthy documents, established separately in the United States, Great Britain, Europe, and Japan, provide standards and specifications for all facets of pharmaceutical materials, including their testing, packaging, and storage. With globalization of the pharmaceutical industry a need arose to harmonize these separate standards, and the International Conference on Harmonisation (ICH) was formed. This committee has the task of harmonizing the requirements for pharmaceutical products among the participating members: The United States, Europe, and Japan. While the process has been a slow and arduous one, much progress has been made. This chapter will discuss the individual pharmacopeia, which are still in effect, and briefly summarize the content of the various guidelines.

3.2. USP/NF

3.2.1. Introduction

The *United States Pharmacopeia* and *National Formulary* (USP/NF) is the largest and most comprehensive of the national compendia, which also include

Analytical Chemistry in a GMP Environment. Edited by J. M. Miller and J. B. Crowther
ISBN 0-471-31431-5 © 2000 John Wiley & Sons, Inc.

the *European Pharmacopeia*, *British Pharmacopeia*, and *Japanese Pharmacopeia*. The purpose of these compendia is similar to that stated in the constitution of the United States Pharmacopeial Convention (USPC), the governing body for the USP/NF:

> To provide authoritative standards and specifications for materials and substances and their preparations that are used in the practice of the healing arts; ... establish titles, definitions, descriptions, and standards for identity, quality, strength, purity, packaging, and labeling, and also, where practicable, bioavailability, stability, procedures for proper handling and storage, and methods for their examination and formulas for their manufacture or preparation [1].

Historically, the USP came into being in 1820 as a result of the efforts of Dr. Lyman Spalding. Dr. Spalding presented a plan to the Medical Society of the County of New York in which delegates representing medical societies and schools from throughout the United States would meet in a national convention to compile a national pharmacopeia. The plan was approved and the first *US Pharmacopeia* was published on December 15, 1820. It contained 272 pages, listed 215 drugs, and was written in Latin and English [2]. The USP did much to promote uniformity in drugs and drug formulations.

The *National Formulary* (NF) arose with the formation of the American Pharmaceutical Association in 1852. At this time the only authoritative and generally recognized compilation of drug standards was the USP. Because it had been published to serve as a guide to the medical profession, its scope was limited to drugs selected by representatives of the medical profession as having the greatest therapeutic value. This selectivity continued until 1975 and was the primary driver in the establishment of the *National Formulary*, which sought to include a large number of drugs that also had substantial acceptance and significant therapeutic value. In July 1974 the NF was acquired by the USPC. Although now contained within a single volume, they continued as two separate official compendia. With the resolution by the USPC that public standards should be set for all drugs, the scope of the USP and NF changed drastically, and the USP was limited to coverage of drug substances and dosage forms, while the NF was limited to pharmaceutic ingredients. Although the goal was to include monographs for all pharmaceutic ingredients, precedence was given to those used in the formulation of drugs with approved New Drug Applications (NDAs) which were widely used and not topically administered. With each revision the number of monographs in the NF continues to grow, and recent emphasis has been placed on international harmonization.

It is important to note that when the first Pure Food and Drug Act was adopted by Congress in 1906, the USP and NF were designated as the official sources of standards of strength, quality, and purity in medicinal products recognized therein and sold in interstate commerce for medicinal use. Congress and the states have continued and expanded the recognition of the USP and NF as "official compendia" today [3]. References to the USP and NF occur in many statutes and regulating articles used in medical practice. The most significant of these is the recognition of the official compendia in the Federal Food, Drug, and Cosmetic Act enacted in the 1970s.

Today's USP/NF, substantially more comprehensive, continues to be the responsibility of the United States Pharmacopoeial Convention, whose membership includes representatives from colleges and medical schools in the United States, from state medical societies, from chemical, pharmaceutical, dental, and hospital societies, and from governmental organizations [4]. Proposed changes may be submitted by any individual and are routed for public review and comment in the Pharmacopeial Forum before becoming effective. Supplements with revisions, changes, and new entries are issued biannually.

3.2.2. Organization/Overview

3.2.2.1. Introduction. The USP/NF is organized into several sections. These include the following:

USP 24

- Introductory information on the people involved, the preamble and information on admissions to the USP
- General Notices and Requirements
- Official Monographs for USP 24
- General Chapters
- Reagents
- Tables
- Nutritional Supplements
  - Monographs
  - General Chapters

NF 19

- Introductory information on the people involved, the preamble, and information on admissions to the NF
- Tables
- General Notices and Requirements
- Official Monographs for NF 19
- General Chapters
- Reagents

The contents of these sections (omitting introductory information) will be reviewed in this chapter.

3.2.2.2. General Notices and Requirements. The General Notices and Requirements sections of the USP and NF provide summaries of the basic guidelines for the interpretation and application of the standards, tests, assays, and other specifications in the USP/NF. Any exceptions made to the general notices are noted in the individual monographs or general test chapter and take precedence. If no specific information is given to the contrary, the General Notices apply.

3.2.2.3. Official Monographs. The official monographs contained in the USP and NF set forth the standards with which drug substances, nutrients, pharmaceutical ingredients, drug products, nutritional supplements, or finished devices contained therein must comply. The monographs are separated into two sections, USP and NF, and are listed alphabetically in each section. A typical monograph specifies chemical structure, formula, nomenclature, USP reference standards (where applicable), potency, purity, packaging and storage, and testing requirements. Methods and required result ranges are given.

3.2.2.4. General Chapters. Each General Chapter is assigned a number that appears in brackets next to the chapter name. Chapters including general requirements for tests and assays are numbered from ⟨1⟩ to ⟨999⟩. Informational chapters are numbered from ⟨1000⟩ to ⟨1999⟩. Chapters pertaining to nutritional supplements are numbered ⟨2000⟩ [5]. General chapters from the USP are equally applicable to the items listed in the NF. General chapters are frequently referred to within official monographs.

3.2.2.5. Reagents. The reagent section of the USP contains the following:

1. General test methods that can be used to examine reagents for compliance of the reagent with the specifications of the individual reagent and are to be used unless otherwise specified in such specifications.
2. Specifications for individual reagents (Reagent Specifications).
3. Indicators and Indicator Test Papers required in the Pharmacopeial test and assays. The necessary solutions of indicators are listed among Test Solutions (TS).
4. Solutions, including buffer solutions, calorimetric solutions, indicator/test solutions, and volumetric solutions and their preparation.

3.2.2.6. Tables. The Tables section contains reference tables pertaining to the following:

1. Containers for dispensing capsules and tablets
2. The description and relative solubility of USP and NF articles
3. The approximate solubility's of USP and NF articles
4. Atomic weights

5. Alcoholometric table
6. Thermometric equivalents

3.2.3. USP/NF and The FDA

Reference to the *United States Pharmacopeia* and *National Formulary* in the Federal Food, Drug, and Cosmetic Act is a significant recognition of their role in establishing nationally binding standards for drugs and drug products. This act empowers the Food and Drug Administration (FDA) to enforce the law using certain defined aspects of the compendia. Most commonly recognized are the USP and NF standards for determining the identity, strength, quality, and purity of the articles and specifications for packaging and labeling.

There has been a close working relationship between the Pharmacopeial Convention and the FDA as the USP and NF have undergone revision in recent years. Both individual FDA scientists and FDA laboratories and centers have interacted closely with the USP/NF during the revision process. During the latest revision, formal liaison efforts were conducted primarily through the Compendia Operations Branch in the Center for Drug Evaluation and Research of the FDA. This branch serves as the official contact point with the Center and is responsible for obtaining and coordinating agency comments on proposals, which appear in the Pharmacopeial Forum. It also coordinates the Joint Compendia Monograph Evaluation and Development project. The formal liaison has allowed FDA to contribute many suggestions for improvement to the USP/NF and has resulted in a greater degree of consistency between FDA and compendia requirements. In addition, the FDA Division of Drug Analysis in St. Louis, Missouri has consistently participated in the revision process. This laboratory has performed extensive development and review of tests and assays. During the latest revision cycle, it served as the primary FDA participant in the evaluation of established and proposed new USP reference standards [6].

3.2.4. FDA Requirements for Regulatory Submissions/ Field Inspections

As a general statement, the FDA looks for compliance to compendia or regulatory requirements as a minimum acceptable quality level. The USP/NF provides much guidance on these minimum standards, but each manufacturer must rationally assess the requirements of his process and product.

The General Notices section of the USP/NF contains considerable information, which clarifies compliance requirements. Tolerances and the basis for their establishment are discussed, including overages for analytical error, manufacturing variability, and deterioration. Compendium standards are the minimum acceptable quality standards for compendial products. During inspection, FDA often looks to see that the manufacturer has established tighter in-house product release specifications for assay, impurities, content uniformity, dissolution, endotoxins, particulate matter, pH, and microbial limits than are required by the compendia. This is particularly true in the case of the dissolution test and limit. The tests specified in USP/NF monographs are regulatory tests by virtue of appearing in the compendia. For this reason, the manufacturer must be able to perform these tests without interference from product excipients. It is also true that the compendial test must be able to be performed in a stand-alone regulatory laboratory, thereby precluding the use of placebos to correct for interference.

CGMPs provide guidance to FDA inspectors and reviewers in the area of parametric release. It states the following [7]:

> Data derived from manufacturing process validation studies and from in-process testing controls may provide greater assurance that a batch meets a particular monograph requirement than analytical data derived from an examination of finished units drawn from that batch. On the basis of such assurances, the manufacturer in judging compliance of the batch with the *Pharmacopeia* standards may omit the analytical procedures in the monograph.

This policy has long been accepted by FDA compliance managers.

The USP includes a discussion of method validation in the General Chapters section. FDA inspectors rely heavily upon this section in establishing minimum requirements for method validation. Of note is that the specificity is called out as a requirement for method validation. Correspondingly, it is specified that general procedures, such as titration methods, must also be validated to verify their accuracy when used for a new product or raw material [8]. With the introduction of the ICH guidelines, these also form the basis for method validation requirements. These guidelines will be discussed in greater detail later in this chapter.

3.2.5. Analysis of Excipients/Raw Materials/Drug Substance/ Drug Product

Excipients and raw materials used in the manufacture of pharmaceutical products fall within the scope of pharmaceutic ingredients, the monographs for which appear in the NF. For the purpose of the NF, an excipient is defined as "any component, other than the active substance(s), intentionally added to the formulation of a dosage form" [9]. It is not defined as being inert.

The monographs for drug substances and dosage forms fall within the scope of the USP. The monographs in both the NF and USP are similarly formatted, although the increased complexity of the items covered in the USP tends to make these monographs somewhat longer and more complicated.

3.2.6. "Meets USP" Labeling

The composition of USP/NF articles is defined to characterize the item as closely as possible by providing tests and stating limits, either in the monograph definition explicitly or by reference to required labeling. Many excipients may also have uses outside the pharmaceutical area and may, therefore, have specifications that differ from the usual pharmaceutical tests and standards. Many companies may manufacture or test a given drug substance or drug product. Only items tested by the methods specified in the current NF and meeting the specifications contained therein may be labeled "Meets USP."

3.2.7. Methodology

The methods included in the USP and NF have been previously submitted and circulated for public review before inclusion in the compendia. These methods remain official until a revision or alternate method is submitted, reviewed, and selected for admission into the formulary. Methods and revisions may be submitted by anyone. However, they are typically submitted by members of a pharmaceutical company that makes or uses the subject article extensively. Because this is a time- and resource-intensive activity, often monographs remain unchanged for many years. As analytical technology advances at its current rapid pace, an individual laboratory may wish to use a more advanced or automated method than that included in the USP/NF. This may be done without loss of the "Meets USP" labeling if a carefully controlled comparison study of the two methods is performed. That is, a sample must be tested by both methods, and comparable results must be obtained. This duplicated testing should be performed on at least three individual samples and must be carefully documented. It is important to be aware of the regulatory status of the method (i.e., has a specific method been filed in the original NDA?). If so, notification of the FDA authority may be necessary. In any case, the documentation of the method comparison must be available for FDA field inspectors if requested. All USP/NF specifications must continue to be met.

3.2.8. Accept/Reject Criteria

The accept/reject criteria for each article are provided in the individual monograph for that item. The monographs typically contain an introductory section in larger type specifying the required potency. Description of the packaging, storage, labeling, and USP reference standard follows. After this, several tests and the methods by which they are to be performed are included. The methodology may be provided within the monograph or may reference a USP general chapter. It is important to be certain that the article you are testing corresponds to the title of the monograph. This is particularly true when dealing with polymers, where the chain length is specified and of drugs that may have several dosage forms. The article being tested must conform to all monograph requirements to be acceptable. The content uniformity and dissolution specifications for drug products are provided within the general chapter, rather than the individual monograph. For dissolution, the time and Q value are provided in the monograph. Failure to meet even one specification prevents the item from being labeled USP or NF and results in rejection. Specifications within the monographs typically take the form of "not more than" the specified amount. This allows the test to pass if the value obtained matches the limit. A test result below this value is also considered passing. For some tests (e.g., heavy metals) a single value is given.

3.2.9. Validation

The methods provided in the official monographs have been validated by the laboratory submitting the monograph. This validation has been performed with material produced or used by this laboratory and on equipment contained in the laboratory. It is important for all compendial methods that each individual laboratory perform a scaled-back validation of the method, or, as some call it, a verification of the method. Chapter 15 of this text details the requirements for method validation. This chapter will not go into detail regarding each aspect of validation.

3.3. *EUROPEAN PHARMACOPEIA, BRITISH PHARMACOPEIA,* AND *JAPANESE PHARMACOPEIA*

The *European Pharmacopeia* (EP), *British Pharmacopeia* (BP), and *Japanese Pharmacopeia* (JP) are the three largest and most influential official compendia outside of the United States. The EP is the one of the world's youngest pharmacopeias, being on the order of 30 years old. The BP has been in existence since 1864, when the pharmacopeias for London, Edinburgh, and Dublin merged.

3.3.1. EP, Third Edition

The parties signing the convention on the elaboration of a *European Pharmacopeia* have agreed to "undertake the necessary measures to ensure that the monographs of the *European Pharmacopeia* shall become the official standards applicable within their respective territories" [10]. The material contained therein serves a parallel purpose for the signing countries as the USP/NF does for the United States. Again, all requirements must be met to have acceptable material.

The EP is organized as follows:

1. Introductory material
2. General notices
3. Explanation on use of the text
4. Analytical methods—general comments on
5. Apparatus
6. Physical and physicochemical methods

7. Identification
8. Limit tests
9. Assays
10. Biological tests
11. Biological assays
12. Methods in pharmacognosy
13. Pharmaceutical technical procedures
14. Containers
15. Reagents/solutions
16. Vaccines
17. Statistical analysis
18. Monographs
19. Dosage forms

The monographs follow a format virtually identical to that of the USP/NF. However, individual monograph requirements may vary slightly, and analysts performing testing for internationally distributed product must be cognizant of the differences. Frequently, multiple testing must be performed unless the original regulatory filings for each country specify that only one pharmacopeia is to be used. It should also be noted that where a monograph for a specific product may not be included in the compendia, any requirements put forth in general chapters pertaining to the dosage form (e.g., content uniformity for tablets) must be met.

3.3.2. BP

The *British Pharmacopeia* is unlike the USP/NF and EP in that it is stated to contain "a publicly available statement concerning the quality of a product that is expected to be demonstrable at any time during its accepted shelf-life; it does not provide a collection of minimum standards with which a manufacturer must comply before release of a product" [11]. The BP is a legally enforceable document throughout the United Kingdom. A marketed product in the United Kingdom may at any time during its shelf life be challenged independently by the methods of the BP. Compliance with the compendial limits is required and if any dispute as to product quality should arise, the methods of the pharmacopeia also are authoritative. Needless to say, it behooves a manufacturer wishing to sell a product in United Kingdom to ensure that the BP specifications will be met when the product is tested using BP methods throughout its lifetime. Again, all specifications must be met to yield an acceptable product.

As the European Union harmonizes, the EP continues to grow in significance for the United Kingdom. The BP reproduces edited versions of almost all of the monographs currently contained in the EP. It should be noted that they are not reproduced in their entirety and that the original EP should be consulted for testing. Additionally, the two compendia differ slightly in focus. The EP concentrates effort on the preparation of monographs for medicinal and pharmaceutical substances combined with general requirements for dosage forms. The BP, on the other hand, focuses on monographs for specific dosage forms.

The BP is organized as follows:

Volume I

1. Introductory material
2. General notices
3. Monographs on medicinal and pharmaceutical substances
4. Infrared reference spectra for many of the included materials

Volume II

1. Monographs on medicinal and pharmaceutical formulated preparations
2. Blood products
3. Immunological products
4. Radiopharmaceutical preparations
5. Surgical materials

Reagents and solutions, test methods, biological tests, and containers are covered in appendices to the text.

The monographs follow a format virtually identical to that of the USP/NF. However, individual monograph requirements may vary slightly, and analysts performing testing for internationally distributed product must be cognizant of the differences. Frequently, multiple testing must be performed to meet filing requirements in individual countries unless the original regulatory filings specify that only one pharmacopeia will be used.

3.3.3. JP, Thirteenth Edition

The *Japanese Pharmacopeia* (JP) was established in accordance with Article 41 of the Pharmaceutical Affairs Law (Law No. 145, 1960), and as such is legally enforceable as it applies to marketed drugs listed therein. As with the USP/NF, EP, and BP, the JP has the characteristics of an "official standard for the description and quality of drugs which are generally recognized to be medically significant from the viewpoint of medical treatment, that its role should be to specify the quality standards not only of drugs which are filed in it, but also of drugs in principle, as well as test, and that at the same time, it should help to ensure compatibility regarding quality of drugs with the rest of the world" [12]. Note that as with the EP, the general concepts, such as dissolution and content uniformity, are expected to extend beyond the drugs listed in the pharmacopeia.

The text itself covers the (a) drugs, which are most important from the point of view of health care, and (b) medical treatment based on demand, frequency of use, and clinical results.

The JP is organized as follows:

1. General introductory material
2. General notices
3. General rules for preparations
4. General tests, processes, and apparatus
5. Monographs on drugs in Part I and general notices
6. General rules for crude drugs (derived from natural sources and minerals)
7. General rules for preparations
8. General tests, processes, and apparatus
9. Monographs on drugs in Part II
10. General information

The monographs follow a format virtually identical to that of the USP/NF. However, individual monograph requirements may vary slightly, and analysts performing testing for internationally distributed product must be cognizant of the differences.

3.4. ICH GUIDELINE

3.4.1. Introduction/Role of the Guidelines

As the pharmaceutical industry becomes increasingly global in nature, the need to harmonize the requirements for the three major regulatory communities is apparent. The International Conference on Harmonisation (ICH) was created to address this need. The committee includes representatives from industry, academia, and the regulatory bodies of the United States, the European Community, and Japan [13]. Over the past several years, significant progress has been made in the harmonization of requirements across many areas important to the pharmaceutical analytical chemist. These areas include the following:

- Analytical method validation
  - Terminology
  - Methodology
- Stability
  - Primary document
  - New dosage forms
  - Photostability

Table 3.1. Review Stages of ICH Documentation

Step 1:	Draft document
Step 2:	Open for public review and comment
Step 3:	Comments incorporated and new draft prepared
Step 4:	Near final signed draft; recommendation by the three regulatory authorities
Step 5:	Published in the *Federal Register* and in international regulatory publications; final; generally, official day of publication unless otherwise noted (e.g., Stability Guideline official 1/1/98 by agreement)

- Impurities
  - Drug substance
  - Residual solvents
  - Drug product
- Specifications

Currently these harmonized documents are in various stages of review and approval. The status of these documents as of November 1997 are noted below. Table 3.1 contains the definitions of the various review stages.

3.4.2. Summary of the Guidelines [14]

3.4.2.1. ICH Guideline on Validation of Analytical Procedures: Definitions and Terminology; Step 5, 3/1/95. The ICH guideline on the validation of analytical methods is intended for registration applications within the United States, European Union, and Japan [15]. It applies to identification tests, control of impurities, and assay procedures. The Definitions and Terminology section is intended as a collection of terms, not specifics on how to perform each procedure. The types of analytical procedures, which must be validated, are as follows:

- Identification tests
- Quantitative test for level of impurities
- Quantitative tests for the active compound in drug substance or drug product or other selected components in the drug product (e.g., preservatives)

Other tests, such as dissolution, particle size, and so on, are not included in this guideline. Identification tests are intended to ensure the identity of an analyte in a sample—for example, by comparison to a property of the analyte with that of a reference compound (e.g., spectrum, chromatographic retention time, etc.). Impurity testing may be accomplished by either quantitative test methods or limit tests, and the validation requirements vary according to the type of test. Validation characteristics will also vary with the objective of the analytical procedure. Assay procedures measure the amount of analyte present in a given sample. In the ICH guideline, the assay represents a quantitative measurement of the major component(s) or other selected components of a drug product.

Required validation characteristics are listed by type of method in Table 3.2. Exceptions to these parameters are made on a case-by-case basis and should be documented. Robustness, while not included, should be evaluated at an appropriate time during analytical development and is often included during method transfer. Revalidation may be necessary after changes in drug substance synthesis, changes in drug product composition, and changes in the analytical procedure. The degree of revalidation required depends on the nature of the changes made. A glossary of terms as used in the ICH document is provided in Appendix II.

3.4.2.2. ICH Guideline on Validation of Analytical Procedures: Methodology; Step 5, 5/19/97.

This guidance addresses the question of how to consider the validation characteristics for each analytical procedure [16]. It allows the overall capabilities of several analytical procedures to be investigated in combination and provides indication of the data to be included in a regulatory application. The goal of method validation is to demonstrate that a method is suitable for its intended application. Validation approaches other than those contained in the ICH guideline may be used, provided that the analyst has chosen the validation procedure most suited for his/her products and can support the choice on a scientific basis. All relevant data and formulae used in calculations should be contained in the regulatory submissions. The ICH guidance presents validation requirements individually, but acknowledges that they may be considered simultaneously. Well-characterized reference materials suitable for their intended use should be used. Only USP reference standards are qualified for monograph purposes. If other standards are used, they may need additional qualification. Biological and biotechnology products may be approached differently.

The requirements for validating the various characteristics of a method are given in Chapter 15, Section 15.7.3. The typical validation protocol will include the following:

- Method robustness
- Linearity over 3 preparations, conducted at five analyte levels for both the drug substance and the finished dosage form for accuracy
- Six replicates at 100% drug substance for repeatability
- Measurement of the quantitation level at five levels plus six replicates for precision
- Detection level with six blank runs for baseline noise

System suitability testing is an integral part of many procedures. It is based on the concept that equipment, electronics, analytical operations, and samples

Table 3.2. Validation Characteristics Required as a Function of Method Type

Identification Tests	Impurity Tests	Evaluate? Quantitative	Evaluate? Limit	Assay, Dissolution (Measurement Only), Content/Potency
Specificity	Accuracy	Yes	No	Accuracy
	Precision			Precision
	Repeatability	Yes	No	Repeatability
	Intermediate precision[a]	Yes	No	Intermediate precision[d]
	Specificity[b]	Yes	Yes	Specificity[e]
	Detection limit[c]	No	Yes	
	Quantitation limit	Yes	Yes	
	Linearity	Yes	No	Linearity
	Range	Yes	No	Range

[a] Where reproducibitiy has been performed, intermediate precision is not needed.
[b] Lack of specificity for one analytical procedure can be compensated for by other(s) supporting procedures.
[c] May be needed in some cases.
[d] Where reproducibility has been performed, intermediate precision is not needed.
[e] Lack of specificity for one analytical procedure can be compensated for by other(s) supporting procedures.
Source: Reference 14.

to be analyzed constitute an integral system that can be evaluated as such. The specific parameters depend on the type of procedure. The pharmacopeia can be consulted for additional information.

3.4.2.3. Stability Testing of New Drug Substances and Products; Step 5, 9/22/94. The goal of the ICH stability guideline was to exemplify the core stability data package required for new drug substances and products in the United States, the European Union, and Japan such that data generated in any one the regions is mutually acceptable in the other two. In this spirit, specific requirements for sampling, testing of particular dosage forms or packaging, and so on, are not covered. The guideline applies to the information required for the registration applications of new molecular entities and drug products, but not to abbreviated or abridged applications, clinical trial applications, and so on. The test conditions were selected based on the climatic conditions in the three areas so that test data provides evidence on the variation in quality with time under the influence of a variety of representative environmental factors. These data in turn allow recommended storage conditions and shelf lives to be established.

Drug Substance. The primary stability studies for the drug substance show that it will remain within specification during the retest period. Long-term (12-month) and accelerated testing are performed on at least three batches. Batches can be manufactured at a minimum of pilot scale, but should use the same synthetic route and a method of manufacture that simulates the final process to be used at manufacturing scale. In addition, supporting stability on laboratory-scale batches may be submitted. The quality of the batches placed on stability should be representative of the quality of (a) material used in preclinical and clinical studies and (b) material to be made at a manufacturing scale. The first three batches made post approval should also be placed on long-term stability using the product registration protocol. Testing should cover physical, chemical, and microbiological properties susceptible to change during storage and likely to affect product quality, safety, and/or efficacy. Validated stability-indicating methods should be used. The number of replicates to be run depends on the results of validation studies, and limits should be derived from material used in preclinical and clinical studies, including both individual and total upper limits for impurities and degradation products. The length of the studies and the storage conditions should cover storage, shipment, and subsequent use, although use of the same conditions as for the drug product will facilitate comparative review and assessment. Other conditions should be included as scientifically justified. Temperature-sensitive drugs should be stored at the labeled long-term storage temperature, and accelerated testing should be conducted at 15°C above the designated long-term storage temperature with appropriate relative humidity conditions.

At the time of regulatory submission, a minimum of 12 months at 25°C $\pm$ 2°C/60% RH $\pm$ 5% and /60% RH $\pm$ 5 and 6 months at 40°C $\pm$ 2°C and /75% RH $\pm$ 5% is required. If significant changes are noted at the elevated temperature, additional testing at an intermediate condition, such as 30°C $\pm$ 2°C/60% RH $\pm$ 5%, should be conducted. The registration application should include a minimum of 6 months of data from a 12-month study at the intermediate condition. Significant change at 40°C and 75% RH is defined as failure to meet specification. Long-term-testing should be continued to cover all retest periods. Normally, testing under long-term conditions is performed every 3 months for the first year, every 6 months for the second year, and then annually. Containers employed in the long-term stability study should be the same or simulate actual packaging used for storage and distribution. As the application is pending review, accumulated stability data should be submitted. Accelerated or intermediate temperature data may be used to support shipping conditions and evaluate the effect of short-term excursions outside the label storage conditions.

Because long-term stability is used to establish appropriate retest periods, it should be noted that the degree of interbatch variability affects the confidence that a future batch will remain within specifications for the entire retest period. As a rule, determination of the time at which the 95% one-sided confidence limit for the mean degradation curve intersects the acceptable lower specification limit is acceptable, combining data into one overall estimate to account for variability. Before combining the data, apply appropriate statistical tests (e.g., p test) to be sure it is allowable. If inappropriate to combine data, the retest period may depend on the minimum time a batch is actually measured to remain in specification. The nature of the degradation relationship determines the need for transformation of the data for linear regression analysis. This relationship can generally be fitted to a linear, quadratic, or cubic function on an arithmetic or logarithmic scale. Use statistical methods to test the goodness of fit of the data on all batches and combined batches, where appropriate, to the assumed degradation curve. If the data show little degradation or variability, a retest period can be justified without statistical analysis and a limited extrapolation of real-time data may be undertaken when supported by the accelerated data. Any extrapolation must be justified, because it assumes that the same mechanism of degradation will continue beyond the observed data; this evaluation should include assay, degradation products, and any other appropriate attributes.

The storage temperature range should be based on the stability data and used in accordance with the national or regional requirements. Specific labeling requirements should be stated, particularly for drugs that cannot freeze; terms such as ambient and room temperature are to be avoided.

Drug Product. The stability program for the drug product should be based on knowledge of the drug substance and experience from experimental and clinical formulations. Unless specifically noted in this section, the requirements for drug substances also apply to drug products. Accelerated and long-term data should be provided on three batches of the same formulation and dosage form in the containers and closure proposed for marketing. This revision of the ICH guideline specifies only solid oral dosage forms, and it states that two of the three batches placed on stability should be at least pilot scale, but that a third may be smaller—for example, 25,000–50,000 tablets or capsules. As with drug substance, at least 12 months of long-term stability data should be submitted at the time of regulatory filing. When possible, manufacture stability batches of the finished product using identifiably different batches of drug substance. Data on laboratory-scale batches of drug product is not acceptable as primary stability data, but may be submitted as supportive information, as may data on associated formulations or packaging. If required, preservative efficacy testing and assays on stored samples should be performed to determine content and efficacy of antimicrobial preservatives. Differences between release and shelf-life specifications for antimicrobial preservatives should be supported by preservative efficacy testing. Limits for tests such as dissolution and particle size require reference to results of bioavailability and clinical batches.

Storage at high relative humidity is important for solid oral dosage forms, but is not necessary for products such as solutions, suspension, and so on, stored in containers designed to provide a permanent water barrier. Low relative humidity (10–20%) is appropriate for products of high water content stored in semipermeable containers. Testing of unprotected drug product can be a useful part of stress testing and package evaluation, as can studies in related packaging materials. If a product needs to be reconstituted or diluted, stability in the final form should also be addressed.

An annex to the ICH stability guideline (step 5, 5/9/97) applies to the owner of the original regulatory application regarding stability of new dosage forms after the original submission. This new dosage form is defined as a drug product, but is a different pharmaceutical product type even though it contains the same active substance as included in the existing approved drug product (e.g., different route of administration, new specific functionality/delivery system, or different dosage form of same route of administration). The guidance in the parent guideline is to be followed, but a reduced stability database at time of submission may be acceptable in certain justified cases.

Photostability Testing of New Drug Substances and Products; Step 5, 5/16/97. Light testing should be an integral part of stress testing. This guideline is an annex to the stability testing guideline. The intrinsic photostability of new drug substances and drug products should be evaluated (normally in one batch). The study may need to be repeated if changes are made to the formulation or packaging, depending on the photostability characteristics and type of change. The guideline primarily addresses photostability information for registration applications and does not cover photostability under conditions of use. If appropriate, studies should contain information on the drug substance and exposed drug product outside of the immediate package. If necessary, it should also include tests on drug product in immediate package and in the marketing package. When changes are made to the drug product, the extent of testing is determined by the acceptability of the change as justified by the applicant. The labeling requirements for photolabile drug substances and drug products are established by national and/or regional requirements.

For drug substances, both forced degradation testing and confirmatory testing are normally performed. Forced testing is done to evaluate overall photosensitivity and for the elucidation of degradation pathways. The substance may be tested "as is" or in simple solutions or suspension in order to validate analytical procedures. The samples should be held in chemically inert and transparent containers with solid substances spread to a thickness of not more than 3 mm. The studies should be ended if extensive degradation is observed. For photostable compounds the study should be terminated after an appropriate exposure level. The experimental design and the exposure levels should be designed based on the chemical properties of the drug substance. If degradation products are found during forced testing, they need not be examined further if they are not also found during confirmatory testing. Confirmatory studies are done to provide information necessary for handling, packaging, and labeling. Normally, one batch is tested during development for forced degradation, and confirmation is done on a single other batch if the drug is clearly photostable or photolabile. If results from the confirmatory testing are equivocal, test up to two additional batches. Changes in the physical as well as chemical properties of the drug substance should be considered, and samples should be cooled or placed in sealed containers to minimize sublimation, evaporation, or melting.

Drug product is normally tested sequentially as exposed product, in the immediate package and in the marketing pack. Testing should continue along this sequence until it is determined that the product is adequately protected from light. Exposed drug product should be placed to maximize surface exposed, and testing in the immediate package should be done with the package placed horizontally or transversely to the light source. The number of batches to be tested is as with drug substance. It may be appropriate to test products that will be exposed to light for long periods of time during use—for example, infusion liquids, under in-use conditions. Again, physical as well as chemical properties should be considered after exposure. Dark controls should always be tested concomitantly.

3.4.2.4. Guideline on Impurities in New Drug Substances; Step 5, 1/4/96.
The objective of this guidance is to address the information that is required for regulatory submission on the content and qualification of impurities in synthesized drug substances not previously registered in a region or member state. It does not cover products used in clinical research stages of development, biological or biotechnological products, radiopharmaceuticals, fermentation and semisynthetic products derived therefrom, herbal products, crude products of animal or plant origin, or extraneous contaminants. The chemistry aspects include classification and identification of impurities, report generation, setting

of specifications, and a brief discussion of analytical procedures. Safety aspects include specific guidance for qualifying impurities not present in batches used in safety and clinical studies and/or impurity levels substantially higher than in those batches. Thresholds are defined.

Impurities in the new drug substance may be classified as indicated below. Analytical procedures used include documented evidence of validation and suitability for the detection and quantitation of the impurities. Differences between the methods used during development and those proposed for the commercial product should be discussed. Organic impurity levels can be measured by a number of techniques, including those that compare an analytical response with the drug substance itself. When the response factors of the impurities are not close to that of the drug substance, a correction factor may be used unless the impurities are being overestimated. Any analytical assumptions made should be discussed.

In reporting impurity contents, provide results for all batches used for clinical, safety, and stability studies, as well as for batches representing the proposed commercial process. Report the content of individual and total impurities with the analytical procedures indicated. Results should be provided in tabular format. For all batches included, provide the information listed below.

Required Batch Information

- Batch identity, strength, and size
- Date of manufacture
- Site of manufacture
- Manufacturing process
- Impurity content: individual and total
- Use of batch
- Link with each safety or clinical study in which batch was used
- Reference to analytical procedure(s) used

Identify impurities by code number or an appropriate descriptor such as retention time. Levels of impurities present, but below the validated limit of quantitation, need not be reported. Chromatograms or the equivalent from representative batches showing the location of the observed impurities should be provided. These chromatograms showing separation and detectability of impurities along with any other impurity tests routinely performed can serve as the representative impurity profiles.

To set impurity limits, which are required, use of stability studies, chemical development data, and routine batch analyses is appropriate. Impurities included in the specifications are referred to as specified impurities. Specified impurities may be identified or unidentified and should be individually listed in the specifications. Present a rationale for the inclusion or exclusion of impurities. Discuss, in addition, impurity profiles observed in the safety and clinical development batches as well as that of the proposed commercial process. Include specific identified impurities and recurring unidentified impurities estimated to be at or above 0.1%. For toxic or extremely potent impurities the detection and quantification limits of these substances must be at levels commensurate with the levels at which they must be controlled. Unidentified impurities must be referred to in the specifications by a descriptor, such as a letter designation or the retention time. The procedures used and the assumptions made in establishing acceptable levels for these impurities should be clearly described. In general, a specification limit of not more than 0.1% for any unspecified impurity should be included. Limits are appropriately set no higher than the level that can be justified by safety data and no lower than the level achievable by the manufacturing process and analytical capability. Where there is no safety concern, limits should be based on (a) data generated on actual batches of the drug substance, allowing sufficient latitude to deal with normal process and analytical variation, and (b) stability characteristics. Drug substance specifications should include organic and inorganic impurities and residual solvents. Mass balance in the test sample may be derived from the summation of the assay value and the impurity levels and need not add to exactly 100% due to analytical error. For nonspecific assay, values such as titration, summation of impurity, and assay values may be misleading.

The qualification of impurities in the drug substance is accomplished by the same procedures described in the next section on the drug product, where it is highly critical, and will not be repeated here. See Table 3.3 for appropriate qualification levels.

3.4.2.5. Guideline on Residual Solvents; Step 4, 7/17/97. The objective of this guidance is to address the organic volatile chemicals used or produced in the synthesis of drug substances or excipients or in the preparation of drug products. It does not cover solvents used as excipients or solvates. In general, residual levels should never be higher than can be supported by safety data. Solvents known to cause unacceptable toxicity (class 1) should be avoided; use of solvents associated with less severe toxicity (class 2) should be limited; and minimally toxic solvents (class 3) should be used whenever practical. Test for residual solvents when production or purification processes are known to result in the presence of such solvents. For drug products in lieu of testing directly for residual solvents, the cumulative method may be used. In this method, residual solvent levels are calculated from the known levels in the ingredients used to produce the drug product. Testing of the drug product is not required if the calculated level is below the level recommended by the guideline. If the calculated level is above the recommended level, test the drug product to see if levels have been reduced during processing. Always test the drug product routinely if class 1 or 2 solvents have been used in the manufacture of purification of the drug substance, excipients, or the drug product. The testing of drug substances,

Table 3.3. Threshold Values for New Drug Substances and Drug Products

Maximum Daily Dose[a]	Threshold[b]
Thresholds for Qualification *of Impurities in New Drug Substances*	
≤2 g/day	0.1% or 1 mg TDI,[c] whichever is lower
>2 g/day	0.05%
Thresholds for Reporting *of Degradation Products in New Drug Products*	
≤1 g	0.1%
>1 g	0.05%
Thresholds for Identification *of Degradation Products in New Drug Products*	
<1 mg	1.0% or 5 μg TDI, whichever is lower
1 mg to 10 mg	0.5% or 20 μg TDI, whichever is lower
10 mg to 2 g	0.2% or 2 mg TDI, whichever is lower
>2 g	0.1%
Thresholds for Qualification *of Degradation Products in New Drug Products*	
<10 mg	1.0% or 50 μg TDI, whichever is lower
10 mg to 100 mg	0.5% or 200 μg TDI, whichever is lower
100 mg to 2 g	0.2% or 2 mg TDI, whichever is lower
>2 g	0.1%

[a] The amount of drug substance administered per day.

[b] The threshold is based on percent of the drug substance.

[c] Total daily intake.

products, or excipients for residual solvents does not apply to material used during clinical research. The guideline does apply to all dosage forms and routes of administration. Solvent classification schemes and methods for describing acceptable solvent limits are given in Table 3.4. The analytical methods of choice must be validated procedures such as gas chromatography with or without headspace analysis. For class 3 solvents, nonspecific procedures such as loss on drying may be used.

3.4.2.6. Guideline on Impurities in New Drug Products; Step 5, 5/19/97.

The objective of this guidance is to address the information that is required for regulatory submission on the content and quantitation of impurities in new drug products composed of chemically synthesized new drug substances. The guidance applies only to degradation products or reaction products of the active ingredient with an excipients and/or container closure system. It does not cover impurities in excipients, clinical research stages of development, bio-

Table 3.4. Classification of Residual Solvents

Classification of Solvents by Risk Assessment:

Solvent Classification	Description of Attributes
Class 1	To be avoided—known human carcinogens, strongly suspected human carcinogens, environmental hazards.
Class 2	Solvents to be limited—nongenotoxic animal carcinogens or possible causative agents of other irreversible toxicity such as neurotoxicity or teratogenicity, solvents suspected of other significant but reversible effects.
Class 3	Solvents with toxic potential—solvents with low toxic potential to humans, no health-based exposure limit is needed. Have PDEs[a] of 50 mg or more per day.

Examples of Solvents of Varying Classes:

Solvent Classification	Solvent	Limit (ppm)[b]
Class 1 (complete list)	Benzene	2
	Carbon tetrachloride	4
	1,2-Dichloroethane	5
	1,1-Dichloroethane	8
	1,1,1-Trichloroethane	1500
Class 2 (partial list)	Acetonitrile	410 (PDE 4.1 mg/day)
	Chloroform	60 (PDE 0.6 mg/day)
	Formamide	220 (PDE 2.2 mg/day)
	Hexane	290 (PDE 2.9 mg/day)
	Methanol	3000 (PDE 30.0 mg/day)
	Toluene	890 (PDE 8.9 mg/day)
	Dichloromethane	600 (PDE 6.0 mg/day)
Class 3 (partial list)	Acetic acid	No limits
	Heptane anisole	
	Pentane isopropyl acetate	
	2-Butanol	
	1-Propanol	
	Butyl acetate	
	Cumene	
	Methyl acetate	
	Ethyl acetate	
	Formic acid	
	Methylethyl ketone	
	Tetrahydrofuran	

[a] PDE = permitted daily exposure or pharmaceutically acceptable intake of residue solvent.

[b] *Options for describing limits for Class 2 solvents:*

Option 1: Limits from Table 2 in guideline can be used. Limits are calculated assuming 10 g product mass daily: Concentration (ppm) = (1000 × PDE)/(dose), where the PDE is in mg/day and dose is in g/day.

Option 2: PDE used with known daily dose and above equation. Can be applied by adding amounts of residual solvents present in each of the components of the drug product.

Note: There is no adequate toxicological data for:

* 1,1-Diethoxypropane
* 1,10-Dimethoxymethane
* 2,2-Dimethoxypropane
* Isooctane
* Isopropyl ether
* Methylisopropyl ketone
* Methyltetrahydrofuran
* Petroleum ether
* Trichloroacetic acid
* Trifluoroacetic acid

It is the manufacturer's responsibility to justify levels.

logical or biotechnological products, radiopharmaceuticals, fermentation and semisynthetic products derived therefrom, herbal products, crude products of animal or plant origin, extraneous contaminants, polymorphs, or enantiomeric impurities. In general, impurities present in new drug substances do not need to be monitored in the drug products if they are not also degradation products.

With regard to the analytical procedures, they should be validated and suitable for the detection of degradation products, including demonstration that drug substance impurities do not interfere or are separated from specified and unspecified degradation products. The means of measuring the degradation products must also be described (e.g., against reference standards or area %). Any differences in the analytical procedures used during development and those intended for the commercial product should be discussed.

In the reporting and control of impurities, summarize the degradation products observed during stability studies, discuss potential degradation pathways and impurities arising from interaction with excipients and/or container closure systems, and summarize laboratory studies conducted to detect degradation products. Compare impurity profiles of batches representative of the proposed commercial process and development batches. Provide rationale for exclusion of impurities that are not degradation products. Values such as 0.09% should not be rounded to 0.1% if the analytical procedure is capable of that level of accuracy. Degradation products observed above thresholds (see Table 3.4) during stability studies should be identified, and unsuccessful attempts to identify them should be discussed. If any degradation products are suspected to be unusually potent, to be toxic, or to produce pharmacological effects at levels lower than thresholds, they should also be identified.

Results should be provided in tabular format for all relevant batches used for clinical, safety, and stability testing, as well as for batches representative of the commercial process. For all batches included, provide the information listed below.

Required Batch Information

- Batch identity, strength, and size
- Date of manufacture
- Site of manufacture
- Manufacturing process, where applicable
- Immediate container/closure
- Degradation product content: individual and total
- Use of batch
- Reference to analytical procedure(s) used
- Batch number of the drug substance used
- Storage conditions

All degradation products greater than the reporting threshold must be reported. The reporting threshold is always lower than the identification threshold, and the analytical method should be capable of quantification of results at the reporting threshold. Chromatograms or the equivalent from representative batches showing the location of the observed degradation products and impurities from the drug substance would be provided. Discuss the origin of the impurities attributed to the drug substance.

With regard to the specification limits for the impurities, include limits for degradation products expected to occur under recommended storage conditions. These limits should be established based on stability studies, knowledge of degradation pathways, product development studies, and laboratory studies. Sufficient latitude should be taken to allow for normal manufacturing, analytical, and stability profile variation. Provide rationale for the inclusion or exclusion of impurities.

To qualify impurities, provide a rational for the limits based on safety considerations. Degradation products adequately included in safety or clinical studies are considered qualified. Include information on the actual content of degradation products at the time of use in clinical or safety studies. Degradation products that are significant metabolites present in human or animal studies do not need further qualification. It may be possible to justify a higher level of a degradation product than the level administered in safety studies. This justification should consider the following:

- The amount of degradation product administered in previous safety and clinical studies and found to be safe
- The percentage change in the degradation product
- Other safety factors as relevant

If data are not available to qualify the proposed level, additional studies may be needed. In unusual circumstances, technical factors such as manufacturing capability, low drug to excipient ratios, or the use of excipients that are also crude products of animal or plant origin may be considered as part of the justification for alternate thresholds. Alternatively, literature data may be used to qualify an impurity. Changes in qualitative degradation profile above threshold values requires additional identification and/or qualification, including safety studies comparing the use of drug product with a representative level of degradation product to previously qualified material.

3.4.2.7. Guideline on Specifications: Test Procedures and Acceptance Criteria for New Drug Substances and New Drug Products: Chemical Substances; Step 2, 7/16/97.

The objective of this document is to provide guidance on the setting and justification of specifications and on the selection of test procedures. It covers new drug substances and products of synthetic origin, which have not been previously registered with any regulatory body. The guideline covers both release and shelf-life specifications as filed in the marketing application, as either universal specifications or as ones specific to a dosage form. Solid, liquid, and parenteral dosage forms are addressed.

The general concepts covered include the following:

- Periodic or skip testing
  - Specified tests are performed at release on preselected batches and/or at predetermined intervals.
- Release versus shelf-life acceptance criteria
  - Applies to drug products.
  - More restrictive criteria for release.
- In-process tests
  - Performed during manufacture.
  - Tests used for adjusting process parameters within an operating range are not included in the specifications.
  - Tests performed that are identical to final product tests (e.g., pH) may be used to satisfy specification requirements.
- Design and development considerations
  - Include/exclude tests based on development experience.
  - Exclude tests such as
    - Microbial tests when a product does not support microbial growth.
    - Extractable from product containers.
    - Particle size.
- Parametric release
  - Alternative for routine testing (e.g., sterility).
  - Batch release based on monitoring of specific parameters.
- Alternative methods
  - Methods used to measure an attribute are comparable or superior to the

compendial procedure (e.g., for stable product use of spectrophotometric methods versus chromatographic).
- Pharmacopial tests and acceptance criteria
  - Use where appropriate.
  - Methods and acceptance criteria must be acceptable to regulatory authorities in each region.
- Limited data available at filing
  - Propose revised specifications as experience is gained (e.g., narrowing of ranges).
- Evolving technologies
  - Use when scientifically justified.
- Impact of drug substance on drug product specifications
  - Synthetic impurities controlled in the drug substance need not be controlled in drug product.
  - Impurities known to be degradation products must be tested in drug product.

In the guidance the following concepts apply. Specifications are defined as lists of tests, references to analytical procedures, and appropriate acceptance criteria. The conformance to specifications occurs whether the product is fully, parametrically, or skip tested. This is also true if the product is tested in process in lieu of final testing. All types of testing may be given in the same document. However, when first introduced, each procedure should be fully discussed and the acceptance criteria should be included and justified. Justification data may be drawn from all relevant development, validation, batch analysis, primary, and accelerated stability studies as appropriate. Data from all manufacturing sites should be considered, allowing for a reasonable range in analytical and manufacturing variation. Graphical presentation of test results may facilitate the justification of limits. Where acceptance criteria are approved on limited data, they should be reviewed and modified where indicated. Identical product manufactured at multiple sites should meet the same specifications. Specific guidance for universal tests and criteria as well as dosage form specific tests and criteria are given in Appendix III.

3.5. CONCLUSION

In this time of increasingly complex pharmaceutical technology, the standards and specifications established by the four major pharmaceutical compendia and harmonized by the ICH guidelines assure uniform product quality throughout the world. This assurance is critical in facilitating the mutual recognition of marketing approvals by different international regulatory bodies. Once this mutual recognition is achieved, the pharmaceutical industry will be able to devote increased resources to the development of new products, rather than to redundant testing required by differing compendial requirements.

REFERENCES

1. United States Pharmacopeia Convention, *U.S. Pharmacopeia*, 24th revision/*National Formulary*, 19th edition, Rockville, MD, Constitution and Bylaws, 2000 p. xxvii.

2. United States Pharmacopeia Convention, *U.S. Pharmacopeia*, 24th revision/*National Formulary*, 19th edition, Rockville, MD, History of the USP, 2000, p. xxvi.

3. United States Pharmacopeia Convention, *U.S. Pharmacopeia*, 24th revision/*National Formulary*, 19th edition, Rockville, MD, History of the USP, 2000, p. xxviii.

4. United States Pharmacopeia Convention, *U.S. Pharmacopeia*, 24th revision/*National Formulary*, 19th edition, Rockville, MD, Constitution and Bylaws, 2000, p. xxvii.

5. United States Pharmacopeia Convention, *U.S. Pharmacopeia*, 24th revision/*National Formulary*, 19th edition, Rockville, MD, General Notices, 2000, p. 4.

6. United States Pharmacopeia Convention, *U.S. Pharmacopeia*, 24th revision/*National Formulary*, 19th edition, Rockville, MD, Preamble, 2000, p. 6.

7. Current Good Manufacturing Practice Regulation, *Federal Register*, September 29, 1978, p. 45086.

8. United States Pharmacopeia Convention, *U.S. Pharmacopeia*, 24th revision/*National Formulary*, 19th edition, Rockville, MD, ⟨1225⟩, 2000, p. 2152.

9. United States Pharmacopeia Convention, *U.S. Pharmacopeia*, 24th revision/*National Formulary*, 19th edition, Rockville, MD, Preface to NF, 2000, p. 2400.

10. *European Pharmacopeia*, 3rd edition, Counsel of Europe, Strasbourg; Introduction, 1996, p. 4.

11. *British Pharmacopeia*, British Pharmacopeia Commission, London, Introduction, 1998, p. 21.

12. *Japanese Pharmacopeia*, 13th edition, The Society of Japanese Pharmacopeia, Tokyo; Introduction, 1996, p. 1.

13. *Federal Register* **1995**, *60*(196), Notices, p. 53078.

14. Presentation by J. Boehlert, ICH Seminar for Johnson & Johnson, November 10, 1997.

15. *Federal Register* **1995**, *60*(40), Notices, p. 11260.

16. *Federal Register*, **1997**, *62*(96), Notices, p. 27464.

4

STATISTICS IN THE PHARMACEUTICAL ANALYSIS LABORATORY

Alvin J. Melveger

> When You Measure You Know.
> —Lord Kelvin

The essence of obtaining knowledge about our universe, environment, or analytical samples depends upon making reliable measurements. The measurements usually involve the generation of data expressed as numbers. The knowledge of a sample that we are examining follows from our ability to record and analyze the numbers that we generate. Unless these data are treated in a consistent and correct manner, the knowledge we are seeking will be flawed and perhaps incorrect.

Handling data, and making decisions about their "correctness," is usually accomplished with statistical methods. In the pharmaceutical industry, statistics are applied throughout—in the design of manufacturing processes, in sampling, in the evaluation and control of processes, in setting product specifications and shelf life, and in the analytical laboratory where methods are developed and evaluated and products are analyzed.

The use of statistical techniques is quite pervasive in the pharmaceutical industry. The *United States Pharmacopeia* (USP) has addressed the issue as

Analytical Chemistry in a GMP Environment. Edited by J. M. Miller and J. B. Crowther
ISBN 0-471-31431-5 © 2000 John Wiley & Sons, Inc.

related to biological assays as in the USP general chapter, "Design and Analysis of Biological Assays" [1, Section $\langle 111 \rangle$].

For statistical treatment of analytical method development and data derived from these methods, the USP has published a general information chapter, "Validation of Compendial Methods" [1, Section $\langle 1225 \rangle$]. Although this chapter outlines the statistical tests necessary, it does not present the methodology to be employed.

The USP is presently considering more formal use of statistical methods applied to interpretation of analytical data [2]. This is in response to judge Wolin's decision [3] in the United States vs. Barr pharmaceutical litigation. The judge clearly asked for compendial guidance for understanding analytical data and out of specification (OOS) results.

Statistical methods are useful in determining the best method of sampling incoming raw materials, analysis of data for accuracy and conformance to a specification, and comparing the measured values of product to established specifications [4].

When developing new analytical procedures, it is necessary to evaluate the accuracy, precision, linearty, specificity, limit of detection (LOD), and limit of quantitation (LOQ) of the method and to compare the method to alternative available methods [5].

Determining the shelf life of a pharmaceutical product enables the assignment of expiry dating. Included in the process is determining stability under accelerated conditions and applying statistical and chemical kinetic methods to extrapolate to actual shelf stability under ambient conditions.

This chapter will focus on the basic concepts regarding the generation and treatment of data obtained from analyzing samples in the pharmaceutical laboratory. The treatments described may not be completely rigorous but should be sufficient for most purposes. These will include a review of basic statistical approaches for handling and evaluating data, the variability of a testing method, and the confidence placed on the results. A list of general references is provided at the end of the chapter for more complete discussions of the subjects.

4.1. ERRORS ASSOCIATED WITH MAKING MEASUREMENTS

It should be recognized that variability is inherent in every physical process or measurement. Some factors may be carefully controlled, but even well-designed and controlled laboratory experiments will exhibit variability that may originate from slight fluctuations in environmental conditions such as temperature, pressure, humidity, cosmic radiation, and so on. Statistical methods are aimed at providing objective, quantitative, and reproducible ways of assessing the effects of variability.

There are two types of variability or error, systematic and random. The total error is the sum of the two, but systematic errors can be, and should be, found and eliminated.

4.1.1. Systematic Error

Errors that could be avoided but creep into measurements at times unknown to the analyst are called systematic errors or determinate errors. They may or may not be proportional to the concentration of analyte and may bias repeated or replicate measurements equally. These errors may be unpredictable or constant.

Examples of such errors and possible remedies are given in Table 4.1. In general, statistical methods cannot be used to analyze such errors since they are unpredictable. As a result, they may seriously affect assay results. The good news is that once recognized, these errors may be addressed and rectified, as suggested in Table 4.1.

4.1.2. Random Error

Random errors are inherent in the measurement system and may be due to small changes in environmental factors such as temperature, atmospheric pressure, humidity, cosmic radiation, and so on. Statistics may be applied to these random errors to determine their magnitude in a series of replicate measurements.

Averaging of replicate measurements ensures that random error will be minimized and will tend toward zero. Statistics allows us to analyze replicate measurements to assess the variability or uncertainty in the analysis. A few of these statistical tools will be examined in this chapter, but before describing the treatment of data let us discuss the essence of data which includes treatment of numbers, significant figures, and rounding.

4.2. SIGNIFICANT FIGURES AND ROUNDING

Any number generated in a laboratory measurement has only a finite number of digits associated with the measurement. These digits are determined by the analyst's ability to read a scale or an instrument's electronic or visual output. The number of significant figures associated with a measurement depends on the capability of making such a measurement. For example, an analyst using a laboratory balance that reads out to 0.01 g should not report a result to 0.0001 g.

4.2.1. Number of Significant Figures

The number of significant figures is the sum of all those digits you can read with certainty from your measuring scale plus one extra digit that is estimated. The estimated digit is a careful estimation from the scale. It is therefore the size and nature of the scale and only the ability to read it which determines the number of significant figures.

Consider the following example, shown in Figure 4.1, which is associated with reading the same point using two separate scales having differing numbers

Table 4.1. Systematic Errors

Type	Observed Results[a]	Possible Remedies
Contamination Samples may have foreign material mixed in. At times this impurity is colored and may be easily detected; however, this may not always be apparent.	1	Reject samples that may appear discolored or have foreign inclusion.
Calibration Errors Incorrect standards, outdated standards, or check solutions. Instruments not properly calibrated.	1, 2 OOS	Utilize traceable standards and follow established procedures. Make sure that the standards are certified and not outdated.
Losses/Degradation Samples and/or sample solution could be degrading during storage or analysis. Heat, light, or humidity may be at fault.	1	Minimize exposure of samples to sources of degradation. Examine results of stability studies to understand environmental factors. Method development and validation activities should suggest sample handling conditions and time limitations.
Unrepresentative Sampling A specific analytical sample does not represent the bulk of the material.	1, 2	Set up a sampling plan involving statistical sampling so that the collection of samples will represent the actual bulk material.
Unsuitable Methods A particular method may introduce bias. Incorrect columns in chromatography may not give proper elution of analytes.	1, 2	Use only validated methods.
Incompetence Improperly trained analysts utilizing instrumentation or methods improperly.	1, 2	Ensure that analyst has trained on method and type of samples being assayed.

[a] 1, Low assay; 2, high assay; OOS, out of specification.

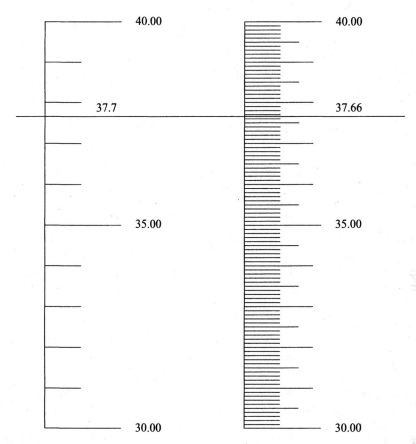

Figure 4.1. Effect of scale divisions on the number of significant figures that can be read.

of graduation marks. Note that the left scale only allows for two digits (i.e., 37) to be read with certainty, and a third is estimated (37.7). This reading has three significant figures. The right scale, which has more graduation lines, allows for three digits (i.e., 37.6) to be read with certainty, and a fourth is estimated (37.66). This reading has four significant figures.

In digital instrumentation, typically the last digit is assumed to be estimated. For instruments such as balances, pH meters, and calculators, the same rules apply. Only report significant figures that are known with certainty based on a known standard, for example. Then provide an additional figure from the readout as an estimate. Traceable standards must be used to verify the instrument's capability for displaying the certain and the estimated digits. When using volumetric glassware or weights, it is necessary to know the classification and tolerances of these so that the correct number of significant figures for weights and volumes may be reported. Tolerances for weight and volume standards are reported in ASTM standard specifications E 617-78 [6] and E 694-79 [7], and some NIST volume tolerances are included in the next chapter (Table 5.2).

It is important to understand the number of significant figures the instrument or device is capable of generating and not to express more significant figures than this number. Specifications for materials must be consistent with the capabilities of the instrument as to the number of significant figures that may be reported.

4.2.1.1. Significant Figure Conventions. There are a number of conventions to be followed in determining the amount of significant figures in a reported number:

1. The digits 1 through 9 are significant.
2. Zeros between nonzero digits are significant; for example, 1.084 has four significant figures.
3. Zeros to the left of nonzero digits are not significant. These zeros only show the position of the decimal point; for example, 0.084 has two significant figures. This could also have been expressed as 8.4×10^{-2} or 84×10^{-3} or 0.00084×10^{2}.
4. Zeros that fall at the end of a number are not significant unless they are marked as significant by having a line placed over them or by some other convention. 1600 has two significant figures and $16\overline{00}$ has four.
5. Zeros to the right of the decimal point are always significant. 25.00 has four significant figures but 25. only has two.

When reporting a value, do not automatically add zeros after the decimal point unless the actual measurement has given those figures as significant. If an instrument is incapable of giving more than a certain number of significant figures, do not assume that addition of zeros after a decimal point is acceptable. Use correct judgment in reporting data calculated with an electronic calculator; it probably has more significant figures than is justified. The correct number of significant figures should be specified in the analytical method being used.

4.2.2. Rounding

After making a laboratory measurement consistent with the rules for significant figures, we may then be asked to employ numbers in calculations. The question arises as to methods for handling the significant figures in our calculations and in the reported final result. There is a tendency to report more significant figures than is actually generated in the measurement. For example, if we were asked to determine the density of a metal by measuring its weight and the volume of liquid it displaced, we would proceed to substitute our measurements in the formula

$$\text{Density} = \frac{\text{Weight (g)}}{\text{Volume (mL)}} \tag{4.1}$$

For a measured weight of 15.125 g and displaced volume of 6.1 mL the computed density read from a calculator is 15.125/6.1 or 2.4795081 g/mL. Clearly, this computed result has many more significant figures compared to the individual weight or volume measurements. The question is, How do we make the calculated result consistent with the measurements, and at what stage of the calculation do we make an adjustment?

The process of adjusting the result to the correct number of significant figures is called rounding. The following rules are followed in the rounding process. There are other conventions used in the rounding procedure, but we will limit ourselves to the following:

1. To round a number, look at all digits beyond the last decimal place desired and raise this last place value by one if the number following the last place is 5 or greater (see example below).
2. For whole numbers, if the series of numbers after the desired place is exactly halfway up (5, 500, 500,000, etc.), then raise the desired digit by one.

The following examples demonstrate the rounding process. In the previous calculation of density, the computed value was 2.4795081. The measurement of volume had only two significant figures (one after the decimal). The final reported density may therefore not have more than two significant figures! To round we look at the computed value and examine the numbers after the first decimal place. For 2.4795081 the value 795,081 is more than halfway to 1,000,000; therefore the required rounding is to report the result as 2.5 (add 1 to the 4), consistent with the number of significant figures of the least known measurement (volume).

It is important to know at what point in our calculation to round. Consider the following guidelines:

1. In addition or subtraction, the result is rounded off to the smallest number of *decimal places* to be found in the numbers that were added or subtracted. The result can have no more decimal places than the number with the least such places.
2. In multiplication or division the result is rounded off to the smallest number of *significant figures* to be found in the numbers that were multiplied or divided.
3. In a chain of calculations, rounding off is not done until the final result has been obtained. The number of significant figures in the final result is

Table 4.2. Some Statistical Definitions

Name	Symbol	Formula
1. Mean (average)	$\bar{X}$	$\bar{X} = \dfrac{\sum X_i}{n}$
2. Average mean deviation	D	$\dfrac{\sum \lvert X_i - \bar{X} \rvert}{n}$
3. % Relative mean deviation	—	$\dfrac{D}{\bar{X}} \times 100$
4. Standard deviation	σ	$\sqrt{\dfrac{\sum (X_i - \bar{X})^2}{n - 1}}$
5. Variance	σ^2	—
6. Relative standard deviation	RSD	$\dfrac{\sigma}{\bar{x}}$
Coefficient of variation	CV	
7. Standard error	SE	$\dfrac{\sigma}{\sqrt{n}}$
Standard deviation of the mean	σ_M	
8. Confidence limit	C.L.	$\bar{X} \pm \dfrac{t\sigma}{\sqrt{n}}$

the same as the least number of significant figures found in the numbers that were used in the chain of calculations.

4. At least one more place beyond the last figure required in a specification will be carried through the calculations to determine the reported values.
5. Rounding will not be performed until the average final value is calculated.
6. Numerical results will be reported to the same number of figures shown for the test requirement in the specification.
7. The number of figures shown in a specification requirement must be consistent with the number of significant figures the test method or equipment can generate.

4.3. SOME DEFINITIONS*

Before proceeding to a discussion of the uses of statistics in handling laboratory data, it is necessary to summarize the terms and symbols to be used. The most important ones are listed in Table 4.2 [8–10].

4.3.1. Accuracy

Accuracy describes how close a measurement is to the correct or true value. It is not unusual for the accuracy to be indeterminate because an incoming sample is

* See Chapter 3 and Appendix II for ICH definitions.

not identified with a true value. In fact the purpose of a lab analysis of a sample is to determine what that value is and if it agrees with a specification.

Accuracy may be assessed from running traceable standards such as from the National Institute of Standards and Technology (NIST), USP, or internally generated samples that have been well-characterized. Standards are available from USP and NIST and are provided as certified standards with the appropriate assay or other parameter given. In developing new methods, accuracy is assessed by determining the agreement of the analytical procedure with the known standard value. For new drug substances, one may use internally developed standards that have been well-characterized by appropriate techniques for purity levels and assay values.

4.3.2. Precision

Precision is a measure of the reproducibility of a measurement. It expresses in a quantitative fashion the degree of scatter of repetitive measurements on the same sample. Good precision implies that the analyst is producing a consistent result each time the same measurement is made on the same sample. Precision does not address how close this measurement is to the true value. It is possible to have high precision but poor accuracy. This may come about, for example, in a situation where there is a systematic error causing a bias. If a sample degrades 50% during a measurement, the reported result could show very little scatter in the data but would yield an assay only one-half the true value. An incorrectly calibrated instrument could also give rise to a situation of poor accuracy but good precision.

Poor precision is influenced by random errors that are part of any measurement. Running replicate analyses will minimize the effect of these random errors. A distinction may also be made for within-run precision and between-run precision. The repeatability, or within-run precision, is a measure of the precision of replicate runs determined for the same sample preparations and glassware within a short time span. The reproducibility or between-run precision is determined for replicated runs at different times. This precision is generally not as good as the repeatability.

4.3.3. Absolute Error

When the true value of a measurement is known, then the absolute error is expressed as

$$\text{Absolute error} = \text{Measured value} - \text{True value} \qquad (4.2)$$

This is usually expressed as a positive number having the units of the measured property. For example, if a standard has a known value of 24.2 g of a particular constituent and we measure it in our laboratory and get the value as 24.5 g, the absolute error of the measurement is 0.3 g.

4.3.4. Relative Error

Relative error compares the magnitude of the absolute error to the size of a measurement and is usually expressed as a percent. It is defined as

$$\text{Relative error} = \left(\frac{\text{Absolute error}}{\text{True value}}\right) \times 100 \qquad (4.3)$$

In the example just given, the relative error is $(0.3/24.2)100 = 1.2\%$. In this case the relative error is a measure of the accuracy of our measurement because we are comparing it to a known standard value.

At times an analyst may be biasing a reading, thereby introducing error. For example, in reading a burette it is found that no matter what volume the meniscus is at, an analyst cannot read the scale to better than 0.02 mL. This absolute error of reading occurs at any point along the burette scale.

The relative uncertainty of this measurement is very dependent on the magnitude of the volume being read. Therefore for a burette reading of 30 mL the relative error is $(0.02/30) \times 100 = 0.07\%$. If the burette reading had been at 10 mL, the relative error would be $(0.02/10) \times 100 = 0.2\%$. It can be seen that by reducing the scale reading from 30 to 10 we have increased the relative error in reading the burette a factor of almost threefold.

In making laboratory measurements it is therefore important to choose sample sizes that allow for minimizing relative error. For example, do not perform a titration in which only very small volumes are dispensed. In chromatography or absorption spectroscopy, make sure that area counts or integrated band absorptions are large enough so that noise levels only contribute a small relative error to the measurement.

4.3.5. Mean

The average of any number of repeated measurements is given by $\bar{X}$.

$$\bar{X} = \frac{\sum X_i}{n} \qquad (4.4)$$

where $\bar{X}$ = mean, X_i = value of single measurement, and n = total number of measurements. This is simply the sum of each measurement divided by the number of measurements.

Consider the following separate measurement values for an assay: 11, 12, 10, 8, 9. The mean is calculated as

$$\bar{x} = \frac{11 + 12 + 10 + 8 + 9}{5} = 10 \qquad (4.5)$$

4.3.6. Average Deviation

The average deviation is a measure of how close each measurement is to the mean (use positive or absolute numbers). Deviation is defined as the difference between any given measurement and the mean of the measurements.

For the previous example in which the mean was 10 we have the following:

Measurement Value (X_i)	Deviation $(X_i - \bar{X})$
11	1
12	2
10	0
8	2
9	1
Sum	6

The average or mean deviation is calculated as

$$D = \frac{\sum |X_i - \bar{X}|}{n} = \frac{1 + 2 + 0 + 2 + 1}{5} = 1.2 \tag{4.6}$$

where | | represents the absolute value. A convenient way to express this precision is to use the % relative mean deviation:

$$\% \text{ Relative mean deviation} = \left(\frac{\text{Average mean deviation}}{\text{Mean}}\right) \times 100 \tag{4.7}$$

For the example given, the relative mean deviation is 12%.

4.3.7. Standard Deviation

Another important measure of the precision of a series of measurements is the standard deviation, σ. The meaning of this will be discussed in the next section, but the conventional formula for a series of n measurements is

$$\sigma = \sqrt{\frac{\sum (X_i - \bar{X})^2}{n - 1}} \tag{4.8}$$

For the previous example we have the following:

X_i	$\bar{X}$	$|X_i - \bar{X}|$	$(X_i - \bar{X})^2$
11	10	1	1
12	10	2	4
10	10	0	0
8	10	2	4
9	10	1	1
Sum		6	10

$$\sigma = \sqrt{\frac{1 + 4 + 0 + 4 + 1}{4}} = \sqrt{\frac{10}{4}} = 1.6 \tag{4.9}$$

The precision of the set of measurements may then be expressed as the average ± an uncertainty range. For example:

$$\bar{X} \pm \sigma \qquad (4.10)$$

Or, as in the case of our example, 10 ± 1.6. The ± indicates that the deviation is above and below the mean and equally spaced. The actual range is therefore $8.4 - 11.6$.

4.3.8. Relative Standard Deviation

Usually the standard deviation is expressed relative to the mean and defined as

$$\% \text{ Relative standard deviation (RSD)} = \left(\frac{\text{Standard deviation}}{\text{Mean}}\right) \times 100 \quad (4.11)$$

In our example the %RSD is 16%. The RSD is also referred to as the coefficient of variation (CV). During the validation process of an analytical method, limits are placed on the magnitude of the allowed RSD or CV. Requirements for good precision will permit RSD of only a few percent. For example, the USP requires an RSD of 2% or less in any assay analysis.

4.3.9. Comparison of Precision and Accuracy

It should be stressed that precision and accuracy may fall in opposite directions for a series of measurements. It is possible to have high or low precision coupled with high or low accuracy. Ideally we strive to make our measurements highly precise and highly accurate. To illustrate, consider the following four sets of replicate measurements of a standard having a true value of 20. These different sets of data could have been generated by different analysts or on different instruments.

    A. 15, 25, 10, 30
    B. 19, 20, 21, 22
    C. 25, 24, 26, 23
    D. 25, 30, 35, 45

We can determine the accuracy and precision of these sets of measurements using the calculated values of relative error of the mean compared to the known true value (accuracy) and the relative mean deviation (precision). (See Table 4.3.) These calculations illustrate that the four sets of measurements may be categorized as to their precision and accuracy:

Table 4.3. Comparison of the Accuracy and Precision of the Four Sets of Data

Data Set	$\bar{X}$	Accuracy		Precision	
		Mean Deviation from True Value (20)	% Relative Mean Deviation from True Value	Average Mean Deviation, D	% Relative Mean Deviation
A	20	0	0	7.5	37.5
B	21	1	5	1.0	4.8
C	24.5	4.5	22.5	1.0	4.1
D	33.8	13.8	69	6.4	18.9

A. Poor precision and good accuracy

B. Good precision and good accuracy

C. Good precision and poor accuracy

D. Poor precision and poor accuracy

If one were to visualize the results as shots at a target, the results would look like those depicted in Figure 4.2. Obviously, set B is the best and is always our objective. Set C could represent a situation in which there is a determinate or systematic error; for example, the given sight could be misaligned, causing the projectile to miss the bull's eye. In such a situation, the cause of the determinant errors should be found and corrected. Notice that sample set A represents measurements that are not precise but are accurate. This is not a desirable situation. The high accuracy may be a fortuitous occurrence in a set of measurements showing poor reproducibility. The poor precision informs us that the measurement is not under proper control. Clearly, set D is not desirable and indicates serious problems with the method.

The analyst must pay attention to how the measurements are being made to evaluate situations of poor accuracy and precision. Attention must be paid to proper instrumental calibration, correct preparation of standards and working solutions, and noninterference from impurities or excipients.

4.3.10. Standard Error

Statistical theory also predicts that the uncertainty of a series of measurements will decrease as the number of measurements increases. The uncertainty decreases in proportion to

$$\frac{1}{\sqrt{n}} \tag{4.12}$$

where n is the number of measurements. For an infinite number of measurements the uncertainty tends to zero.

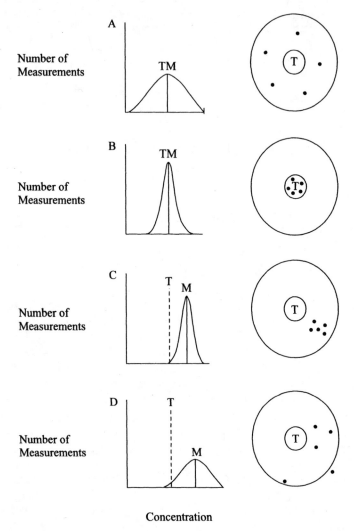

Concentration

Figure 4.2. Illustrations of the four possible combinations of precision and accuracy. T equals target value and M equals mean value.

The standard error is an additional estimate of precision, and for large sample sizes it is referred to as the standard error of the mean. This standard error is calculated by dividing the standard deviation by the square root of the number of replicate measurements:

$$SE = \frac{\sigma}{\sqrt{n}} \tag{4.13}$$

This measure of precision implies that increasing the number of replicates will decrease the error in a square root fashion. Assuming a standard deviation of 1 for a series of measurements, the change in standard error for 4, 16, and 64 measurements would be proportionate as the ratios 1/2, 1/4, 1/8, or 4:2:1. Therefore 64 measurements would only give us a fourfold decrease in uncertainty compared to 4 measurements. The extra work in running such a large number of replicates may not be worth the effort in terms of decreasing the uncertainty.

4.4. NORMAL DISTRIBUTION OF REPEATED MEASUREMENTS

We have discussed some methods for estimating the precision of repeated or replicate measurements. The standard deviation, σ, is an important indicator of precision or repeatability. This parameter is derived from statistical theory and has the following implications.

When we perform a chemical analysis, perhaps on a standard, random error will cause repeated measurements to give scattered results around the true value (assuming no systematic error is present). If we were to repeat the same measurement many times the measured values of X would take on the appearance of the curves in Figure 4.3 and Figure 4.4, where the Y axis represents the number of times a particular value of X is found [8–11].

This distribution curve comes about due to random errors associated in making the repeated measurements and is symmetrical around a value of X, labeled μ, which is the true value of the standard or the mean value for any set of repeated measurements. The more measurements that are made, the closer the value μ is to the mean or average value. This curve is called a normal or Gaussian distribution and is rigorously correct for an infinite number of repeated measurements. In practice one can develop such distribution curves with many less than an infinite number of measurements.

Statistical probability theory uses this normal distribution to predict the probability that a given measurement will fall into a certain range of X values. The standard deviation becomes the measuring yardstick of data spreading as illustrated in Figure 4.4. It is predicted that for repeated measurements on the same specimen, any measurement of X would have a value in the range of $(\mu \pm \sigma)$ 68.3% of the time, $(\mu \pm 2\sigma)$ 95.5% of the time, and $(\mu \pm 3\sigma)$ 99.7% of the time [11]. As the measured points deviate more and more from the true value (large spread of the curve), the probability of finding a measurement within this larger range increases. This implies that the uncertainty of finding a value in this expanded range goes down. The standard deviation, σ, defines the narrowness of the distribution curve. For low standard deviation the curve narrows and the repeated measurements cluster about the average value.

We express the confidence limits of a series of measurements in terms of the standard deviation. Thus a 68% confidence interval refers to measurements that fall in the range of the mean $\pm 1\sigma$. The 95% confidence interval includes the mean $\pm 2\sigma$ and the 99.7% confidence limit includes the range of the mean $\pm 3\sigma$. The greater the confidence interval, the greater the expectation (probability) that a value will fall into the range.

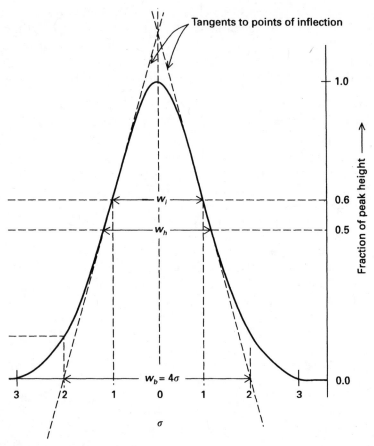

Figure 4.3. A normal distribution of repeated measurements. (From H. M. McNair and J. M. Miller, *Basic Gas Chromatography*, Wiley, New York, 1998, p. 37. Reprinted by permission of John Wiley & Sons, Inc.)

4.5. STUDENT *t* TEST

There is a useful statistical tool for analyzing data which allows us to deal with a limited number of replicate measurements rather than infinite numbers. This tool was elucidated by an Irish chemist, W. Gosset, working at the Guinness Brewery in Dublin in the early 1900s. The statistic is referred to as the Student *t* or more usually just *t*. This tool allows us to determine the uncertainty interval of the mean for a series of replicates. The calculation may be performed at any desired level of confidence. The *t* test enables calculation of the agreement of a series of measurements compared to a known standard (accuracy) and allows us to compare the results of different analytical methods, instruments, and operators for the level of agreement.

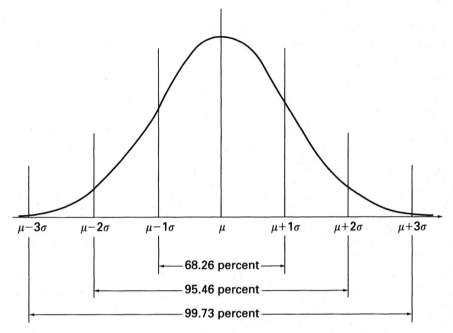

Figure 4.4. Illustration of standard deviations on a normal distribution.

The confidence interval about the mean of a series of measurements is expressed as

$$\bar{X} \pm \frac{t\sigma}{\sqrt{n}} \tag{4.14}$$

where n is the number of replicates, and t is read from a table (see Table 4.4). Examination of Table 4.4 shows that values of t are available at different levels of confidence and for different numbers of replicates [11, 12]. The degrees of freedom are actually the number of replicates minus 1. Five replicates would have four degrees of freedom.

4.5.1. Applications of *t* Test

4.5.1.1. Confidence Limits. The confidence interval (C.I.) or confidence limit (C.L.) was given in equation 4.14. To see how it is used, consider the following example.

Calculate the confidence limits for a set of five measurements of the moisture level in a solid sample with the following individual measurements: 11.6%, 10.9%, 12.0%, 11.7%, 11.5%.

$$\text{C.I.} = \bar{X} \pm \frac{t\sigma}{\sqrt{n}} \tag{4.15}$$

Table 4.4. Table of *t* Values

Degrees of Freedom	Confidence Level (%)						
	50	90	95	98	99	99.5	99.9
1	1.000	6.314	12.706	31.821	63.657	127.32	636.619
2	0.816	2.920	4.303	6.965	9.925	14.089	31.598
3	0.765	2.353	3.182	4.541	5.841	7.453	12.924
4	0.741	2.132	2.776	3.747	4.604	5.598	8.610
5	0.727	2.015	2.571	3.365	4.032	4.773	6.869
6	0.718	1.943	2.447	3.143	3.707	4.317	5.959
7	0.711	1.895	2.365	2.998	3.500	4.029	5.408
8	0.706	1.860	2.306	2.896	3.355	3.832	5.041
9	0.703	1.833	2.262	2.821	3.250	3.690	4.781
10	0.700	1.812	2.228	2.764	3.169	3.581	4.587
15	0.691	1.753	2.131	2.602	2.947	3.252	4.073
20	0.687	1.725	2.086	2.528	2.845	3.153	3.850
25	0.684	1.708	2.068	2.485	2.787	3.078	3.725
30	0.683	1.697	2.042	2.457	2.750	3.030	3.646
40	0.681	1.684	2.021	2.423	2.704	2.971	3.551
60	0.679	1.671	2.000	2.390	2.660	2.915	3.460
120	0.677	1.658	1.980	2.358	2.617	2.860	3.373

Calculating the mean gives

$$\bar{X} = \frac{11.6 + 10.9 + 12.0 + 11.7 + 11.5}{5} = 11.5 \qquad (4.16)$$

Calculating the standard deviation gives

$$\sigma = \sqrt{\frac{(0.2)^2 + (0.6)^2 + (0.5)^2 + (0.1)^2}{4}} = 0.4 \qquad (4.17)$$

The confidence limits may be calculated for any desired confidence by choosing the appropriate t value from Table 4.4. At 90% confidence, $t = 2.132$ for four degrees of freedom (five replicates). To calculate the confidence interval at 90% confidence, substitute as shown:

$$\text{C.I.} = 11.5 \pm \frac{(2.132)(0.4)}{\sqrt{5}} = 11.5 \pm 0.5 \qquad (4.18)$$

Expressed another way, the confidence interval at 90% confidence for our data is 11.0–12.0% water in our sample.

For 50% confidence the confidence interval computes as 11.5 ± 0.1. Please note that at lower confidence levels the uncertainty goes down and the precision is expected to be tighter. The probability is therefore decreased for a given measurement to fall into this narrower interval. In most analytical chemistry work we don't impose specification requirements below 90–95% confidence.

4.5.1.2. Comparison of a Set of Measurements to a Standard. The results of replicate measurements on a standard may be used to assess the accuracy of the measurement, to determine how close our average value is to the known value. For a set of measurements on a standard buffer having a pH value of 10.5 the following data were obtained in five replicate measurements:

$$10.7, \ 10.5, \ 10.9, \ 10.6, \ 10.8$$

Do these data agree with the standard value?

We apply the t test and impose the requirement that agreement must be at the 95% confidence level. Substituting into our equation for confidence limits we obtain the following:

$$\bar{X} = 10.7$$

$$\sigma = 0.2$$

$$t = 2.776$$

so

$$\text{C.I.} = 10.7 \pm \frac{(2.776)(0.2)}{\sqrt{5}} = 10.7 \pm 0.2 \tag{4.19}$$

The uncertainty interval is therefore the pH range of 10.5 to 10.9. The standard pH value of 10.5 is within our uncertainty interval, and we may conclude that our measurements are accurate at 95% confidence. If we had imposed stricter requirements for agreement such as 68% or 50% confidence, our data would not have been considered accurate.

4.6. PROPAGATION OF UNCERTAINTY (ERRORS)

Because we have determined methods for calculating the uncertainty interval for data sets, how do we treat these uncertainties in calculations? The rules differ when dealing with addition or subtraction and multiplication or division.

4.6.1. Addition and Subtraction of Uncertainties

If the uncertainty of a given mean value is known, then the sum and difference are calculated as shown. Calculate the sum of the mean values obtained for three separate samples and consider the uncertainty.

Sample	Mean Value ($\bar{X}$)	Uncertainty (U)
1	1.76	$\pm 0.03\ (U_1)$
2	1.89	$\pm 0.02\ (U_2)$
3	−0.59	$\pm 0.02\ (U_3)$
Sum	3.06	$\pm 0.04\ (U_4)$

Uncertainties are given as absolute and labeled U_1, U_2, and U_3. The means are treated as in normal addition or subtraction, and in this case the sum is simply $1.76 + 1.89 - 0.59 = 3.06$ The question is the method for calculating the sum of the uncertainties labeled U_4, it is not simply the sum of the uncertainties. The following formula should be applied when adding or subtracting uncertainties:

$$U_4 = \sqrt{(U_1)^2 + (U_2)^2 + (U_3)^2} \tag{4.20}$$

For the example given we obtain

$$U_4 = \sqrt{(0.03)^2 + (0.02)^2 + (0.02)^2} = 0.04 \tag{4.21}$$

The final result is therefore expressed as 3.06 ± 0.04.

4.6.2. Multiplication or Division of Uncertainties

To carry out a series of multiplication or divisions, one must first convert the absolute uncertainties to % relative uncertainties. These % values are then treated as in the previous example, but the end result is an uncertainty expressed as % relative uncertainty that must be reconverted to an absolute uncertainty. Consider the following example. Calculate the value of the following, including the uncertainty of the calculated result:

$$\frac{(1.76 \pm 0.03)(1.89 \pm 0.02)}{0.59 \pm 0.02} \tag{4.22}$$

Excluding the uncertainties the numbers calculate as

$$\frac{(1.76)(1.89)}{0.59} = 5.6 \tag{4.23}$$

The uncertainty U_4 is calculated as

$$\%U_4 = \sqrt{(\%U_1)^2 + (\%U_2)^2 + (\%U_3)^2} \tag{4.24}$$

The individual % uncertainties are then calculated:

$$\%U_1 = \frac{(0.03)100}{1.76} = 1.7 \quad \%U_2 = \frac{(0.02)100}{1.89} = 1.1$$

$$\%U_3 = \frac{(0.02)100}{0.59} = 3.4$$

(4.25)

$$\%U_4 = \sqrt{2.89 + 1.21 + 11.56} = 3.96 = 4.0\% \tag{4.26}$$

The final result is $5.6 \pm 4.0\%$, which is now converted to an absolute uncertainty of 5.6 ± 0.2. Note that the final result can have no more than two significant figures because 0.59 used in the calculation has only two.

4.7. REJECTION OF OUTLIERS

Before going into this section it should be cautioned that no data are to be discarded in any analyses unless there are clear-cut standard operating procedures to allow this in your laboratory.

The following is presented to make you aware that at times we generate data points in replicate measurements which do not seem to fit in with the other measurements. If this should occur in a series of four or more readings from the same sample, then it is possible using a valid statistical technique to check if a particular value is an outlier. In an important court decision [3], the *United States v. Barr Laboratories, Inc.*, the presiding judge, Wolin, decided that outlier tests were permitted in microbiological assays but not in chemical analyses. It is not clear how binding this will be on the pharamaceutical industry in general.

Proceed as follows. Check all calculations, instrument settings, and calibrations to ensure that these were not in error. If a source of error cannot be identified, then a suspected outlying data point may be tested by a statistically valid test. An example of such a test is the Dixon Q test [13]. This is only applied to a series of four or more replicates where a value appears out of line. A term Q is defined as follows:

$$Q = \frac{\text{Gap}}{\text{Range}} \tag{4.27}$$

where gap = difference between a questionable value and its nearest value and range = difference between the highest and lowest values. The differences are expressed as positive numbers. Once the Q is calculated, it is compared to a table of critical Q values determined for a given confidence level. An example of such is Table 4.5 [14] for 95% confidence.

If the calculated Q from our experimental data is greater than the critical Q from the table, then the questionable point may be discarded. Consider the following experimental values determined in a laboratory analysis:

$$12.73, 12.36, 12.95, 12.90, 12.85$$

Table 4.5. Q-Test Table

Number of Data Points	Q at 95% Confidence
4	0.831
5	0.717
6	0.621
7	0.570
8	0.524
9	0.492
10	0.464

Derived from: Reference 14.

Examination of these data might lead us to suspect the 12.36 value as being out of line.

Applying the Q test gives us

$$Q = \frac{\text{Gap}}{\text{Range}} = \frac{12.73 - 12.36}{12.95 - 12.36} = \frac{0.37}{0.59} = 0.63 \qquad (4.28)$$

Examination of the table of critical Q values for five observations shows the critical value of Q as 0.717. Because our calculated Q is not greater than the Q from the table, the questionable point 12.36 may not be rejected.

4.8. LINEAR REGRESSION ANALYSIS

In the analytical laboratory we encounter situations where we generate calibration or standard curves that are expected to be linear; that is, the points fall on a straight line whose slope and y intercept may be calculated.

This linear curve fitting is encountered in all forms of analysis including spectroscopy and chromatography where a range of concentrations is used to examine the response of an instrument. The data generated can be tested statistically to examine how closely the data points fit a straight-line relationship or for that matter any other mathematical function.

In order to check whether a group of data points fit a straight line we make the assumption that the data fit the general linear relationship:

$$y = mx + b \qquad (4.29)$$

where y is predicted from a known value of x (concentration), m is the slope of the line, and b is the y intercept of the line.

When a standard curve is developed, known sample concentrations are made up and the instrument response (area counts, absorption) is measured. The goodness of fit of these data points to a straight line may be estimated by determining a correlation coefficient, r. The closeness of the r value to 1 confirms that the assumption of linearity is correct. Usually the data are plotted and a visual or computer estimate of the best straight line is drawn through the data.

It is, however, possible to numerically calculate the best straight line from the data. This calculated line is called a regression line and is based upon minimization of the sum of squares of the residuals or deviations of data points from the straight line. It is beyond the scope of this book to go into the actual calculations that are usually generated by a computer program associated with the analytical instrument.

The value of r may range from -1 to $+1$. A perfect linear correlation would have an r value of $+1$. For r equal to -1 a perfect negative correlation would exist and the slope would be negative. For r equal to zero there is no correlation between the instrument response y and the concentration x. Normally, laboratory standardizations give r values between 0.9 and 1.0. We usually don't deal with the negative slope and we report r^2 values between 0 and 1. Values of r better than 0.99 are readily obtainable in normal laboratory operations.

It is possible to use statistics to determine the standard deviation and confidence limits associated with the slope and intercept. When a standard curve is used to calculate the concentration of an unknown sample, it is also possible to determine the errors associated with the use of the regression line at any level of confidence. These calculations are usually included in the statistical software programs supplied with analytical instruments and in common spreadsheet programs.

At times the response of an instrument as a function of concentration may only be linear over a limited concentration range. Deviation from linearity is measured by the r or r^2. In many cases these values are high, and trending away from linerarity is masked and not apparent.

Another technique for assessing deviations from linerarity include calculating the residuals that are the deviations in Y value (signal) of each data point away from the linear regression line. For any measured value of y, the plotted point will have no residuals if the point is on the line. Residuals are characterized as $+$ or $-$ deviations from the line, and the numerical sum will add up to zero.

The trending of the residuals may be an indication of the deviation from linearity. For a linear plot, the signs of the residuals ($+$ or $-$) should be randomly distributed along the regression line. However, if the residuals appear in groups of $+$ or $-$, deviations from linearity are indicated. Examples are given in Figure 4.5.

4.9. QUALITY ASSURANCE/CONTROL

Any process may be continuously monitored against specific requirements that ensure that the process is producing material according to expected requirements. The weight of tablets or the volume of a liquid fill are examples of vari-

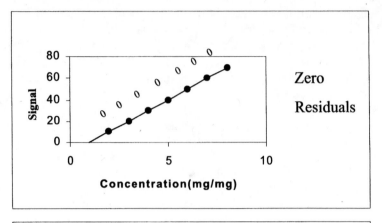

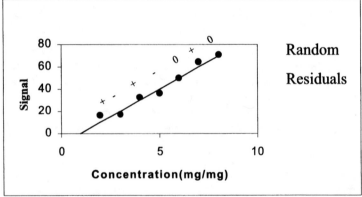

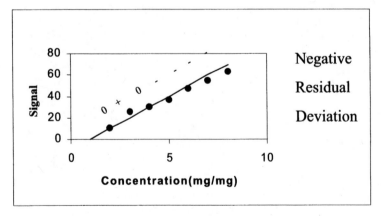

Figure 4.5. Trending of residuals in a linear plot.

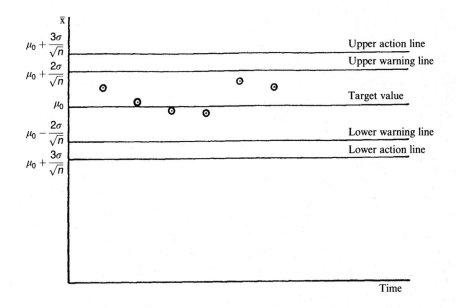

Figure 4.6. A typical (Shewhart) control chart. (Reprinted with permission from J. C. Miller and J. N. Miller, *Statistics for Analytical Chemistry*, third edition, Prentice-Hall, Upper Saddle River, NJ, 1993, Figure 4.2.)

ables that could be monitored. These parameters, when monitored on a continuous basis, allow for examination of drifting or trending and for corrective action to be taken before the process gives out of specification product. Such data charts are called control charts and may take on different formats.

A convenient control chart model (Shewhart) is shown in Figure 4.6. The value of a particular attribute is plotted on the y axis as a function of time. Typically we establish a target value for a process; for example, the fill volume may be targeted at 10 mL. The actual value may vary for each bottle fill due to random error and perhaps errors in the filling operation. The control chart is set up so that the average fill volume is monitored with time over many runs. We expect the measured fill volume (sampled by QA) to cluster about the target value (which we call μ_0) and have a standard deviation of σ. According to our previous discussion, it is expected that the 95% uncertainty range will be represented by

$$\mu_0 \pm \frac{2\sigma}{\sqrt{n}} \tag{4.30}$$

The 99.7% uncertainty range will be

$$\mu_0 \pm \frac{3\sigma}{\sqrt{n}} \tag{4.31}$$

The lines on the chart represent these uncertainty limits. In practice the fill volume would be monitored to ensure that the value hovers around the target value within its standard deviation.

The 95% control line may be designated as an alarm line indicating that data outside this limit indicate that the process may be drifting out of control. The 99.7% uncertainty limit is usually considered an action limit that requires examination of the process and that an appropriate adjustment be made.

There are other forms of control charts, and one can use any uncertainty level for alarm and action limits. In some cases it might be necessary to estab-lish an action limit at 95% confidence.

The control line may be calculated at any level of confidence according to the methods described in this chapter using the *t* test. These control charts provide a visual record of the important attributes. A continuous drift upward or downward might indicate the influence of systematic error. Machinery parts could be wearing, incoming materials may have different levels of impurities, and so on.

4.10. CONCLUSION

This chapter has dealt with basic concepts relating to the treatment of numbers and data relating to working in an analytical laboratory. Many of the topics have not been treated in depth; should you desire a fuller treatment, you may consult some of the references listed below particularly reference 8. Remember that there is no substitute for doing accurate and precise work. Statistics may tell you how you are doing but will not correct for sloppy laboratory practices.

REFERENCES

1. United States Pharmacopeia Convention, *United States Pharmacopeia 24/NF 19*, Rockville, MD, 2000.
2. R. G. Whitfield, D. W. Hughes, T. P. Layoff, R. R. Cox, G. E. Gressett, P. J. Jimenez, J. Andersen, R. R. Peck, and S. Schneipp, Interpretation and Treatment of Analytical Data, *Pharmacopeial Forum* **1998**, 24, 7051.
3. *United States v. Barr Laboratories, Inc., 812 Federal Supp. 458(1)*, NJ, 1993.
4. Z. Deyl (editor), *Quality Control in Pharmaceutical Analysis: Separation Methods*, Elsevier, New York, 1997.
5. Christopher M. Roley and Thomas W. Rosanshe (editors), *Development and Vali-dation of Analytical Methods Progress in Pharmaceutical and Biomedical Analysis*, Vol. 3, Pergamon Press, Elmsford, NY, 1996.
6. ASTM, Vol. 14.02.
7. ASTM, Vol. 14.02.
8. J. C. Miller and J. N. Miller, *Statistics for Analytical Chemistry*, 3rd edition, Ellis Horwood PTR Prentice-Hall, Englewood Cliffs, NJ, 1993.

9. F. Mosteller, S. E. Fienberg. R. E. K. Rourke, *Beginning Statistics with Data Analysis*, Addison-Wesley, Reading, MA, 1983.

10. J. T. McClave and F. H. Dietrich II, *A First Course in Statistics*, 4th edition, Macmillan, New York, 1992.

11. W. J. Youden, *Experimentation and Measurement*, NIST Special Publication 672, 1991.

12. *Standard Guide for Statistical Procedures to Use in Developing and Applying ASTM Test Methods*, ASTM E 1488-92, Vol. 14.02.

13. *Standard Practice for Dealing with Outlying Observations*, ASTM E 178-94, Vol. 14.02.

14. P. King. *J. Am. Stat. Assoc.* **1958**, *48*, 531.

5

BASIC ANALYTICAL OPERATIONS AND SOLUTION CHEMISTRY

Nicholas H. Snow and Wyatt R. Murphy, Jr.

Basic laboratory operations are fundamental to performing analytical methods. None of the instrumental techniques described later in this text can be operated effectively without proper pretreatment of samples, preparation of buffers and mobile phases, and handling of laboratory equipment. Chemical equilibrium provides the basis for many of the operations that analytical chemists perform, including chromatography, dissolution, buffer preparation, titration, extraction, pH measurements, and potentiometry. Thus, a study of chemical equilibrium is central to the understanding of most analytical methods. This chapter describes several of the basic operations that are commonly employed in analytical methods. Equilibria provide a common theme throughout most of these techniques.

5.1. ANALYTICAL REAGENTS

Most instrumental analysis methods are performed on solutions containing the analyte(s) of interest. There are many grades and types of solvents available, and the analyst should ensure that the solvents being used are appropriate for the instrumental method being employed. Table 5.1 shows available grades of

Analytical Chemistry in a GMP Environment. Edited by J. M. Miller and J. B. Crowther
ISBN 0-471-31431-5   © 2000 John Wiley & Sons, Inc.

Table 5.1. Grades and Classifications of Solvents

Grade	Comments
Technical/laboratory	Reasonable but unknown purity; use where no official specification exists.
Certified	Certified by whom? Check lot analysis to be sure purity is acceptable.
Certified ACS	Reagents that meet or exceed purity guidelines established by the American Chemical Society.
USP, NF, FCC, EP, BP	Reagents that meet or exceed purity guidelines established by the *United States Pharmacopeia, National Formulary, Food Chemicals Codex, European Pharmacopeia,* and/or *British Pharmacopeia*
HPLC, GC	Reagents specially purified and filtered for chromatography applications; check lot analysis for purity
Spectranalyzed, spectronic	Reagents that have been purified for use in spectrometry; should contain little or no absorbing impurities; check lot analysis for purity.
Biotechnology, electronic, etc.	Reagents that have been purified for specific applications. Check lot analysis for purity.
Optima, gold label, etc.	Most vendors offer extremely pure solvents. Check individual method requirements.

common solvents available from various vendors. For most spectrophotometric methods, the "American Chemical Society (ACS) reagent" grade is usually sufficient for solvents. For high-performance liquid chromatography (HPLC), only filtered HPLC grade solvents (including water) should be used. Be aware that each vendor prepares HPLC solvents to slightly different specifications and that this may cause variation in results. For high-resolution trace gas chromatography (GC) analysis, solvent cleanliness is especially important, because minor impurities in solvents cause major interferences at part-per-billion and lower levels. For these analyses, only the highest grades of solvents should be used, and even these may require further distillation or drying.

Solvent bottles should be handled with extreme care. Secondary containment (placing the bottle in a second container such as a tray or sleeve when in use) is recommended to avoid hazardous spillage. To avoid contamination, never leave a solvent bottle uncovered in the laboratory for more that a few minutes. All solvents are hygroscopic to an extent, and they will readily absorb water or traces of other contaminants from the laboratory air. Also, unless you are using an automatic pipettor that is sealed onto the bottle, never pipet directly from reagent stock bottles. Pour approximately the necessary amount into a clean beaker or Erlenmeyer flask and pipet from that. Never return unused portions of solvent aliquots to the stock bottle. These should be treated as waste.

When obtaining solids, similar cautions should be applied. Never place a spatula or other delivery tool directly into a stock bottle. Pour the approximate amount of solid needed from the bottle into another container such as a clean beaker or the bottle cap. Obtain the necessary material from that and treat the remaining solid as waste. Never return unused solid aliquots to the stock bottle. Like solvents, solids can be hygroscopic, so the bottles should remain capped at all times when not in immediate use.

5.2. SAMPLING

Obtaining and preparing samples is at the beginning of any analytical method and provides one of the major opportunities for inaccuracies and errors to be introduced. While this work is not "glamorous," none of the instrumental technologies described in this text can readily overcome the errors that can result from poor sampling and sample handling. Basic techniques, such as handling beakers and washing bottles, filters, pipets, and other simple glassware, are very important here and will make the difference between accurate, precise analysis and technique error-prone analysis. The reference section at the end of the chapter can be consulted for further details. Especially useful are videos, such as the one produced by the American Chemical Society [1], and the WWW site produced at the University of Wisconsin [2].

5.2.1. Obtaining a Representative Sample

In an analytical method, the sampling protocol is often one of the most easily questioned portions of the procedure. Sampling protocols should be designed to ensure that the small samples obtained are representative of the larger population of material. The statistical reasons for this are discussed in Chapter 4. In considering basic laboratory operations when obtaining samples, the major issue is contamination, which should be avoided. Generally, samples should be handled as little as possible and exposed to the environment as little as possible. Small samples of solids should be handled with clean tongs, forceps, or spatula; avoid handling them directly with bare or gloved hands, because these both contain small amounts of contaminants, such as finger oils, talc, placticizers, or polymeric materials. Samples should be sealed in an appropriate clean container immediately following collection, especially those that might absorb water or other contaminants from the atmosphere. When sampling liquids, it is best to pour the sample from the original container into the sampling container, rather than to transfer it with a pipet. This avoids any possible contamination of the sample and original material from the pipetting technique.

5.2.2. Preparing Samples for Analytical Methods

Almost all samples will require preparation before being subjected to instrumental or other analysis. These techniques may include grinding, weighing, filtering, dissolving, extraction, and drying. In the course of performing any of these methods, the analyst should take care to follow standard operating procedures every time, and in a consistent manner. Note that each time a sample is handled, error is introduced into the final quantitative result. When developing methods, it is best to reduce the number of sample handling steps as much as possible. Due to the complex sample preparation methods and the small amounts of analyte involved, the results of trace environmental analyses often have standard deviations on repetitive analysis results of over 20%.

5.2.3. Weighing and Balances

Weighing samples is one of the most common laboratory operations. If not performed carefully, large errors can result. Solid samples should always be handled with tongs or a spatula, never with your hands. Liquid samples should be transferred using pipets and should be transferred from the storage bottle into a smaller container, before the final sample is aliquotted for weighing. An accurate weighing procedure typically involves tare weighing the empty container, adding the appropriate weight of sample, recording the total weight, and calculating the weight of sample by the difference. Note that this involves a subtraction. The experimental error in each weighing step will be additive to generate the uncertainty in the reported sample weight, as described in Chapter 4. Although modern balances have digital readouts that can appear exact, this uncertainty may become important when weighing very small sample quantities.

A modern electronic balance must be calibrated according to an established schedule. Standard weights that are certified by the National Institute of Standards and Technology should be used. Two types of electronic balances are commonly employed: a single-pan balance and a top-loading balance. A single-pan balance, shown in Figure 5.1, measures the force generated by the object being weighed as it is resisted by a servomotor. The system contains a zeroing device, such as an optical vane or lamp and photodetector. As force is applied to the pan, the null detector signals a drop in current that is compensated by the servomotor until the pan is returned to its original position. The applied correction current is related to the weight of the object. Calibration is performed by placing known weights on the balance pan and adjusting the balance electronics. In a top-loading balance, shown in Figure 5.2, a solid parallelogram load constraint is attached to the balance pan. This prevents torsional forces related to off-center loading. Top-loading balances usually are less affected by off-center loading. Schnoover [3] describes the operation of many types of analytical balances.

There are several common errors associated with weighing. First, samples that may absorb water from the atmosphere must be kept in closed containers, both before and during weighing. Glass vessels should not be wiped with a dry cloth, because this may cause a buildup of static electricity and particulates that may cause errors. The sample and container should be weighed at the same temperature as the balance. Typically, this means allowing samples that have been heated or cooled to equilibrate to room temperature. Electronic balances

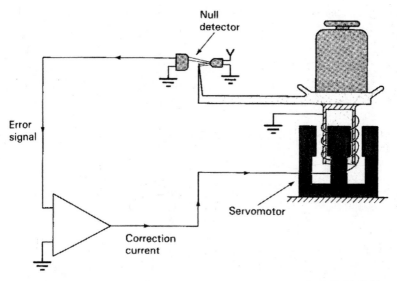

Figure 5.1. Single pan analytical balance. (Reprinted with permission from R. M. Schoonover, *Anal. Chem.*, **1982**, *54*, 373A. Copyright 1982, American Chemical Society.)

may be subject to errors from (a) magnetized samples, (b) electromagnetic radiation from other equipment, and (c) dust. Small errors can also result from off-center sample loading and from buoyancy of samples. Johnson and Wells [4] summarize errors that are common in the weighing of samples in detail.

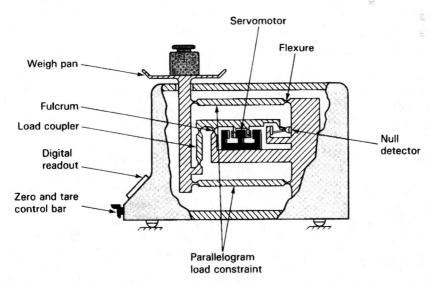

Figure 5.2. Top-loading analytical balance. (Reprinted with permission from R. M. Schoonover, *Anal. Chem.*, **1982**, *54*, 373A. Copyright 1982, American Chemical Society.)

Table 5.2. Tolerances of Volumetric Glassware, as Defined by NIST

Glassware Type	Class A Tolerance ($\pm$mL, %)	Class B Tolerance ($\pm$mL, %)
Volumetric flask, 10 mL	0.02, 0.20	0.04, 0.40
Volumetric flask, 25 mL	0.03, 0.12	0.06, 0.24
Volumetric flask, 50 mL	0.05, 0.10	0.10, 0.20
Volumetric flask, 100 mL	0.08, 0.08	0.16, 0.16
Volumetric flask, 250 mL	0.12, 0.048	0.24, 0.096
Volumetric pipet, 1 mL	0.006, 0.60	0.012, 1.20
Volumetric pipet, 5 mL	0.01, 0.20	0.20, 0.40
Volumetric pipet, 10 mL	0.02, 0.20	0.04, 0.40
Volumetric pipet, 25 mL	0.03, 0.12	0.06, 0.24
Volumetric pipet, 50 mL	0.05, 0.10	0.10, 0.20

5.2.4. Volumetric Glassware

There are many types of glassware in use the analytical laboratory. You will commonly employ beakers, several types of flasks, graduated cylinders, pipets, and burets. Its markings give the purpose of any glassware item. Typically, the markings include the volume, "TC" for glassware designated "to contain" such as a volumetric flask, or "TD" for glassware designated "to deliver" such as a buret, the temperature at which it is calibrated, the class of the glassware (A, B, or nothing), the stopper type (if appropriate), and the drainage rate for pipets. In general, items such as beakers, Erlenmeyer flasks, and other glassware with rudimentary volume markings should not be used for measuring volumes. These should only be used as reaction vessels and for making or storing solutions or reagents. Graduated cylinders should be read by placing them on the bench and kneeling down so that your eyes are level with the meniscus; *do not* hold a graduated cylinder in your hand, up in the air, to read the volume. Always read the volume of aqueous solutions from the bottom of the meniscus.

Volumetric flasks and pipets require special care for proper use. They are calibrated "to contain" the specified volume of water within specified tolerances at a given temperature. Table 5.2 gives the tolerances, as defined by the National Institute of Standards and Technology (NIST), of class A and B volumetric flasks and pipets. Class A glassware is manufactured to more exact tolerances. You should also note that the larger flasks provide smaller relative errors in the volume. Avoid using glassware that has no indication about class. Changes in ambient laboratory temperature and the use of nonaquoeus solvents may also cause small errors. Finally, volumetric flasks with glass stoppers are matched and should be kept and used together. Using an unmatched flask and stopper may result in leakage when inverting the flask to mix a solution.

To dissolve a solute in a volumetric flask, first weigh the solute on weighing paper or in an appropriate weighing container. Quantitatively transfer the solute to the volumetric flask, using a funnel to avoid depositing solute in the ground-glass area at the neck. Rinse the weighing container and funnel with the solvent. Fill the flask about halfway with solvent and swirl to dissolve the solute. Dilute to the mark, stopper the flask, and invert at least 10 times, while holding the stopper in place, to ensure complete mixing. Excessive heat from hot plates or ovens, solutions such as HF and strong bases that etch glass, volatile drying solvents (such as acetone), compressed air for drying, and paper towels should be avoided, because these are all sources of damage or contamination. Volumetric flasks should be rinsed immediately following use and then dried on a drying rack. Abrasive detergents should avoided.

There are several types of volumetric pipets, the choice of which depends upon the precision required. Transfer pipets, which are designed to deliver a single, specific volume provide the greatest precision, with up to five significant digits. Measuring or seriological pipets (those with graduations) can provide up to three significant digits, while syringe pipets and automatic pipets (these latter devices deliver small volumes, usually for chromatography) provide two significant digits. You should examine the markings on the pipet to determine whether it is class A or class B and whether it is a "blow-out" or non-blow-out design. Also note that small errors (1–2%) result from using nonaqueous solutions.

To use a volumetric pipet, the pipet should first be inspected carefully. If it has cracks at either end, discard it, or if appears dirty, clean it. It should be rinsed at least twice in the sample solution before use. The solution should be drawn above the mark using an appropriate pipet bulb. The best pipet bulbs are those with the conical beslow insert, which allows easy insertion of the pipet into the bulb. Once the pipet bulb is removed, flow is best controlled using a forefinger. Do not try to control the flow using the pipet bulb, because liquid can easily be sucked into the bulb and because the flow is very difficult to control this way. While touching the tip to glass, lower the solution level to the mark; the bottom of the meniscus should reach the top of the mark. Wipe excess liquid gently from the glass, with a downward motion; then deliver the aliquot, without wiping again. After the flow stops, wait a few seconds, then remove the pipet. Most volumetric transfer pipets are designed so that you do not expel the small amount of liquid remaining in the tip. Running a continuous stream of deionized water through them while inverted best cleans pipets. Pipet cleaning baths are available for this purpose.

5.2.5. Filtering

Filtering is another very common operation. As a pharmaceutical laboratory analyst, you will most likely be filtering large volumes of solvents for HPLC mobile phases and small volumes of analytical samples prior to instrumental analysis. There are a wide variety of filtering devices and filter media. In most analytical applications, a vacuum filter, usually using 47-mm-diameter filter membrane disks, will be used for large volumes, while syringe filters will be used for small volumes. Choosing the proper filter media is critical, because some are reactive toward acids, bases, or solvents, while others may retain certain analytes, especially proteins. Be sure to check the manufacturer's specifications for the filter device and media to ensure compatibility with your samples. Filter media are generally available both sterile and nonsterile, depending upon the application. The *United States Pharmacopeia* (USP) provides little guidance or specification on filter media, stating merely: "liquid [should] be filtered through suitable filter paper or equivalent device until filtrate is clear"; however, specific methods often have requirements [5]. A summary of common filter media, applications, and compatibility is given in Table 5.3 [6]. Also, filtering techniques using classical equipment, such as filter papers, glass funnels, and Buchner funnels, are described in detail in most undergraduate laboratory texts, as listed in the reference section.

5.3. CHEMICAL EQUILIBRIUM

The fundamental principles of chemical equilibrium and the drive of chemical systems to equilibrium provide a model for explaining much of the behavior of chemical systems encountered by the laboratory analyst. The study of equilibrium constants is crucial to understanding operations such as extraction, dissolution, chromatography, pH and pH measurement, and buffer preparation. Throughout this text, equilibrium will appear in many areas as a common theme in nearly all analytical methods. A system that is at equilibrium is undergoing no net changes; the rates of the forward and reverse reactions are constant and equal. Analytical extractions and acid–base chemistry will be used in this chapter as practical examples and consequences of the need for laboratory analysts to be familiar with equilibrium calculations.

5.3.1. Equilibrium Constants

The equilibrium constant is a measure of the extent to which a chemical reaction or process proceeds from reactants to products. For a hypothetical reaction,

$$aA + bB \rightleftarrows cC + dD \tag{5.1}$$

the equilibrium constant expression is obtained from the concentrations of the reactants and products present at equilibrium and the coefficients in the balanced chemical equation and is given by

$$K = \frac{[C]^c [D]^d}{[A]^a [B]^b} \tag{5.2}$$

The concentrations of pure substances such as water are not written into equilibrium expressions. These unchanging concentrations are incorporated into the value of the equilibrium constant.

A large value for an equilibrium constant indicates a reaction that proceeds

Table 5.3. Filter Media, Applications, and Compatibility

Filter Material	Major Applications	Compatibility
Nylon	General use; HPLC solvents	Resists most common HPLC solvents; limited resistance to acids, bases, halogenated hydrocarbons, aldehydes, strong oxidizing agents
Polyvinylidinefluoride	HPLC sample preparation	Hydrophobic; low water breakthrough; resists most common HPLC solvents; low protein binding
Polytetrafluoroethylene	Nonaqueous solvents	Hydrophobic; requires high pressure to pass water; must be wetted with alcohol to pass water
Polyethersulfone	Protein containing solutions where minimal adsorption of protein desired	Durable; inert; thermally stable; low protein adsorption
Polypropylene	High flow applications	High flow; solvent- and water-compatible
Cellulose acetate	Biological solutions	Low surfactants; low extractables; low protein binding; high capacity
Borosilicate glass	Retention of particles and microorganisms	High flow; high capacity; low extractables; low adsorption; good for rapid, repetitive analyses

Source: Reference [6].

mostly to products, while a small value indicates a reaction that mostly stays as reactants. In laboratory analysis, you will use equilibrium constants extensively. Table 5.4 gives a summary of common situations where equilibrium constants are used in laboratory analysis. Note that there are many "names" for equilibrium constants, depending upon the specific technique.

A more complete treatment of concentration is based on the concept of activity, which is an effective concentration more indicative of the behavior of a species. Under thermodynamically ideal circumstances, a solute is surrounded by solvent and is functionally remote from any other solute molecules. We can then treat the solute as an independent entity. This condition exists to a good approximation when the solute is in low concentration. Under higher concentrations (10^{-3} M and greater) and at interfaces, the solute is not fully solvated and experiences the effects of being close to nonsolvent molecule species, such as a surface or other solutes. The nonideal conditions result in the apparent concentration of the solute being lower than the actual concentration. This is expressed mathematically as $a = \gamma c$, where a is activity, γ is the activity coefficient, and c is the actual concentration. Nonideal conditions lead to values of γ less than 1. Table 5.5 lists selected activity coefficients as a function of concentration for a +1 cation (K^+) and a +2 cation (Ca^{2+}). All of the values are less than 1 and they decrease as the concentration increases, Ca^{2+} faster than K^+. For further information, consult one of the basic texts on quantitative analysis listed at the end of this chapter.

It should be noted that, in practice, relatively few chemical systems used by lab analysts achieve equilibrium. Imagine a GC column. If the molecules are constantly moving, how can they be at equilibrium and how can equilibrium expressions be used to describe the behavior in a column? This is addressed in detail in Chapter 7.

5.3.2. Le Chatelier's Principle

If a system at equilibrium changes in such a way that it is no longer at equilibrium, the system will shift in such a way as to restore the equilibrium. This is a brief statement of Le Chatelier's principle, which provides an explanation for many forces that drive chemical reactions. This also indicates that systems not at equilibrium will shift in order to achieve equilibrium. For example, suppose the hypothetical reaction system described above (equation 5.1) is at equilibrium. Then 50% of product C is removed. The system will shift reactant and product concentrations to restore the equilibrium. More products will be formed until the concentrations again give the equilibrium values.

A practical example is observed in a GC column, which is described in detail in Chapters 7 and 8. When a sample is vaporized in the inlet, all of the analyte molecules are in the vapor state and are transferred onto the column. When they encounter the stationary phase, most of the molecules are dissolved in the stationary-phase liquid and stop moving. The system will attempt to establish vapor pressure equilibrium at this point. The few molecules that remain in the

Table 5.4. Equilibrium Constants

Technique	Typical Reaction	Equilibrium Constant Expression	Equilibrium Constant Name
Extraction	$A(aq) = A(org)$	$K_d = [A(org)]/[A(aq)]$	Distribution constant
Dissolving (ionic solids)	$A_xB_y(s) = xA^+(aq) + yB^-(aq)$	$K_{sp} = [A^+(aq)]^x[B^-(aq)]^y$	Solubility product
Chromatography	$A(mp) = A(sp)$	$K_p = [A(sp)]/[A(mp)]$	Partition coefficient
Dissolving (acids)	$HA(aq) + H_2O(l) = H_3O^+(aq) + A^-$	$K_a = [H_3O^+][A^-]/[HA]$	Acid dissociation constant
Dissolving (bases)	$B^-(aq) + H_2O(l) = BH(aq) + OH^-(aq)$	$K_b = [BH][OH^-]/[B^-]$	Base dissociation constant
Water	$2H_2O(l) = H_3O^+(aq) + OH^-(aq)$	$K_w = [H_3O^+][OH^-]$	Water ionization constant

Table 5.5. Activity Coefficients of K^+ and Ca^{2+}

Concentration (M)	$\gamma_+(K^+)$ (in KF)	γ_+ (Ca^{2+}) (in KCl)
0.01	0.903	0.675
0.1	0.774	0.269
1.0	0.646	0.263

Source: A. Evans, *Potentiometry and Ion Selective Electrodes*, Wiley, New York, 1991.

vapor phase are moved down the column by the carrier gas, where they will attempt to establish a new equilibrium. At the site of the initial equilibrium, more molecules must evaporate in order to return that system to equilibrium. Thus, the movement of molecules along a GC column can be pictured as a succession of steps that are driving to reach equilibrium. It is this drive, along with the differences between equilibrium constants of analytes between the stationary and mobile phases, that allows separation to occur in GC. If you have ever observed fractional distillation column heating up, then you have seen a similar example.

5.3.3. Equilibrium as a Basis for Sample Pretreatment

Chemical equilibrium provides the basis for much of the sample pretreatment that must be performed prior to instrumental analysis. Perhaps the most common sample treatment method is liquid–liquid extraction (LLE), which is often employed for sample cleanup. Other types of extraction performed by laboratory analysis for specific methods include solid-phase extraction (SPE), solid-phase microextraction (SPME), and supercritical fluid extraction (SFE). Equilibrium principles also guide practice in dissolution and influence many instrumental techniques described throughout this text.

5.3.3.1. Liquid–Liquid Extraction. In LLE, analytes are partitioned between two immiscible liquid phases, typically water and an organic solvent. In most applications, the analyte is dissolved in an aqueous phase and it is desired to transfer it to an organic phase. The equilibrium reaction can be described as

$$A(aq) \rightleftarrows A(org) \tag{5.3}$$

and the equilibrium constant expression can be described as

$$K_d = \frac{[A(org)]}{[A(aq)]} \tag{5.4}$$

K_d is called a distribution constant or a partition coefficient and is easily estimated from solubility data. For example, the solubility of cocaine in water is

0.00167 g/ml and the solubility in ether is 0.286 g/ml [7]. The distribution constant, calculated as a ratio of the two solubilities, is equal to 171. Thus, in designing extraction and methods, it is important to know as much solubility information about the compounds of interest as possible.

Attempts have been made to correlate distribution constants obtained by liquid–liquid extraction with HPLC partition coefficients. The extraction solvents most often used are 1-octanol and water, and a large number of K_d values have been reported. In a recent paper by Hanna et al. [8], partition coefficients for drugs obtained by HPLC, MECC, and octanol–water extraction are compared, and the compilations by Hansch [9, 10] provide representative octanol–water partition coefficients (log P) values for a wide range of analytes. Octanol–water partition coefficients have been shown to correlate well with the strength of retention on reversed phase HPLC columns.

Common LLE procedures are well-described in most organic chemistry laboratory textbooks, several of which are listed in the reference section. LLE suffers from the difficulty that large quantities of organic solvents are often used and may present a safety or disposal issue.

5.3.3.2. Solid-Phase Extraction.
As an alternative to LLE, which commonly requires large volumes of organic solvents, solid-phase extraction (SPE) has been employed in many clinical, pharmaceutical, and environmental applications. In SPE, analytes are partitioned between the solution (typically aqueous) phase and a solid-phase sorbent, such as an ion exchange resin, silica, or a reversed-phase (C_8 or C_{18}) bonded-phase packing. (See Chapter 9 for a description of chromatographic bonded phases.) The aqueous solution is passed through the packing, allowing the analytes to be sorbed by the stationary phase. Following a washing step, analytes are eluted using an organic solvent. The sorbent in SPE may be contained within a cartridge, a syringe filter, or a disk, and it can be employed in both manual and automated configurations. Majors and Raynie [11] recently reviewed SPE and sample preparation equipment and methods.

5.3.3.3. Solid-Phase Microextraction.
Solid-phase microextraction (SPME) is a solvent-free alternative to LLE and SPE methods. In SPME, the organic-containing aqueous solution is equilibrated in contact with a fused silica fiber that is coated with a stationary organic phase. Analytes partition between the aqueous and organic phases. In this method, the analyte is significantly concentrated as it passes into the organic phase, due to generally large partition coefficients and the very small volume of the fiber coating. The coated fiber is contained within a syringe needle, as shown in Figure 5.3. Following equilibration, the coated fiber is removed from the aqueous solution and transferred directly to the inlet of a gas chromatograph, where the analytes are thermally desorbed into a capillary GC column.

SPME has been mostly applied in trace environmental [12, 13] and clinical [14, 15] analysis, where the organic–water partition coefficients are usually

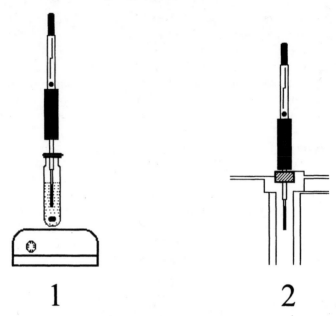

Figure 5.3. SPME apparatus.

very high (10^3 or greater). The SPME–GC analysis of a part-per-trillion-level hydrocarbon sample, extracted from water, is shown in Figure 5.4. SPME has several advantages including no solvent usage, ease of use, and straightforward automation. The theory and practice of SPME are thoroughly described in a textbook by Pawliszyn [16].

5.3.3.4. Soxhlet and Supercritical Fluid Extraction. The extraction methods described above are effective for aqueous samples. If the sample is nonaqueous (solid, for example), it is often difficult to extract using an organic solvent, because the kinetics of liquids passing into solids to dissolve the analyte molecules are often very slow. The classical method for doing this is Soxhlet extraction, which involves the continuous exposure of the solid sample to a stream of hot solvent. A typical apparatus for Soxhlet extraction is shown in Figure 5.5, and Soxhlet extraction is described in the text by Kebbekus and Mitra [17]. Soxhlet methods are often very slow and suffer from poor recovery due to kinetic effects and thermal degradation. The solubility data that would be used to choose an appropriate organic solvent for liquid–liquid extraction can also be used to choose a solvent for Soxhlet extraction.

In supercritical fluid extraction (SFE), a gas, typically carbon dioxide, is heated and compressed above its critical point (32°C and 70 atm for carbon dioxide) to form a supercritical fluid. These fluids have been described as "dense gases" and have diffusion properties similar to gases, allowing them to easily penetrate to the interior of solids, with dissolving power intermediate

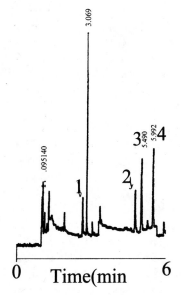

400 part per trillion
1. Toluene
2. Ethylbenzene
3. o-xylene
4. p-xylene
extraction: 3 mL, 30 min
desorption: 220°C, 5 min
Column: 30m x 0.32mm x 1 μm SPB-1,
Temperature program: 40°C (5min),
10°C/min.
Detector: FID, 250°C.

Figure 5.4. SPME-GC analysis of trace hydrocarbons from water.

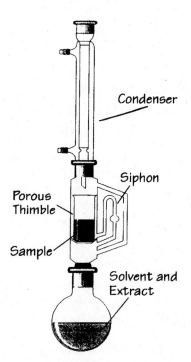

Figure 5.5. Soxhlet extraction apparatus. (Reprinted with permission from B. B. Kebbekus and S. Mitra, *Environmental Chemical Analysis*, Blackie, London, 1999.)

between gases and liquids, enabling them to dissolve a wide variety of analytes. SFE is most commonly employed in environmental analyses involving solid matrices such as soils [18], sludges [19], and smoke particles and, more recently, in the analysis of bulk fat in foods for FDA labeling requirements [20, 21]. The theory and practice of SFE is described in a text by Taylor [22]. SFE has been widely employed in environmental and food analysis; however, acceptance in the pharmaceutical industry has been slow.

5.3.3.5. Conclusions—Equilibrium and Sample Pretreatment. The principles of chemical equilibrium have a profound effect on our ability to sample and remove analytes of interest from the interfering matrix. Some conclusions are as follows: The distribution constant or partition coefficient for the extraction process determines the maximum possible recovery of analyte from an extraction; multiple-step extractions are preferable to single-step extractions, and they ensure maximum recovery; the extraction recovery can depend upon factors such as sample pH, salt content, and temperature, as well as the matrix form (liquid, particulate, nature of the particles, etc.); back-extraction, or washing, can increase the purity of an extract, but will decrease the recovery. Chemical equilibrium is the driving force behind many other processes carried out in aqueous solution, as described below.

5.4. AQUEOUS SOLUTION EQUILIBRIA

5.4.1. Introduction

In this section we will discuss the nature of species that result from the dissolution of compounds in water. When a compound is dissolved, a number of changes have to be considered. First, the solubility of the compound affects the amount that can be dissolved. The compound can dissociate into cations and anions. The compound can react, most commonly via hydrolysis reactions. If other ions are present in the solvent, a new species can precipitate. Each of these processes is governed by solution-phase equilibria, and an understanding of these equilibria will help the laboratory analyst both to quantitatively control these processes and to use them as a means to determine solution species concentration.

Many equilibria that exist in homogeneous aqueous solutions are of importance to the pharmaceutical analyst. The acidities of HPLC mobile phases and the moisture content of tablets are two examples. The equilibria are usually classified by the type of reaction taking place. Acid–base and oxidation–reduction equilibria are two of the most important ones, and they will be presented briefly in this chapter. A basic text on quantitative analysis should be consulted for further information. Some are listed at the end of this chapter.

The reaction for the autoprotolysis of the solvent, water, is given in equation 5.5:

$$2H_2O_{(l)} \rightleftarrows H_3O^+{}_{(aq)} + OH^-{}_{(aq)} \tag{5.5}$$

It produces two ions, hydronium (H_3O^+) and hydroxide (OH^-). Water acts as both a weak acid and a weak base in equation 5.5. In pure water, free of carbon dioxide, the concentrations of hydroxide and hydronium are both 1×10^{-7} M, and the appropriate equilibrium constant expression is

$$K_w = [H^+][OH^-] = 1.00 \times 10^{-14} \text{ M}^2 \tag{5.6}$$

The importance of this equilibrium is that it links the concentration of H_3O^+ and OH^- in *any* aqueous solution, because equation 5.5 must always be satisfied. If one concentration increases, the other must decrease in order for the K_w value to be true.

Other ions in an aqueous solution result from the dissolution of other chemicals such as other acids, bases, salts, oxidizing reagents, reducing agents, and so on. Water promotes the formation of ions, so these chemicals dissociate into ions or they react with water to produce ions. The degree of reaction depends upon the equilibrium constants for the indivdual reactions.

5.4.2. Acids and Bases

There are two definitions of acids and bases in common use. In the more commonly encountered Brønsted–Lowry formalism, an acid is a proton donor (equation 5.7 is an example) and a base is a proton acceptor (equation 5.8 is an example):

$$HCl_{(aq)} + H_2O_{(l)} \rightleftarrows Cl^-{}_{(aq)} + H_3O^+{}_{(aq)} \tag{5.7}$$

$$NH_3 + H_2O \rightleftarrows NH_4^+ + OH^- \tag{5.8}$$

The strength of acids is expressed by the equilibrium constant for the proton donor reaction:

$$HA_{(aq)} + H_2O_{(l)} \rightleftarrows A^-{}_{(aq)} + H_3O^+{}_{(aq)} \tag{5.9}$$

$$K_a = \frac{[A^-][H_3O^+]}{[HA]} \tag{5.10}$$

Acids and bases, like salts, are divided qualitatively into strong and weak categories. Strong acids are those that are better proton donors than H_3O^+, and the equilibrium shown in equation 5.9 lies completely to the right.

Weak acids donate a proton to water only slightly, so the equilibrium in equation 5.9 lies essentially to the left. Similarly, the strength of bases can also be expressed by the values of the K_b (equation 5.12), the equilibrium constant for the dissociation or proton acceptor reaction shown in equation 5.11.

$$B + H_2O \rightleftarrows HB^+ + OH^- \qquad (5.11)$$

$$K_b = \frac{[HB^+][OH^-]}{[B]} \qquad (5.12)$$

5.4.2.1. pH

Logarithmic Relationships. Because most acid–base equilibrium constants span many orders of magnitude, it has proven convenient to express them in logarithmic form. The most important of these expressions is pH, which is conventionally defined as the negative logarithm of the hydronium ion concentration:

$$pH = -\log[H_3O^+] \qquad (5.13)$$

While the absolute concept of pH is complex, it is widely accepted that in practical situations the definition in equation 5.13 will suffice. Note, however, that the USP adopts an operational definition based on electrochemical measurements (to be described later). The logarithmic form of the K_w expression for water (equation 5.6) is given in equation 5.14:

$$pK_w = pH + pOH = 14.000 \qquad (5.14)$$

We can also express the equilibrium constants for the strength of acids and bases in a logarithmic form:

$$pK_a = -\log(K_a) \qquad (5.15)$$

$$pK_b = -\log(K_b) \qquad (5.16)$$

$$K_w = K_a \cdot K_b \qquad (5.17)$$

Buffers. Aqueous buffers are solutions that resist changes in pH and contain both a weak acid and a salt of its conjugate base in roughly equal concentrations. The major species are the acid and its conjugate base. Acetic acid and sodium acetate constitute one such conjugate acid–base pair that can be used to prepare a buffer solution. The acetic acid is consumed by any added base:

$$CH_3COOH + OH^- \rightleftarrows CH_3COO^- + H_2O \qquad (5.18)$$

$$K_{eq} = \frac{1}{K_b} = 1.8 \times 10^9 \ M^{-1} \qquad (5.19)$$

Any added acid reacts with the conjugate base to yield the conjugate acid:

Table 5.6. Buffers for Use in HPLC Separations

Buffer	pK_a	Buffer Range[a]	UV Cutoff[b] (nm)
Trifluoracetic acid	$\gg 2$	1.5–2.5	210 (0.1%)
Phosphoric acid/mono or	2.1	<3.1	<200 (0.1%)
di-K phosphate	7.2	6.2–8.2	
	12.3	11.3–13.3	<200 (10 mM)
Citric acid/tri-K citrate	3.1		
	4.7	2.1–6.4	230 (10 mM)
	5.4		
Formic acid/K-formate	3.8	2.8–4.8	210 (10 mM)
Acetic acid/K-acetate	4.8	3.8–5.8	210 (10 mM)
Mono-/di-K carbonate	6.4	5.4–7.4[c]	<200 (10 mM)
	10.3	9.3–11.3	<200 (10 mM)
Bis-tris propane[e] · HCl	6.8	5.8–7.8	215 (10 mM)
/Bis-tris propane	9.0	8.0–10.0	225 (10 mM)
Tris[d] · HCl/tris	8.3	7.3–9.3	205 (10 mM)
Ammonium chloride	9.2	8.2–10.2	200 (10 mM)
1-Methylpiperidene · HCl/1- Methylpiperdine	10.1	9.1–11.1	215 (10 mM)
Triethylamine–HCl/triethylamine	11.0	10.0–12.0	<200 (10 mM)

[a] pH range allowed with this buffer (conservative estimate).

[b] Absorbance from reference 7.

[c] Requires addition of an acid (e.g., acetic or phosporic).

[d] Tris (hydroxymethyl) aminomethane.

[e] 1.3-bis[Tris(hydroxymethly)methlaminol propane.

Source: Reprinted with permission from L. R. Snyder, J. J. Kirkland, and J. L. Glajch, *Practical HPLC Method Development*, 3rd edition, Wiley, New York, 1997, p. 299.

$$CH_3COO^- + H_3O^+ \rightleftarrows CH_3COOH + H_2O \qquad (5.20)$$

$$K_{eq} = \frac{1}{K_a} = 5.6 \times 10^4 \text{ M}^{-1} \qquad (5.21)$$

Some common buffers, shown in Table 5.6, are those recommended for HPLC due to their optimal UV absorbance. The USP [23] describes the preparation of standard buffers.

Because both members of the conjugate acid–base pair are major species in buffer solutions, the calculations are simplified to some extent. The equilibrium concentrations of the acid and the base are essentially the same as the initial concentrations (because they are both weak). As long as the initial concentrations of the acid and the base differ by no more than a factor of 10, the following equation can be used to determine buffer solution pH:

$$pH = pK_a + \log\left\{\frac{[A^-]_{initial}}{[HA]_{initial}}\right\} \qquad (5.22)$$

This is the well-known Henderson–Hasselbalch (or buffer) equation. It can be used to prepare buffers that yield solutions of specific pH ranges, and it requires only the knowledge of the pK_a of the weak acid and the initial concentrations of the acid and base.

Buffer Capacity. There is a limit to the amount of acid or base that a buffer solution can absorb before the pH of the solution begins to change. This limit is defined as the buffer capacity, which is a function of the concentration of the buffer, the volume of the buffer solution, and the pH variation that the system can tolerate. A 0.1 M buffer solution has a higher buffer capacity than an equal volume of a 0.05 M buffer solution. Greater volumes of buffer can absorb more acid or base than smaller volumes. The broader the pH tolerance, the greater the capacity of a given buffer solution.

Useful Buffer pH Range. One observation that can be made from the buffer equation is that the pH of a buffer solution is close to the pK_a of the acid used to prepare the buffer. Another observation is the limitation that the concentrations of the acid and base must not differ by a factor of 10 defines a range over which a given acid–base pair forms an effective buffer. This effective range turns out to be $pH = pK_a \pm 1$. Therefore, a particular weak acid can be used to prepare a buffer solution with a pH one unit above or below its pK_a. The buffer solution must contain both the acid and the salt of the conjugate base, so the cation of the conjugate base must have no acid–base properties of its own. For example, sodium acetate would be a good choice, but ammonium acetate would not. If an appropriate salt of the desired conjugate base is not readily available, the buffer can be prepared by dissolving sufficient acid in water to supply the total amount of anion, then adding base to form the desired amount of conjugate base. Acid can also be added to concentrated solutions of the appropriate base.

5.5. REDUCTION–OXIDATION EQUILIBRIA

5.5.1. Introduction

Reduction–oxidation, or redox, equilibria involve the transfer of electrons between two species. One of the principal advantages of these types of equilibria is that they can be monitored electrically, usually with voltmeters adapted for this purpose. The common pH meter is one such device. Electrical detection makes such measurements ideally suited to computerization, yielding such devices as automated titrators. All electroanalytical methods are based on redox equilibria or related capacitive processes.

Redox equilibria are made up of two coupled reactions: an oxidation, where a species is losing electrons; and a reduction, where a species is gaining electrons. The species undergoing oxidation is referred to as the reducing agent, or reductant, because it is reducing the other species. The species undergoing reduction is referred to as the oxidizing agent, or oxidant, because it is oxidizing the reductant.

Electroanalytical methods are useful for the investigation of many pharmaceutical problems. These methods are particularly valuable for bulk materials and dosage forms of pharmaceuticals. The linear ranges of detection for many electroanalytical techniques are quite wide (typically 1×10^{-5} to 1×10^{-3} M), but the lower limit is not as good as for other techniques. Many electroanalytical techniques do not require standardization, and they offer simpler sample preparation due to the insensitivity of the technique to common excipients. A particular advantage of electrochemical methods is their suitability for continuous monitoring, as in dissolution analysis. Chromatographic methods require extensive robotics and autosamplers to perform such analyses.

Coulometric methods are absolute, because they determine the amount of material electrolyzed from the amount of charge passed through an electrochemical cell. They can be conducted under controlled potential or controlled current conditions. Constant current coulometry is particularly valuable for electrogenerating reagents (coulometric titrations) and can be used for any titration that can be done volumetrically. Many reagents that cannot be used in volumetric titrations, such as halogens, Ag(II) and Cu(I), can be generated by this method. In particular, electrogenerated I_2 is the titrant in Karl Fischer titrations using pumped solutions and biamperometric flow-through detection. Amperometric methods are based on measuring the current passed between two electrodes with a potential applied across them. Potentiometric methods involve the measurement of the equilibrium cell potential. Voltammetric methods are dynamic and based on the observing the current flow as the potential is varied. These methods have been used to quantitatively study many pharmaceuticals, and they have detection limits in the 10^{-4}–10^{-5} M range with solid electrodes; these limits are as low as 10^{-8} M with mercury electrodes. They have a broader applicability due the greater speciation of voltammetric methods. Selectivity can be substantially improved via the use of ion-selective electrodes. The selectivity of all of these methods can be improved through coupling them with a chromatographic separation, such as in electrochemical detectors. We will limit our discussion to potentiometric methods, particularly the determination of pH, but many reviews are available covering the other electroanalytical techniques mentioned above. (See the reference section provided at the end of this chapter for general information on electroanalytical techniques.)

5.5.1.1. Electrochemical Cells. The laboratory analyst is principally interested in electrochemical reactions that occur within a cell, where the electrons pass through an external circuit. Electrochemical cells are best thought of as collections of interfaces between phases, and we are particularly concerned about the transport of charge across these interfaces. An electrolyte solution is usually one of the two phases, and the other is an electrode. Practically speaking, we must consider at least two interfaces consisting of an electrolyte solution

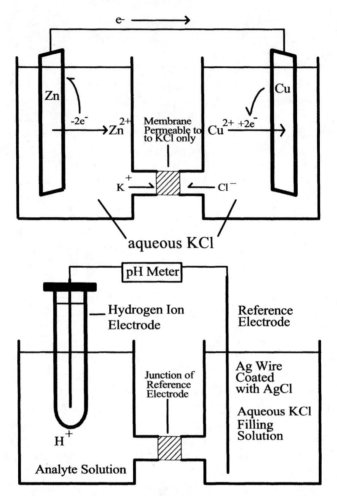

Figure 5.6. Comparison of a Danielle cell (top) with a glass electrode and Ag/AgCl reference electrode in a similar configuration.

in contact with two electrodes. In general, there exists a difference in potential between the interfaces, and the cell potential is the difference in potential between all of the interfaces as measured at the two electrodes. This cell potential results from a number of contributors, including the charging of the surfaces, but we are most interested in the contribution due to the analyte.

It is convenient to divide the redox reaction into two half-cells, separating the oxidation and reduction reactions (Figure 5.6). Thus, for the reaction

$$Zn_{(s)} + Cu^{2+}{}_{(aq)} \rightleftarrows Zn^{2+}{}_{(aq)} + Cu_{(s)} \tag{5.23}$$

the oxidation half-cell is

$$Zn_{(s)} \leftrightarrows Zn^{2+}{}_{(aq)} + 2e^- \qquad (5.24)$$

and the reduction half-cell is

$$Cu^{2+}{}_{(aq)} + 2e^- \rightleftarrows Cu_{(s)} \qquad (5.25)$$

Unlike the overall redox reaction, electrons appear in these half-cell reactions as reactants (for reductions) or products (for oxidation). Such a formal division is useful, because the half-cells can be treated separately and can be combined to form a variety of analytically useful cells.

5.5.1.2. Current and Potential in Electrochemical Cells.
Electrochemical cells are characterized by several quantities associated with the electrons within the reaction. The coulomb (C) is the fundamental unit of electrical charge. The Faraday ($\mathscr{F}$), 9.64846×10^4 C/mol, is the amount of charge due to one mole of electrons. The rate at which electrons flow through an electrochemical cell is the current (I), measured in amperes (amp):

$$1 \text{ amp} = 1 \text{ C s}^{-1} \qquad (5.26)$$

The total amount of charge is given by the product of the current (I) and the time (t) over which the current flowed (charge $= I \times t$). We can obtain the number of electrons transferred in an electrochemical reaction from

$$\text{Moles of electrons} = n = \frac{(I \times t)}{\mathscr{F}} \qquad (5.27)$$

This links the stoichiometry of a redox reaction with a quantity measurable from an electrochemical reaction and is exploited in coulomb techniques.

Potential, E, is related to the thermodynamic driving force of the redox reaction:

$$\Delta G = -n\mathscr{F}E_{cell} \qquad (5.28)$$

where n is the number of electrons in the reaction, $\mathscr{F}$ is the Faraday, and E_{cell} is the measured cell potential.

Electrical potential has the units of voltage, where one volt is one joule of electrical potential energy per coulomb of charge transferred through the circuit. The number of electrons in the redox reaction is denoted by n.

Standard cell potentials ($E°$) are defined for reductions under standard conditions (1.00 M solution concentrations, 25°C). The absolute potential for a half-reaction cannot be measured, so the reduction of hydronium ion to hydrogen gas has been given a defined potential of 0.000 V:

$$2H_3O^+_{(aq, 1.00 \text{ M})} + 2e^- \rightleftarrows H_2 \text{ (g, 1 atm)}, \qquad E° = 0 \text{ V by definition} \qquad (5.29)$$

This half-cell is used in the standard hydrogen electrode (SHE), which consists of a platinized platinum electrode immersed in a 1 M solution of H_3O^+, in contact with 1 atm of hydrogen gas:

$$Pt|H_2(a = 1)|H^+(a = 1, \text{ aqueous}) \tag{5.30}$$

where the | indicates a phase boundary.

Half-cell potentials also have a reference direction in order to define the sign of spontaneity. By convention, half-cell potentials are tabulated for reductions, so electrons are the reactants. The signs of the reduction potentials are such that if the reaction occurs as written (a reduction) relative to the SHE, then the sign of the reduction potential is positive.

It is important to remember that the cell potential is independent of number of electrons involved in the reaction, but the ΔG does depend on the reaction stoichiometry and n. The relationship between the cell potential and the concentration of the species in the cell is given by the Nernst equation:

$$E_{\text{cell}} = E^\circ - \left(\frac{\mathscr{R}T}{n\mathscr{F}}\right) \ln\left[\frac{a(\text{reduced form})}{a(\text{oxidized form})}\right] \tag{5.31}$$

$\mathscr{R}$ is the gas constant and T is the absolute temperature. Measurement of E_{cell} therefore allows us to determine concentrations (under ideal conditions) and is the basis for all potentiometric and volumetric techniques.

5.5.1.3. Electrodes. Potentiometric measurements are carried out in a cell consisting of two electrodes in contact with the analyte solution, which should also contain an electrolyte. Often the laboratory analyst focuses on the reaction occurring at one electrode, such as in pH or ion-selective electrode measurements. In this case, one electrode, termed the reference electrode, is constructed of a series of phases having a constant composition, to be a nonchanging reference. The second electrode, at which the reaction of interest is occurring, is termed the indicating electrode. A variety of reference and indicating electrodes are commonly in use.

Practical Reference Electrodes. Several types of reference electrodes are in common use. Whereas the standard hydrogen electrode is defined as the reference voltage, it is difficult to use in practice. More common reference electrodes are based on calomel:

$$Hg|Hg_2Cl_2|MX(\text{variable concentrations in water}) \tag{5.32}$$

or silver halide:

$$Ag|AgX|MX(\text{variable concentrations in water}) \tag{5.33}$$

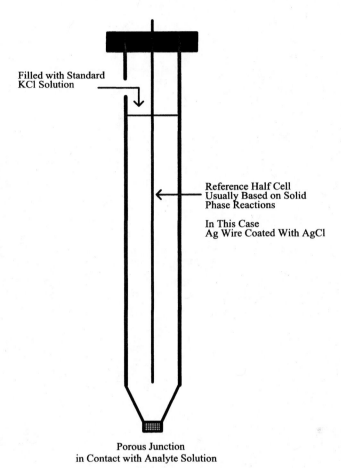

Filled with Standard
KCl Solution

Reference Half Cell
Usually Based on Solid
Phase Reactions

In This Case
Ag Wire Coated With AgCl

Porous Junction
in Contact with Analyte Solution

Figure 5.7. Essential features of an Ag/AgCl reference electrode.

The constant makeup of the reference electrode yields a fixed potential, so the change in the potential of the indicating electrode with respect to the reference electrode is observed.

The common reference electrodes in use, whether based on silver ion or calomel, share a number of physical features. In Figure 5.7, the essential features of these electrodes are shown. The most important is the interface between the cell solution and the analyte solution. This interface is a junction that allows the flow of electrolyte solution to maintain charge balance. The reference electrode junction is the point where the analyte solution and the reference electrode filling solution mix. A potential, due to the differing rates of diffusion of the cations and anions across the junction, results from this mixing, and this potential must be compensated for in order to obtain the most accurate measurements. The junction is constructed of a variety of materials, included fritted glass, Vycor (thirsty glass), ground glass sleeves, or cellulose fiber. The junction potential is minimized by allowing the filling solution to flow into the analyte solution, thereby avoiding the buildup of ions in the junction from back-diffusion. The flow of the electrolyte solution introduces some sample contamination by the electrode filling solution, but this is usually minimal, on the order of microliters per minute.

Ion-Selective Indicating Electrodes. The use of glass and other membranes to selectively detect a variety of species of pharmaceutical interest is well-developed [24–30]. Glass membranes exhibit ion-selective properties and have been used to develop electrodes to sense H_3O^+, Na^+, K^+, and NH_3 [31–36]. Membranes such as Nafion, poly(vinyl chloride), and ionic compound coatings are used for other species, the chemistry of which can be quite complex. Orion Research, Inc. (see General References) lists 18 different species for which commercial electrodes are manufactured to sense, including halides, halogens, ammonia, Cu^{2+}, CN^-, BF_4^-, Pb^{2+}, NO_3^-, NO_2^-, NO_x, O_2, ClO_4^-, K^+, and Na^+. Ion-selective electrodes are widely used to detect pharmaceuticals via reactants immobilized on the membrane.

The generalized Nernst equation for ion-selective electrodes is

$$E(\text{cell}) = E' \pm \left(\frac{\mathscr{R}T}{n\mathscr{F}}\right) \ln a_i \qquad (5.34)$$

or

$$E(\text{cell}) = E' \pm \left(\frac{2.303\mathscr{R}T}{n\mathscr{F}}\right) \log a_i \qquad (5.35)$$

or

$$E(\text{cell}) = E' \pm \left(\frac{0.0591}{n}\right) \log a_i \qquad \text{at 298 K} \qquad (5.36)$$

E' is a constant that incorporates the potential of the reference electrode and the standard potential of the half-cell containing the solution under investigation and the ion-selective electrode. The sign is positive for cations and negative for anions. The common log is often used because it allows the consideration of the response in decades. Practically, if the slope of the $E(\text{cell})$ versus $\log a_i$ lies between 0.0550 and 0.0591, the electrode is taken to have Nernstian response.

The procedures for calibrating ion-selective electrodes are well-established [39–42]. The principal difficulty of such calibrations is the preparation of standard solutions of known *activity*, as opposed to concentration, particularly because the other ionic and neutral species have an effect on the activity of the selected species. In the case of pH standards, typical buffers used for electrode calibration are usually only accurate to ± 0.02 pH, but electrodes can discrimi-

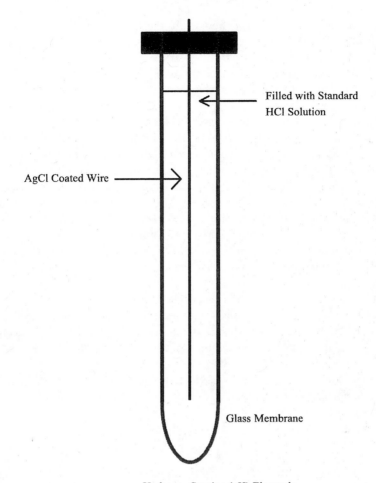

Filled with Standard
HCl Solution

AgCl Coated Wire

Glass Membrane

Hydrogen Sensing (pH) Electrode

Figure 5.8. The common glass electrode.

nate betweens solutions differing by ± 0.002 pH. Also, the composition of the buffers is usually different from the test solutions, so there are some uncompensated junction potentials. Often, ionic strength adjustors (inert electrolytes) are employed to minimize these problems. Because the activity coefficient of the analyte is a function of the total ionic strength, the high concentration of the ionic strength adjustor determines the activity coefficient. In general,

$$E(\text{cell}) = E' \pm \left(\frac{0.0591}{n}\right) \log c_i \gamma_i = E' \pm \left(\frac{0.0591}{n}\right) \log c_i \pm \left(\frac{0.0591}{n}\right) \log \gamma_i$$

$$(5.37)$$

For a given ionic strength, c_i and γ_i are the concentration and activity coefficient of the analyte, n and γ_i are constant, therefore the $0.0591/n \log \gamma_i$ term is a constant and can be incorporated into the E' constant. It is important to remember that the use of ionic strength adjustors does change the pH of the analyte solution, particularly when the solute concentration is low [34, 43].

Glass Electrodes. The common glass pH electrode (Figure 5.8) can be used to illustrate the important features of ion-selective electrodes. The glass membrane usually is prepared from chemically bonded Na_2O and SiO_2, which is low in Al_2O_3 and B_2O_3 [34]. Other materials, such as semiconductors, have been used commercially for pH electrodes, but studies have shown that the glass electrode is superior in rejecting interference from Mg^{2+} and Ca^{2+} [34]. Some controversy has existed over the mechanism of ion-selective electrode response, but the current model is based on the existence of a phase boundary equilibrium between the anionic groups on the glass surface and the ions in the solution [32]. This equilibrium is thought to be responsible for the glass electrode response.

Other Electrodes. While the pH electrode is the most common ion-selective electrode encountered, numerous other materials can be sensed using glass membrane electrodes. Other sensing materials, including coated semiconductors, coated solids, and plastic membranes, can be used to adjust the sensitivity of the electrode to specific materials. Platinum, gold, mercury, and carbon are often used as the working electrodes for redox measurements. The difficulty of using liquid mercury can be circumvented by coating solid electrodes with a film of mercury.

5.5.1.4. Measuring pH

Electrode Configurations (Dual Versus Combined). Potentiometric measurements require at least two electrodes: a reference electrode and a sensing electrode. In pH measurements, the reference electrode is usually calomel or silver/silver ion, with the latter being more common. The sensing electrode normally is a glass electrode. These electrodes can be arranged in one of two configurations: dual electrode, Figure 5.9a; or combination electrode, Figure 5.9b. There are pros and cons to each electrode configuration.

The dual electrode configuration uses two separate electrodes with the reference half-cell housed in one body and the glass electrode housed in another. Separate electrodes allow maintenance procedures specific to each type of electrode and the use of reference electrode filling solutions, junction materials, or glass electrode surfaces that can accommodate a broader range of analyte solutions (see below). If the analyte solutions degrade one of the two electrodes, that electrode can be replaced or rejuvenated separately. The analyte solution may contain ions that are incompatible with the reference electrodes incorporated in typical combination electrodes. The sample may contaminate the reference electrode junction, requiring frequent cleaning. Separate electrodes facili-

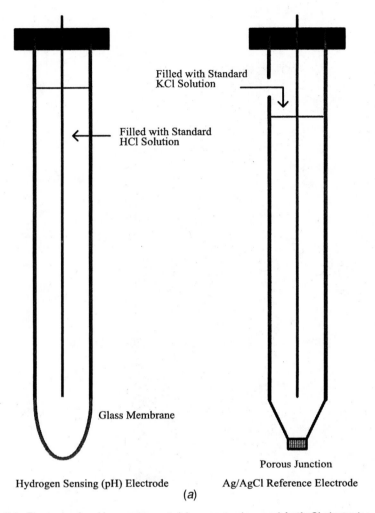

Filled with Standard
KCl Solution

Filled with Standard
HCl Solution

Glass Membrane

Porous Junction

Hydrogen Sensing (pH) Electrode   Ag/AgCl Reference Electrode

(a)

Figure 5.9. Electrodes for pH measurement: (*a*) separate glass and Ag/AgCl electrodes.

tate junction cleaning and/or replacement, particularly due to the larger size of the reference electrode. The disadvantages of the two-electrode system are that larger analyte solution volumes are necessary, the two-electrode system is more difficult to manipulate, and the glass-reference electrode responses are not necessarily matched.

Combination electrodes, where both half-cells are combined within a single body, are far more common for routine measurements. They can be quite small, facilitating pH measurements in very low solution volumes. The performance of the reference and pH sensors can be carefully matched to produce the best response. Combination electrodes are also simpler to maintain when used for routine pH measurements.

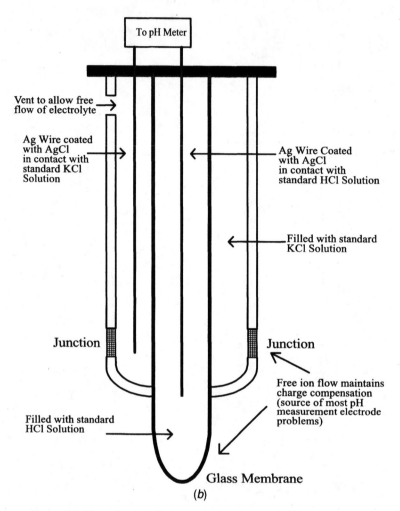

Figure 5.9 Electrodes for pH measurement: (b) the combination electrode.

Standardization. The ion-selective electrode must be calibrated by standards in order to determine the value of the intercept of the Nernst equation (equation 5.36). In pH measurements, standard commercial buffer solutions or solutions prepared from commercial standard dry solids (preferably those certified by the NIST) are normally used to calibrate the electrodes. The USP describes such standardization of pH and indicating electrodes [33]. The salts used to prepare standard buffers are available from NIST or are directly traceable to NIST and can be obtained from commercial sources. Whether prepared from salts or purchased from suppliers, the buffers must be replaced every three months to conform to USP recommendations. If prepared from salts, carbon dioxide-free water must be used and the solutions must be stored in Type I chemically resistant glass bottles.

The best procedure for calibrating the electrode with multiple buffers is readily performed with modern, microprocessor-controlled pH meters. At the very least, buffers of known pH bracketing the sample pH must be used. Temperature must also be considered. Substantial problems can result from temperature differences of the sample, electrode, and calibration standards. The pH of the buffers also change with temperature due to the equilibria associated with acids and bases that compose the buffer. The pH values at various temperatures are normally listed on their packaging. This latter problem is one that is often overlooked.

Potentiometers (pH Meters). Potentiometric measurements are typically made with high-impedance voltmeters. These devices measure potential under conditions of minimal current flow so the equilibrium being measured is not disturbed. These voltmeters are modified to include microprocessors and additional sensing inputs. The microprocessors are used to adjust the slope and intercept of the Nernst equation from data obtained by calibrating the electrode with standard buffers. The meter can also incorporate a printer for data output, or ports to interface with computers. The elements of a potentiometer are often incorporated into other devices for automated control, such as autotitrators and flow analysis systems. An important application of flow analysis is in automated dissolution systems.

The USP has adopted a working definition of pH to facilitate practical comparisons between laboratories. Here, the pH is defined as the value measured by a pH meter using a reference electrode and a pH-indicating electrode standardized with the appropriate NIST buffer solutions [33, 39–42]. This definition does not necessarily correspond to the theoretical definition of pH, but it is close enough for practical purposes.

Temperature Compensation. The measured pH is sensitive to solution temperature, due to the $(\mathcal{R}T/z_iF)$ term [44]. For example, the slope of the E versus pH line is 59.2 mV/pH at 25°C, 64.1 mV/pH at 50°C, and 54.2 mV/pH at 0°C. This change can be compensated for by either (a) manually adjusting the slope of the calibration line using the measured solution temperature or (b) automatically adjusting the meter response using a probe to sense the solution temperature. Slow drift of the electrode response may also be observed, due to the time it takes for the electrode itself to reach thermal equilibrium with the solution. The particular component that must come to equilibrium is the reference electrode, because this half-reaction is usually deep within the body of the electrode.

Troubleshooting. Common problems observed in pH measurements are slow response, drift, unstable readings, and measurement errors. These problems often arise from breakdown of the reference electrode junction conditions, dehydration or contamination of the glass electrode surface, or low solution conductivity. These problems are frequently encountered in the measurement of pH in solutions containing nonaqueous components, extreme pH, or high ionic strength (these systems will be addressed in detail below). Clogged reference junctions can mimic the effect of low solution conductivity, and they often result from solutions containing particles, polymers, or biological materials that precipitate on the junction. Nonaqueous components can also cause precipitation. High junction flow rates or "blow-out" junctions minimize such problems.

Nonaqueous pH Measurement. The control and determination of the pH of nonaqueous and mixed aqueous–nonaqueous solutions is problematic at best. Such solutions yield readings that are unstable and drift, have long response times, and are (by definition) inaccurate. Instability, drift, and long response times are due to low solution conductivity, glass electrode dehydration, and incompatibilities between the reference electrode filling solution and the analyte solution (see above). The low solution conductivity of nonaqueous solutions results from the lack of free ions in the solution to balance the charge between the glass and reference electrode half-cells. Nonaqueous solutions may dissolve salts, but these salts are normally ion-paired. Addition of tetraalkylammonium salts to the analyte solution may alleviate conductivity problems, but the salts must be of high purity to not contaminate the analyte.

Many junction potential problems can be addressed by using separate electrodes. The reference electrode half-cell can then be based on a solution compatible with the analyte solution, such as also described in the section on nonaqueous titration. One common nonaqueous reference electrode half-cell is the $Ag/AgClO_4$ redox pair. Silver perchlorate is being supplanted by silver triflate (triflate = trifluoromethanesulfonate) to avoid the use of perchlorate salts.

Dehydration of the glass electrode can be minimized by frequently rehydrating the electrode in the recommended storage solution. Hydrogen ion activity in nonaqueous media is a complex problem in general. Buffer solutions for standardizing pH electrodes are based on aqueous solutions, and the pH of these solutions is *not* comparable to the pH in nonaqueous solutions. Development of standard buffers in the mixed solvent medium is recommended, but is a formidable problem. Carryover of the nonaqueous component is also a problem, and exhaustive rinsing is necessary to avoid contamination of the storage solutions.

Acid and Alkaline Error. While pH electrodes are quite effective at selectively sensing the concentration of H_3O^+, the selectivity falls off at the pH scale extremes. Solutions of high acid content show positive deviations due to the breakdown of the assumption that hydronium concentration equals hydronium activity ($[H_3O^+] = a_{H_3O^+}$). Generally, this approximation breaks down for all species in concentrations greater than 10^{-3} M. The principal reason for this is that proper solvation of any species requires a number of solvent layers to isolate solute molecules from each other. Protons, as well as other small ions, have substantial solvation requirements because the high charge-to-size ratio polarizes the solvent. Complete solvation of H^+ requires far more water molecules than the approximate formula H_3O^+ implies. Solutions of high acid concentration have insufficient water molecules to properly solvate the proton, so the solution appears to have less acid in it than it actually does (because the pH electrode senses the fully solvated proton). In basic solution, negative deviations are observed due to the low concentration of H_3O^+ and the high concentration of small metal ions used as the cations to balance the charge of the hydroxide. Potentials associated with the reference junction also become important in solutions of high ionic strength because of incompatibilities of the analyte solution and the reference electrode filling solution. These incompatibilities are manifested as slow response and drift of the readings.

Summary of pH Measurement. It is important to remember that pH is a complex phenomena, because of the multiple interactions that take place in solution. Practically, the laboratory analyst has to make decisions regarding the importance of each of the above factors, which may be involved in the analysis. Often, it is more important to develop methods that are consistent from analysis to analysis. The control of the pH of HPLC mobile phases is one example where it is more important to have consistently prepared solutions than it is to have a solution of an absolute pH.

5.5.1.5. Titrations

Definitions. Titration is an analytical technique that measures the concentration of an analyte by the volumetric addition of a reagent solution (the titrant) that reacts quantitatively with the analyte and is covered in USP section ⟨541⟩ [45]. Typically, in a direct titration, an accurately measured volume of analyte solution is placed in a suitable container and the titrant is added with a burette. The titrant concentration or the burette size is adjusted so that 30–95% of the burette volume is delivered in the analysis. The reaction is monitored visually or instrumentally.

Titration analysis requires that a suitable reaction between a titrant and the analyte be developed. Successful titration analyses have been based on acid–base, compleximetric, and reduction–oxidation chemistry. The reaction must be quantitative, fast, and well-behaved. Quantitative reactions will be those with large K_{eq} values. The relative rate of the reaction must be tested empirically. Well-behaved reactions will not be air-sensitive, will use thermally stable and easily standardized reagents, will avoid noxious reagents, and will have few side reactions, among other considerations.

The advantages and disadvantages of titration are summarized in Table 5.7. The major disadvantage in the pharmaceutical lab is the lack of specificity. Very few titration reactions are available for one specific drug, for example.

A variety of methods can be used to monitor the progress of the titration reaction. One of the most convenient methods is the use of colorimetric indicators. Colorimetric indicators are species added to the analyte solution that clearly and abruptly change color when the slightest excess of titrant is added.

Table 5.7. Advantages and Disadvantages of Titrimetric Methods

Advantages	Disadvantages
Great flexibility	Requires relatively large amounts of analyte.
Suitable for a wide variety of analytes	Lacks speciation, especially for compounds with similar structure.
Manual methods need only simple, commonly available instrumentation	Colorimetric indication is operator-dependent.
Capable of excellent precision and accuracy	Manual titrations very sensitive to skill of analyst.
Readily automated	Reagents are often unstable.
Well-established methods available in literature for many analytes	Chemical complications and interferences can be a problem.

The most commonly encountered colorimetric indicator is phenolphthalein, used for the titration of strong acids and bases. The USP describes a number of colorimetric indicators and their uses [46]. Instrumental methods, such as spectroscopy, potentiometry and coulometry can also be used, especially for the analysis of unknowns or where the use of appropriate indicators is not possible.

Electrochemical measurements are a versatile method for detecting endpoints. An indicator electrode sensitive to the species undergoing titration and a suitable reference electrode, both immersed in the analyte solution, can be used to form a galvanic cell. The potential of this cell can then be measured with a pH meter, yielding a plot of pH (or mV in the case of complexometric or redox titration) versus titrant volume. A sigmoid curve is typically obtained and the midpoint of the sharply rising linear portion of the curve represents the endpoint. This type of data analysis works for all acid–base titration; but for asymmetric reactions, in which the number of anions reacting is not equal to the number of cations reacting, the midpoint will not be exactly at the endpoint.

Titration of a Strong Acid with a Strong Base. The titration of a strong acid with a strong base is a straightforward procedure because the reaction is rapid and quantitative. Addition of the base consumes an equimolar amount of the acid, decreasing the concentration of protons in solution. The pH at any point in the titration can be calculated from a limiting yield method. While titration of a strong acid with a strong base is usually followed colorimetrically, the titration can be followed by measuring pH directly with a pH meter and electrode (see below). This procedure is typical of many titrations, and it serves as a general example. A plot of pH versus volume of OH^- added is shown in Figure 5.10 and can be used to illustrate some of the salient features of such a titration. Prior to the addition of base, the pH of the solution is simply that due to the strong acid present. As base is added, the pH slowly rises as the acid is consumed. When the number of moles of added base approaches that of the moles

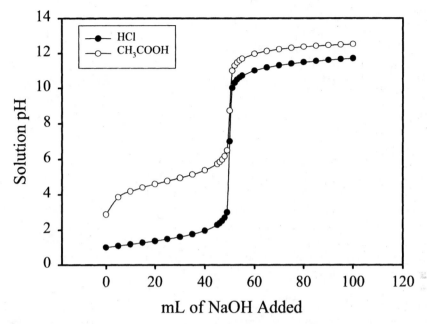

Figure 5.10. Titration curves for 50 mL solutions of 0.1 M HCl and CH₃COOH respectively, titrated with 0.1 M NaOH.

of acid, an abrupt change in pH is observed. This abrupt pH change signals the stoichiometric or *equivalence point*, where the amount of titrant added equals the amount of analyte. At this point the reaction is complete. The *endpoint* is that volume of titrant where the indicator signals that the reaction is complete. The pH at which the endpoint is signaled depends on the indicator, but for the titration of a strong acid with a strong base the pH change at the equivalence point is so large that a wide number of indicators will work. Phenolphthalein is often the preferred indicator, due to its readily observable color change under typical experimental conditions. Figure 5.10 also shows the titration curve for more dilute solutions of the acid and base. Dilute solutions exhibit a smaller pH change at the equivalence point, making the indicator choice more critical.

Titration of a Weak Acid with a Strong Base. The titration of a weak acid with a strong base is more complex than the previous case. Consider the titration of acetic acid with KOH. Before a significant amount of base is added, the major species are water and the acid. The pH of this solution is that of a solution of a weak acid. The pH of the analyte solution is initially quite a bit more basic than for the strong acid case, because only a small amount of H_3O^+ is supplied by the dissociation of the weak acid. Early in the titration, the base added merely lowers the concentration of the weak acid.

As the titration proceeds, the major species become acetic acid and acetate, the latter resulting from the consumption of the acetic acid by the added base. When a solution contains roughly equal concentrations of weak acid and its conjugate base, the solution resists changes in pH because it is buffered. The pH will remain essentially constant until the buffer capacity is reached (see above). The pH now rises sharply, but only over a small range because the majority of the acid has been consumed and the initial pH was more basic to begin with. This small pH range necessitates care in the choice of a colorimetric indicator, because an inappropriate indicator will yield a false endpoint.

One drawback of the manual titration procedure is that it is operator-dependent, requiring some skill in the use of the burette and in judging the indicator color changes. Because careful operators can obtain quite accurate and precise results, this old technique remains in use in the analytical laboratory, but modern automated titrators, such as those used for Karl Fischer analysis for moisture (see below), improve the titration analysis by removing operator involvement and coupling reagent delivery with reaction monitoring, through either optical or electrochemical means.

Back Titration. The residual or back titration is a variant of the direct titration. In this analysis, a known but excess amount of a solution containing a species that reacts completely and quantitatively with the analyte is added to the analyte solution. Thus, all of the analyte is consumed and an excess of the species that consumed the analyte is present. This excess is then titrated with a second titrant in a direct fashion. Because the amount of reacting species added to the analyte is known, the amount of analyte can be determined by difference. This is often done when a suitable titrant for the analyte cannot be conveniently formulated.

Blank Titration. The USP prescribes blank titration, in which the analyte matrix is titrated without analyte present in residual titration [45]. This allows the analyst to determine whether there are any contaminants in the reagents used for the analysis or any complications in the procedure.

Nonaqueous Titration. Many organic-soluble materials are more basic or acidic in organic solvents than in water, which facilitates their quantitation by nonaqueous titration methods. The USP lists a variety of pharmaceutically important compound classes that can be analyzed this way [45]. Organic bases can be titrated with perchloric acid in glacial acetic acid or dioxane solution. The glacial acetic acid system is preferred because the glass–calomel electrode pair behaves as described by theory in this medium. Acids can be titrated with alkali metal alkoxides or tetraalkylammonium hydroxides with methanol–toluene or methanol–benzene as the solvent. The preferred nonaqueous solvent systems are listed in the USP [45]. For these titrations the antimony-sensing electrode, while erratic, must be used, due to the alkaline error of the glass electrode. The tetraalkylammonium hydroxides in methanol–benzene or iso-propyl alcohol are preferred as titrants, because the R_4N^+ salt is more soluble and the glass–calomel or similar glass–silver ion electrode pair can be used. However, the salt bridge solution in the calomel electrode must be replaced with 0.1 M $LiClO_4$ in glacial acetic acid or KCl in methanol. The silver ion reference electrode is the easiest to modify, because the aqueous KCl filling solution can be removed, the electrode interior can be rinsed with distilled water, and then the nonaqueous solvent can be removed. Finally, the new filling solution is placed inside the salt bridge.

The Need for Buffers in Liquid Chromatography Mobile Phases. When analyzing a sample by HPLC, it is particularly important to control the pH of the mobile phase (see Chapter 9). The most important reason for this is that many common samples are weak acids or bases and can exist as both the conjugate acid and conjugate base forms within the normal pH range. If the mobile phase has the same pH as the pK_a as that of a weak acid analyte, then 50% of the analyte is in the acid form, and 50% is in the base form. Small changes in pH around the pK_a result in large changes in the ratio of the acid–base forms, so the pK_a is usually not chosen as the pH of the buffer. The stationary phase also has requirements for pH. If the mobile phase is too basic or acidic, the stationary phase can deteriorate.

The addition of nonaqueous solvents to a mobile phase causes changes in the pH of the overall solution, but the pH of nonaqueous and mixed nonaqueous–aqueous media is difficult to measure. Typically, when an aqueous buffer solution is mixed with a nonaqueous solvent, the pH rises because most organic solvents are Lewis bases. The pH of such a buffer should be adjusted *before* the organic phase is added to the aqueous phase.

5.6. KARL FISCHER TITRATION

Moisture determination is one of the most common and important analyses in the pharmaceutical industry. It is typically necessary in the quality assurance of raw materials and final products. Most organic solvents will dissolve water to an extent and most solids will absorb it. The presence of excessive amounts of water may affect product stability or effectiveness or may affect the chemical properties of a raw material. The need to analyze for moisture in many types of samples goes back to the early parts of this century, before the development of modern chromatographic equipment. The need for simplicity and high throughput makes the use of instrumental techniques such as chromatography difficult. The classical method, developed in the 1930s by Karl Fischer [47], remains the simplest and most common way to determine water content in a wide variety of materials. A detailed description of Karl Fischer titration and applications is provided in the short textbooks produced by Reidel de Haen, one of the major vendors [48, 49].

5.6.1. Karl Fischer Reagents and Reactions

The classical reaction involves an alcohol and an organic base and is given by the following equations:

$$ROH + SO_2 + R'N \rightarrow [R'NH]SO_3R \tag{5.38}$$

$$H_2O + I_2 + [R'NH]SO_3R + 2R'N \rightarrow [R'NH]SO_4R + 2[R'NH]I \tag{5.39}$$

In the first reaction the alcohol (ROH) reacts with sulfur dioxide and the base ($R'N$) to form a reactive complex. In the second reaction, this complex reacts with iodine and water, consuming any moisture present. This may be considered a form of iodometric titration. As may be implied from these reactions, many combinations of organic bases and alcohol have been tried since Fischer's development of the technique, which used 0.20 M I_2, 0.60 M SO_2, and 2.00 M pyridine in ethanol. This example of a one-component reagent was used for decades, but the technique was limited by the safety and convenience concerns of using pyridine. In the 1970s, reagents were developed that substituted basic salts such as sodium acetate and potassium iodide for pyridine, and the safety of Karl Fischer (KF) titration was improved considerably.

In modern KF titration, which is automated, the reagents are purchased in premixed form directly from vendors [50, 51], with the exact composition often being protected by patents. The apparatus includes a buret, titration vessel, and controller. All of the openings are covered with drying agents to prevent water leakage into the system from the surrounding laboratory. The endpoint is detected either by colorimeter for less sensitive applications or by potentiometer if more sensitivity is desired.

For volumetric KF titration, detection limits are now about 50 ppm [52] and about 1 ppm for coulometric analyses [53].

5.6.2. Karl Fischer Titration Procedures

KF titration procedures may be divided into two types: one-component and two-component. In a one-component titration, the titration vessel is filled with solvent, typically methanol and pretitrated with a one-component reagent that contains all of the reactants (except water) to a stable endpoint. This removes any residual water from the system. A known mass of analyte is added to the titration vessel, and titration is performed until a second stable endpoint is reached. In a typical two-component titration, the reagents are separated, with the base and sulfur dioxide in methanol as one component and the methanol iodine solution as the other. In this procedure, the solvent component contains the alcohol and base and is added to the titration vessel. The titrant contains the iodine and is added to the buret. Before addition of the sample, the solvent component is pretitrated to a stable endpoint to remove residual water. The sample is added to the titration vessel and dissolved if possible. It is then titrated with the iodine solution until a stable endpoint is reached.

While the one-component method is convenient, there are several advantages to two-component titration. First, because the reagents are not mixed, they have a longer shelf life. Two-component reagents are also designed with an excess of sulfur dioxide and the amine component, which accelerate the reaction. Remember that titration reactions must be extremely rapid. This leads to easier and more accurate endpoint detection. The two-component method is better for low masses of water because the titrant (iodine) mass is easy to control in very small quantities.

5.6.3. Method Development Issues in Karl Fischer Titration

The reactions shown above are complex; as a result, there are many cautions that must be observed when performing Karl Fischer titration. There are effects related to titration solvents, pH, side reactions, atmospheric moisture, indication methods, sample handling, and safety that must be considered.

The working solvents must ensure that the stoichiometry of the titration reaction is complete and rapid. Thus, the solvent must be able to dissolve the samples, must not interfere with indication, must be maintained at the correct pH, and must be safe to handle. Methanol is the preferred solvent for most Karl Fischer reactions; however, 1-propanol, 1-butanol, or 2-methoxyethanol is often used. Little work has been performed with nonalcoholic solvents, because these have proven unsuitable for most applications. Generally, acidic pH, between 4 and 7, is used to ensure that the starting materials remain in solution and do not react with each other. The specific reagent manufacturer's recommendations should be consulted when choosing the pH conditions. A sluggish or vanishing endpoint is often the result of poorly chosen pH conditions. Side reactions between the reagents, solvents, and matrix components should be avoided. Some of the common problems include the following: The presence of reducing agents will reduce iodine; oxidizing agents will oxidize iodide ions present from the reagents; carbonates, hydroxides, and oxides will often react with pyridinium salts, while carboxylic acids, aldehydes, and ketones can react with methanol. In each of these cases, presence of the matrix components can affect the outcome.

Atmospheric moisture is a particularly insidious problem in water determinations. Remember that water is ubiquitous in the environment and in the laboratory. The apparatus must be kept scrupulously clean and dry. All components can absorb or adsorb moisture. Properly charged drying fixtures must be maintained at all orifices. Check the drying agents often (use indicating materials) to ensure that they are not expended, and regenerate or replace them if in doubt. Be aware that 1 liter of air can contain up to 20 mg of water and that

Table 5.8. Applications of Karl Fischer Titration for Moisture Determination

General Category	Selected Applications
Organic compounds	Hydrocarbons, halogenated hydrocarbons, alcohols, phenols, ethers, aldehydes and ketones, carboxylic acids, esters, salts, amines, diamones, cycloamines, aromatic amines, heterocycles, sulfur compounds, siloxanes, silanols, peroxides
Inorganic compounds	Water of crystallization, hydrates, entrapped moisture, acids, oxides
Foodstuffs and natural products	Carbohydrates, fats and fatty products, diary products, proteins, vegetable products, cellulose, chocolates, bakery products, pastas
Medicines and cosmetic products	Procedures described by pharmacopeias (See USP, EP, or BP for details)
Other products	Mineral oil, paint, lacquer, thinner, plastics, liquid gases

a dried titration vessel can still contain 5 mg of water hydrogen bound to the glass surface.

Table 5.8, summarized from information in reference 49, shows a brief description of many of the applications of Karl Fischer titration. Recent research in Karl Fischer methods has included a thorough study by Margolis [54] of systematic bias in both volumetric and coulometric methods. He observed that bias can occur from many sources, including accuracy of moisture calibration standards, adjustable (incorrectly set parameters) and nonadjustable instrumental bias, solvent composition, titration cell design, and sample composition. He developed a standard set of protocols that can be used to assess the performance of KF instruments.

5.7. OTHER METHODS FOR DETERMINING WATER

5.7.1. Loss on Drying

Loss on drying is another classical method for determining the water content of a variety of analytes. The basic principle is very simple. The moist sample is weighed; then heating, dessication, or application of microwave energy drives off the water and the dried sample is reweighed. The weight difference gives the water content. This analysis can be performed with apparatus as simple as a drying oven and an analytical balance. Care must be taken to ensure that the sample does not reabsorb moisture following drying. Modern thermogravimetric analyzers combine the balance and drying system into a single instrument, using either heat or microwaves to dry the samples. Part-per-million water concentrations can be determined, without sample pretreatment using a combination heater and coulometric detector. Modern systems for thermogravimetric analysis were recently reviewed [55], while other studies [56, 57] show comparative data between thermogravimetric analysis and Karl Fischer titration.

5.7.2. Instrumental Methods

Gas chromatography (GC), near-infared (NIR) spectrometry and nuclear magnetic resonance (NMR) spectrometry have also been applied to the routine determination of moisture. GC usually requires a thick film polar column, such as Carbowax, which limits sample types and thermal conductivity detection, which limits sensitivity. GC also requires that the water either be in a volatile matrix or be extracted into a volatile matrix. Zhou et al. [58] have recently described the use of GC as a reference method for NIR spectrometric determination of moisture. Infared spectrometry can also be used to determine water. Typically, spectra data in the range of $1100-2500$ cm^{-1} are used, and partial least-squares quantitation must be employed. IR has shown advantages over Karl Fischer titration and loss on drying for water content in difficult matrices such as biomolecules [59] and freeze-dried drug products [58]. NMR spectrometry has also been employed for the analysis of water in coals, using deconvolution of the ^{1}H free induction decay signal [60].

5.8. MISCELLANEOUS TECHNIQUES

5.8.1. Differential Scanning Calorimetry and Thermal Analysis

Thermal analysis in the pharmaceutical industry, which includes all methods in which a physical property is measured as a function of temperature, were thoroughly reviewed by Giron [61]. The techniques include differential scanning calorimetry (DSC), thermogravimetric analysis (TGA), which is described earlier in this chapter, thermomechanical analysis (TMA), and moisture analysis (MA), which is also described earlier in this chapter. In DSC, heat flow through a sample is measured as a function of temperature. Any change in the sample, such as melting, boiling, glass transition, or conformational changes, will be observed as a perturbation in the heat flow. This can be performed by placing the sample into an oven and comparing to a blank oven as the sample is heated. If nothing happens to the sample, the two ovens will show the same thermal properties. If a transition occurs in the sample, it will absorb additional heat, as compared to the reference oven. By calibration, the temperature difference between the ovens can be translated to heat flow. Thermomechanical analysis is typically performed on polymers. Expansion of the polymer as it is heated is measured versus temperature. Changes in the polymer are detected using a transducer. Types of samples for which DSC and thermal techniques are appropriate include: raw materials, precursors, intermediates, drugs, and excipients.

REFERENCES

1. *Basic Analytical Techniques*, American Chemical Society, Washington, DC, 1983 (video series).
2. http://www.chem.wisc.edu/
3. R. M. Schoonver, *Anal. Chem.* **1982**, *54*, 973A.
4. B. B. Johnson and J. D. Wells, *J. Chem. Educ.* **1986**, *63*, 86.
5. United States Pharmacopeia Convention, *United States Pharmacopeia 24/NF 19*, Rockville, MD, 2000, p. 9.
6. *1998–1999 Catalog*, Fisher Scientific, Raritan, NJ, 1998, p. 638.
7. *The Merck Index*, 11th edition, Merck and Company, Rahway, NJ, 1989, p. 383.
8. M. Hanna, V. deBiasi, B. Bond, C. Salter, A. Hutt, and P. Camilleri, *Anal. Chem.* **1998**, *70*, 2092–2099.
9. C. Hansch and A. Leo, in *Substituent Constants for Correlation Analysis in Chemistry and Biology*, Wiley, New York, 1979.
10. A. Leo, C. Hansch, D. Elkins, *Chem. Rev.* **1971**, *71*, 525.
11. R. E. Majors and D. E. Raynie, *LC-GC* **1997**, *15*, 1106–1117.
12. D. Potter and J. Pawliszyn, *J. Environ. Sci. Technol.* **1994**, *28*, 298.
13. C. L. Arthur and J. Pawliszyn, *Anal. Chem.* **1990**, *62*, 2145.
14. P. Okeyo, S. M. Rentz, and N. H. Snow, *J. High Resolut. Chromatogr.* 20, 171 (1997).
15. X. P. Lee, T. Kumazawa, K. Sato, and O. Suzuki, *Chromatographia* **1996**, *42*(3), 135.
16. J. Pawliszyn, *Solid Phase Micro-extraction: Theory and Practice*, Wiley, New York, 1997.
17. B. B. Kebbekus and S. Mitra, *Environmental Chemical Analysis*, Blackie Academic Press, London, 1998, p. 194.
18. S. B. Hawthorne, D. J. Miller, and J. J. Langenfeld, *J. Chromatogr. Sci.* **1990**, *28*, 2.
19. S. B. Hawthorne and D. J. Miller, *Anal. Chem.* **1987**, *59*, 1705.
20. N. H. Snow, M. Dunn, and S. Patel, *J. Chem. Educ.* **1997**, *74*, 1108.
21. J. W. King and M. L. Hopper, *J. Assoc. Off. Anal. Chem. Int.* **1992**, *75*, 375.
22. L. T. Taylor, *Supercritical Fluid Extraction*, Wiley, New York, 1996.
23. United States Pharmacopeia Convention, *United States Pharmacopeia 24/NF 19*, Rockville, MD, 2000, p. 2231.
24. S. K. Menon, A. Sathyapalan, Y. K. Agrawal, *Rev. Anal. Chem.* **1997**, *16*, 333.
25. R. Füglein, C. Bräuchle, and N. Hampp, *Anal. Sci.* **1994**, *10*, 959.
26. J. L. F. C. Lima, M. Conceicao, B. S. M. Montenegro, and M. G. F. Sales, *J. Pharm. Sci.* **1997**, *86*, 1234.
27. R. Aubeck, N. Hampp, *Anal. Chim. Acta* **1992**, *256*, 257.
28. B. B. Saad, Z. A. Zahid, S. A. Rahman, M. N. Ahmad, and A. H. Husin, *Analyst* **1992**, *117*, 1319.
29. R. Katasky, P. S. Bates, and D. Parker, *Analyst* **1992**, *117*, 1313.

30. A. Campiglio, *Analyst* **1993**, *188*, 545.

31. G. Nagy, K. Tóth, and E. Pungor, *Anal. Lett.* **1993**, *26*, 1391

32. F. G. K. Baucke, *Fresenius J. Anal. Chem.* **1994**, *349*, 582.

33. United States Pharmacopeia Convention, *United States Pharmacopeia 24/NF 19*, ⟨791⟩, Rockville, MD, 2000, pp. 1977–1978 and 2231–2232.

34. P. M. Pooler, M. L. Wahl, A. B. Rabinowitz, and C. S. Owen, *Anal. Biochem.* **1998**, *256*, 238.

35. K. Kreuer, *Sensors and Actuators* **1990**, *B1*, 286.

36. E. Pungor, *Electroanalysis* **1996**, *8*, 348.

37. E. Pungor, *Microchemical J.* **1997**, *57*, 251.

38. E. Pungor, *Fresenius J. Anal. Chem.* **1997**, *357*, 184.

39. R. P. Buck and V. V. Cosofret, *Pure Appl. Chem.* **1993**, *65*, 1849.

40. E. Bakker, *Electroanalysis* **1997**, *9*, 7.

41. P. Spitzer, R. Eberhardt, I. Schmidt, and U. Sudmeier, *Fresenius J. Anal. Chem.* **1996**, *356*, 178.

42. R. Naumann, Ch. Alexander-Weber, and F. G. K. Baucke, *Fresenius J. Anal. Chem.* **1994**, *349*, 603.

43. R. C. Metcalf, D. V. Peck, and L. J. Arent, *Analyst* **1990**, *115*, 899.

44. M. S. Frant, *Am. Lab.* **1995**, 18.

45. United States Pharmacopeia Convention, *United States Pharmacopeia 24/NF 19*, 541, Rockville, MD, 2000, pp. 1882–1884.

46. United States Pharmacopeia Convention, *United States Pharmacopeia 24/NF 19*, Rockville, MD, 2000, pp. 2229–2240.

47. K. Fischer, *Angew. Chem.* **1935**, *48*, 394 (in German).

48. E. Scholz, *Karl Fischer Titration Determination of Water: Chemical Laboratory Practice*, Springer-Verlag, Berlin, 1984.

49. *HYDRANAL® Manual*, Riedel de Haen, Seelze, Netherlands, 1995.

50. *1998–1999 Catalog*, Fisher Scientific, Raritan, NJ, 1998, p. 1821.

51. *HYDRANAL® Manual*, Riedel de Haen, Seelze, Netherlands, 1995, p. 123.

52. A. Coates, *Lab. Equip. Dig.* **1990**, *28*, 23.

53. *HYDRANAL® Manual*, Riedel de Haen, Seelze, Netherlands, 1995, p. 29.

54. S. A. Margolis, *Anal. Chem.* **1997**, *69*, 4864–4871.

55. P. Butler, *Lab. Equip. Dig.* **1992**, *30*(11), 33–34.

56. M. L. Bostian, D. L. Fish, N. B. Webb, and J. A. Arey, *J. Assoc. Off. Anal. Chem.* **1985**, *68*, 876–880.

57. W. R. Windham, F. E. Barton II, and J. A. Robertson, *J. Assoc. Off. Anal. Chem.* **1988**, *71*, 256–262.

58. X. Zhou, P. A. Hines, K. C. White, and M. W. Borer, *Anal. Chem.* **1998**, *70*, 390–394.

59. T. Iwaoka, F. Tabata, S. and Tsutsumi, *Appl. Spectrosc.* **1994**, *48*, 818–821.

60. R. Graebert and D. Michel, *Fuel* **1990**, *69*, 826–829.

61. D. Giron, *Acta Pharm. Jugosl.* **1990**, *40*, 95–157.

General References on Laboratory Techniques and Analytical Procedures

J. B. Umland and J. M. Bellama, *General Chemistry*, 3rd edition, Brooks Cole, Pacific Grove, CA, 1999. (A good General Chemistry text is an important part of any chemist's bookshelf.)

D. L. Pavia, G. L. Lampman, and G. S. Kriz, *Introduction to Organic Laboratory Techniques*, 3rd edition, Saunders, Fort Worth, TX 1989. (The section on laboratory techniques provides excellent descriptions of many fundamentals, such as weighing, filtering, extracting, and so on.)

D. C. Harris, *Quantitative Chemical Analysis*, 5th edition, W. H. Freeman, San Francisco, 1999.

D. A. Skoog, D. M. West, and F. J. Holler, *Fundamentals of Analytical Chemistry*, 7th edition, Saunders, New York, 1996.

D. A. Skoog, F. J. Holler, and T. A. Nieman, *Instrumental Analysis*, 5th edition, Saunders, New York, 1998.

General References Related to Ion-Selective Electrodes and Potentiometry

R. G. Bates, *Determination of pH: Theory and Practice*, Wiley, New York, 1973.

R. A. Durst, *Standard Reference Materials: Standardization of pH Measurements*, NIST, SP260 53, revised edition, 1988, PB88-217472.

Orion 1998 Laboratory Products and Electrochemistry Handbook, Orion Research, Inc.

D. R. Lide (editor), *CRC Handbook of Chemistry and Physics*, 75th edition, Boca Raton, FL, CRC Press, 1994.

A. Evans, *Potentiometry and Ion Selective Electrodes*, Wiley, New York, 1991.

R. G. Bates, *Crit. Rev. Anal. Chem.* **1981**, 10, 247.

A. V. Covington, *Ion Selective Electrode Methodology*, Vols. 1 and 2, CRC Press, Boca Raton, FL, 1979.

J. Koryta and K. Štulík, *Ion Selective Electrodes*, Cambridge University Press, New York, 1983.

P. T. Kissinger and W. R. Heineman, (Editors), *Laboratory Techniques in Electroanalytical Chemistry*, Marcel Dekker, New York, 1996.

6

SPECTROSCOPY

Perlette Abuaf and Alvin J. Melveger

Of the spectroscopic techniques used in the pharmaceutical industry for identification and/or quantitation, the more common ones are ultraviolet–visible (UV-VIS) spectrometry, infrared (IR) spectrometry, mass spectrometry (MS), and nuclear magnetic resonance (NMR). These spectroscopic techniques can be used as spectrophotometric tools to obtain the spectrum of the samples being analyzed, or they can be used as detectors in chromatography giving rise to, in certain cases, the hyphenated techniques such as GC-MS, LC-MS, LC-NMR, and so on. This chapter will deal primarily with the instruments operated in the UV-VIS and IR regions of the electromagnetic spectrum.

6.1. THE ELECTROMAGNETIC SPECTRUM

The electromagnetic spectrum includes different forms of radiation such as radio, NMR, radar, microwave, IR, UV-VIS, X-ray, and cosmic rays. The UV-VIS radiation makes up only a very small section of the electromagnetic spectrum (see Figure 6.1). That part of the ultraviolet (UV) region not absorbed by air extends from 200 to 380 nm. Below 200 nm, the atmosphere starts absorbing, and that region is appropriately called the vacuum UV region.

6.2. WAVE-PARTICLE DUALITY

Electromagnetic radiation is a combination of oscillating electric and magnetic fields traveling through space with a wavelike motion. Because radiation behaves

Analytical Chemistry in a GMP Environment. Edited by J. M. Miller and J. B. Crowther
ISBN 0-471-31431-5 © 2000 John Wiley & Sons, Inc.

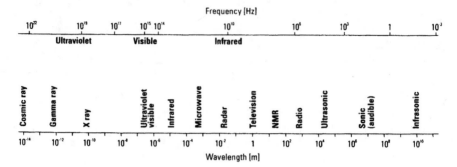

Figure 6.1. The electromagnetic spectrum.

as a wave, it has wave characteristics. Yet, electromagnetic radiation is also composed of discrete energy particles called photons, which manifest photo-electric effects. Therefore electromagnetic radiation is said to have dualistic properties: those of wave and of particle.

6.2.1. Wave Parameters

Because electromagnetic radiation acts as a wave, it has wave properties such as wavelength and frequency. Wavelength (λ) is the distance from one point to the next of the corresponding phase between two adjacent waves, measured along the line of propagation of the wave. Wavelength can be expressed in any unit of distance, such as micrometer, nanometer, or angstrom; however, in the UV-VIS region it is usually expressed in nanometers (1 nm $= 10^{-9}$ m).

Another wave parameter is frequency (ν), which is the number of waves that pass a given point in a given period of time; it has the units of reciprocal time (e.g., $\sec^{-1}$). Wavelength and frequency are related by the following equation:

$$\lambda\nu = v \tag{6.1}$$

where v is the velocity of the radiation. In the special case where radiation travels in vacuum, the following equation holds true:

$$\lambda\nu = c \tag{6.2}$$

where c is the velocity of light, with a value of 2.99×10^{10} cm/sec.

Because frequency can sometimes be a very large number, absorption can be reported in wave numbers ($\bar{\nu}$) whose units are number of waves per centimeter (cm^{-1}). The relationship between $\bar{\nu}$ and λ is given by the following equation:

$$\bar{\nu} = 10^7/\lambda \tag{6.3}$$

where λ is measured in nanometers. We note that wave number is inversely proportional to wavelength; the shorter the wavelength, the higher the wave number or the frequency.

For example, consider converting a wavelength of 20 µm to wave number:

$$\lambda = 20 \text{ µm} = 20 \times 10^3 \text{ nm} = 2 \times 10^4 \text{ nm} \tag{6.4}$$

$$\bar{v} = \frac{10^7}{2 \times 10^4} = 0.5 \times 10^3 \text{ cm}^{-1} = 5 \times 10^2 \text{ cm}^{-1} \tag{6.5}$$

The UV-VIS region of 200–780 nm corresponds to 50,000–12,800 cm^{-1}.

6.2.2. Particle Parameters

The kinetic energy associated with electromagnetic radiation is given by

$$E = hv = \frac{hc}{\lambda} \tag{6.6}$$

where E is kinetic energy (in ergs), h is Planck's constant (6.62×10^{-27} erg sec), c is the speed of light (3×10^{10} cm/sec), v is the frequency, and λ is the wavelength. It follows from equation 6.6 that radiation with shorter wavelength (higher frequency) has higher energy, and vice versa. For example, visible light is relatively harmless, while ultraviolet or X-rays cause burns.

When radiation interacts with a substance, in addition to transmittance and absorption there are other processes that occur. These may be reflection, refraction, diffraction, scattering, fluorescence, phosphorescence, and so on. When measuring UV-VIS spectra we are interested only in absorbance. However, there is no direct way of measuring the absorbed radiation; we get that information indirectly by measuring the transmittance with the assumption that all other kinds of processes are minimal.

6.3. TRANSITIONS AND ENERGIES

Radiation is a form of energy. The interaction between a molecule and radiation causes the molecule to move from its ground energy state E_0 to a higher one such as the next excited state E_1. When a molecule absorbs light, its energy content increases by ΔE, where $\Delta E = hv$. This energy difference is exactly equal to the energy of the photon that has been absorbed by the molecule. Radiant energy can be absorbed or emitted only in discrete quantities called quanta. The quantum is characteristic of each absorbing species. If the photon's energy does

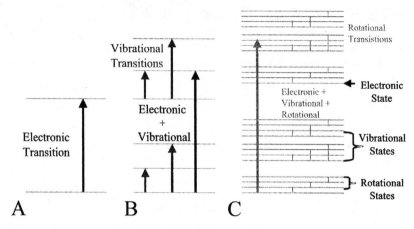

Figure 6.2. Energy transitions

not precisely match the energy difference between two energy states of the molecule, the molecule cannot absorb that radiation.

The total energy of a molecule is equal to the sum of its electronic, vibrational, and rotational energies. Electronic energy is associated with electronic orbital changes within the atom or molecule. Vibrational and rotational energies are associated with molecular vibrations and rotations. Electronic energy (A in Figure 6.2) is higher in energy than vibrational energy (B), which in turn is higher than rotational energy (C). Energy differences between rotational energy states are much smaller than energy differences between electronic energy states. Energy differences between vibrational energy states are intermediate between rotational and electronic energy states.

Radiation in the UV and visible regions of the electromagnetic spectrum affects electronic transitions, which require high energy. The energy absorbed in an electronic transition, ΔE, may cause the excitation of electrons from an occupied molecular orbital to the next-higher-energy orbital, thus exciting the molecule from the ground state to the excited state. In theory it is expected that a single electronic transition will give rise to a single discrete sharp line in the absorption spectrum. However, this is only true of molecules in the gaseous state or of atoms where transitions with different energies are suppressed. In the case of molecules in solution, we observe broad absorption bands in UV spectra because the different types of energy transitions (electronic, vibrational, and rotational) are interrelated, and also because of solvent–solute interactions. Both rotational and vibrational energy states are superimposed on the electronic states. When electronic transitions overlap with vibrational and rotational transitions, the results are wide absorption bands or band envelopes that spread out over not a single wavelength but over a range of wavelengths.

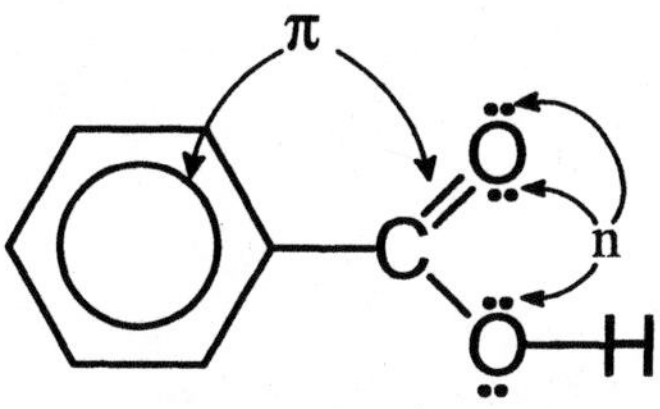

Figure 6.3. Types of electrons for electronic transitions.

6.4. ULTRAVIOLET/VISIBLE SPECTROSCOPY

6.4.1. Electron Type

The type of electrons forming the bonds of a molecule determines whether the molecule will have characteristic UV absorption. For example, sigma (σ) electrons are involved in the formation of single bonds such as C–H bonds. Saturated compounds have single bonds throughout the molecule. In general, the sigma electrons are held tightly, and therefore a great deal of energy is required to excite them. These kinds of compounds have no UV absorbance; they are transparent in the UV region. Higher energy, corresponding to the short-wavelength UV in the far-UV region is needed to excite sigma electrons.

Electrons involved in the formation of double bonds such as C=C or C=O bonds involve both sigma (σ) and pi (π) electrons. Nonbonding electrons are called *n* electrons; they are encountered in molecules containing atoms like nitrogen or oxygen that have unshared electrons. Consider, for example, the electrons in benzoic acid as shown in Figure 6.3. Compared to the sigma electrons, the *n* and π electrons are more loosely held. In the UV region most of the observed absorption bands are due to the transitions involving *n* and π electrons.

The electronic transitions in the UV region can be *n* to π^* where an electron from an unshared pair moves to an unstable antibonding pi (π^*) orbital—that is, from the *n* orbital to the antibonding π^* orbital. Another kind of electronic transition in the UV region could be due to π to π^* excitation, where an electron from a stable (bonding) π orbital moves to an unstable antibonding pi (π^*) orbital—that is, from the bonding π orbital to the antibonding π^* orbital.

6.4.2. Chromophores

A chromophore is an unsaturated group usually containing a π bond. When a chromophore is introduced into a saturated hydrocarbon (which exhibits no UV-VIS absorbance), it causes the new compound to have an absorption band between 185 and 1000 nm. For example, *n*-octane does not have a UV absorption because it is a saturated hydrocarbon. However, octyl nitrite has a strong absorption at 230 nm due to the presence of the nitrite group, which is substituted for a hydrogen atom. Consequently the nitrite group is classified as a

Table 6.1. Sample Chromophoric Groups[a]

Chromophore	Example	λ_{max} (nm)	ε_{max}	Solvent
>C=C<	Ethylene	171	15,530	Vapor
	1-Octene	177	12,600	Heptane
−C≡C−	2-Octyne	178	10,000	Heptane
		196	ca 2,100	Heptane
		223	160	Heptane
>C=O	Acetaldehyde	160	20,000	Vapor
		180	10,000	Vapor
		290	17	Hexane
	Acetone	166	16,000	Vapor
		189	900	Hexane
		279	15	Hexane
−CO₂H	Acetic acid	208	32	Ethanol
−COCl	Acetyl chloride	220	100	Hexane
−CONH₂	Acetamide	178	9,500	Hexane
		220	63	Water
−CO₂R	Ethyl acetate	211	57	Ethanol
−NO₂	Nitromethane	201	5,000	Methanol
		274	17	Methanol
−ONO₂	Butylnitrate	270	17	Ethanol
−ONO	Butylnitrite	220	14,500	Hexane
		356	87	Hexane
−NO	Nitrosobutane	300	100	Ether
		665	20	Ether
>C−N	neo-Pentylidene n-butylamine	235	100	Ethanol
−C≡N	Acetonitrile	167	weak	Vapor
−N₃	Azidoacetic ester	285	20	Ethanol
=N2	Diazomethane	~410	3	Vapor
	Diazoacetic ester	249	10,050	Ethanol
		378	16	Ethanol
−N=N−	Azomethane	338	4	Ethanol

[a] Data in the table were selected mostly from Vols. I, II, and IV of *Organic Electronic Spectral Data*, Interscience, New York, 1946–1959.

chromophore. Other examples of chromophores would be molecular groups such as nitrate, nitro, nitrile, sulfoxide, carbonyl, amide, alkene, and so on. Table 6.1 lists characteristics of sample chromophoric groups.

Even though the absorption intensity of the different chromophoric groups varies widely from one group to the other, it should be noted that it does not vary so much among members of the same class of chromophoric group. For example, carboxylic acids (RCOOH) have absorption bands that vary between 204 to 210 nm, with their molar absorptivities ranging from 40 to 75 on going from acetic acid to stearic acid (the R group changes from C_1 to C_{18}).

Table 6.2. Effect of Conjugation: Ultraviolet Absorption of Some Polyene Aldehydes, $CH_3-(CH=CH)_n-CHO$

n	λ_{max} (nm)	ε_{max}
1	217	15,650
2	270	27,000
3	312	40,000
4	343	40,000
5	370	57,000
6	393	65,000
7	415	63,000

6.4.3. Conjugation and Spectral Shifts

If there are two or more chromophores in a single molecule, their relative positions affect the UV spectrum. For example, 1-hexene has molar absorptivity of 10,000 at 180 nm. When a second ethylene group is introduced the resulting molecule, 1,5-hexadiene, retains its λ_{max} of 180 nm, and its absorptivity doubles to 20,000. However, when the two-ethylene groups are repositioned in the molecule so that they are in conjugation with each other as in 2,4-hexadiene, λ_{max} shifts to longer wavelengths (227 nm) and molar absorptivity increases to 22,500. This example shows that when two chromophores are separated from each other by more than one carbon atom, they do not interact, there is no band effect, and the absorptivity is a simple summation of the absorption of each of the two chromophores. When the two chromophores are adjacent to each other, forming a conjugated system, there is an interaction that causes the absorption maximum to shift to longer wavelengths and to increase its intensity.

Conjugated chromophores usually cause the absorption maximum to shift to longer wavelengths; this effect is called bathochromic shift or a red shift. Conjugated chromophores also increase the absorption intensity; this effect is known as hyperchromic effect. Shift of absorption to a shorter wavelength due to solvent effects or substitution in the molecule is called hypsochromic shift or a blue shift. Hypochromic effect is a decrease in absorption intensity.

In the case of extended conjugation, as the number of conjugated chromophores increase, the wavelength and intensity also increase. For example, in the polyunsaturated aldehydes listed in Table 6.2, we note that each additional double bond added to the conjugated system shifts the λ_{max} to longer wavelengths and gives rise to an increase in intensity, producing a bathochromic shift and hyperchromic effect.

6.4.3.1. Auxochromes. An auxochrome is a saturated group with no absorption of its own, which when introduced to a chromophoric system, due to its nonbonded electrons, causes a bathochromic shift and hyperchromic effect. The hydroxyl group is such an example. Alcohols, even though they have the hydroxyl group, are transparent to UV and therefore are commonly used as

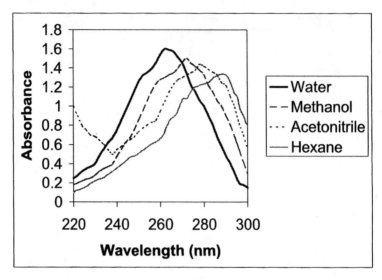

Figure 6.4. Solvent effects on absorption spectra of acetone.

solvents in UV. However, when the hydroxyl group is introduced into a chromophoric system, it will cause a bathochromic shift. For example, when a hydroxyl group is introduced into benzene, the absorptivity increases from 200 at 256 nm (for benzene in hexane) to 1450 at 270 nm (for phenol in water). Other examples of auxochromes are alkyl, halogen, and amino groups.

6.4.3.2. Solvent and pH Effects. UV spectra of compounds can be determined either in the vapor phase or in solution. In the latter case, the solvents used to dissolve the compounds can affect their absorption band maxima. Consequently, the solvent used should always be specified when reporting the λ_{max} of a compound. Figure 6.4 shows how increasing the solvent polarity affects the λ_{max} and the intensity of acetone. We note that the more polar the solvent, the shorter the λ_{max} (hypsochromic shift) and the higher the intensity (hyperchromic effect).

In addition to solvents, the pH of the solution also affects the absorption maximum and intensity of compounds. Changing the pH of a solution generally results in shifting the equilibrium between two different forms of a compound, which in turn affects its UV spectrum. These equilibrium shifts could be between keto–enol tautomers or Bronsted acid–base pairs. Figure 6.5 shows the drastic effect of changing the pH from acidic to basic for metharbital.

6.5. INFRARED SPECTROSCOPY

The absorption of IR radiation comes about from the vibration and rotation of atoms engaged in chemical bonding within molecules. The wavelength or wave

Metharbital, 10 μg/mL; (a) in 0.1M HCL
(b) in 0.1 M NaOH

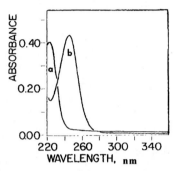

Figure 6.5. pH Effect on metharbital spectrum. (Reprinted with permission from A. L. Hayden, O. R. Sammul, G. B. Selzer and J. Carol, *J. Ass. Off. Anal. Chem.* **1962**, *45*, 883, copyright 1962, by AOAC International.)

number of the absorbed radiation is characteristic of the chemical entities that are absorbing the radiation. Because of these characteristic absorptions, an IR spectrum of a material is a collection of the absorption frequencies arising from the chemical bonds in a molecule. An IR spectrum so obtained is referred to as a fingerprint spectrum because it is expected that any given molecule will have a unique spectrum of absorption intensities at different frequencies.

We will use the convention of expressing frequencies as wave numbers ($\bar{v}$) having units of cm^{-1}. Other conventions refer to wavelengths, usually in microns (μm). The region of the electromagnetic spectrum usually referred to as mid-infrared or fingerprint infrared runs from ~200 cm^{-1} to 4000 cm^{-1} (50 μm to 2.5 μm).

Examples of vibrational motions giving rise to IR absorption may be shown for a molecule such as CO_2. The arrows indicate the direction of motion of the atoms:

← →	→ ← →	↑ ↑	+ − +
O=C=O	O=C=O	O=C=O	O=C=O
			↓
Symmetric stretch	Asymmetric stretch	Deformation	Deformation
(inactive)	(2350 cm^{-1})	(667 cm^{-1})	(667 cm^{-1})

These vibrations are called fundamentals, and their number may be calculated as follows:

- For nonlinear molecules of n atoms there are $(3n - 6)$ fundamental vibrations.
- For linear molecules of n atoms there are $(3n - 5)$ fundamental vibrations.

As the number of atoms in a molecule increases, the complexity of the IR spectrum increases accordingly.

Because CO_2 is a linear molecule, it is expected to have $3(3) - 5 = 4$ fundamental vibrations, which are depicted above. The actual IR spectrum of CO_2 shows only two fundamental vibrations. However, there are theoretical and practical factors that may reduce the number of observed vibrational frequencies; these include the following:

1. Symmetrical stretches are inactive in IR because there is no change in dipole moment during the vibration, a requirement for IR absorption. The symmetric stretch in CO_2 is therefore not observed in the spectrum.

2. Highly symmetric molecules may give rise to multiple vibrations at the same frequency. These so-called degenerate vibrations cannot be distinguished from one another. This is the case for the deformation vibrations at 667 cm^{-1} in CO_2. These are observed as a single absorption band rather than two separate bands.

3. Vibrations that may be very weak and only absorb tiny amounts of IR radiation that are below the detection limit of the spectrophotometer.

4. Fundamental bands that are close to one another and cannot be resolved by the spectrophotometer.

5. Fundamental bands may be beyond the range of the spectrophotometer. Typical instruments only respond from ~400 to 4000 cm^{-1}. There are, however, fundamentals that are below 400 cm^{-1}, especially in molecules containing heavy atoms such as halogens.

A nonlinear molecule such as H_2O is predicted to have three fundamental vibrations. These are in fact all observed in the IR because dipole moment changes are associated with these vibrations.

6.5.1. Group Frequencies

Because individual groupings of atoms in a molecule vibrate separately, these functional groups will have characteristic frequencies. Thus a carbonyl (−C=O) group may show an IR absorption in a certain spectral region independent of the other groups in the molecule. These characteristic absorptions are called group frequencies and may be used to identify the presence of specific functional groups in a molecule. To aid in this process, tabulations of group frequencies have been made by examining the IR spectra of numerous compounds containing such groups. Examples of some important group frequencies are given in Table 6.3. This information is useful in constructing the chemical makeup of a molecule from its IR spectrum.

These group frequencies tend to span a frequency range for each chemical grouping and depend to a large degree on the surrounding chemical environment of the molecule. By studying large numbers of similar compounds, it is

Table 6.3. Some Important IR Group Frequencies

Functional Group	Approximate Frequency (cm^{-1})	Vibration Type
–OH	3600	O–H stretch
–N–H	3350	N–H stretch
=C–H	3020	C–H stretch alkenes, aromatics
≡C–H	2960	C–H stretch alkanes
–C≡C–	2050	C≡C stretch
–C=C–	1650	C=C stretch
–C=O	1700	C=O stretch
–C–O–C–	1100	C–O stretch ethers
–C–Cl	650	C–Cl stretch

possible to associate specific frequencies with functional groups in specific environments. Figure 6.6 depicts an IR spectrum of a fatty acid ester showing the assignment of observed frequencies to specific groups in the molecule.

Such information is tabulated in correlation tables such as shown in Table 6.4 for the carbonyl C=O stretching vibration. The horizontal lines show the expected frequency ranges for different types of aldehydes and ketones. The entire range of these carbonyl vibrations is between 1600 and 1800 cm^{-1}.

Such information is available in computer databases and is useful for identifying compounds by comparing their group frequencies to the stored information. This is useful for synthetic chemists synthesizing compounds and record-

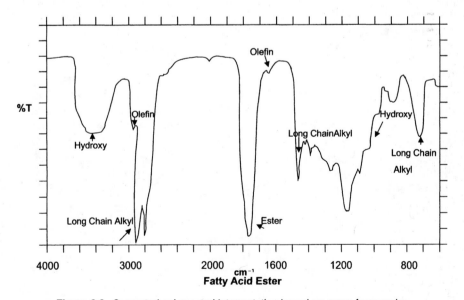

Figure 6.6. Computerized spectral interpretation based on group frequencies.

Table 6.4. IR Correlation Table for Carbonyl Groups

	Frequency of Carbonyl Vibration (cm^{-1})		
Grouping	1800	1700	1600
Aldehydes			
H–C=O			
• Saturated aliphatic and side-chain aromatic		‒‒‒	
• Aryl and αβ-unsaturated		‒‒‒‒‒	
Ketones			
–C=O			
• Saturated open chain		‒‒‒	
• αβ-Unsaturated			‒‒‒
• Monoaryl		‒‒‒	
• α-Diketones	‒‒‒‒‒‒‒‒‒‒‒‒‒‒		
• 2 C=O in a ring			‒‒‒‒‒‒‒‒

ing the corresponding IR spectra. A computer search may reveal the presence or absence of particular group frequencies in the synthesized compounds.

6.5.2. Fingerprinting

As indicated earlier, an IR spectrum of a molecule represents a unique finger-print of the substance. As with fingerprints of people, where the matching of an individual's print with that in a database may serve to identify that person, IR spectra may be used in a similar fashion. The spectrum of a material may be compared point by point to a database of spectra created from a universe of compounds. A computer search algorithm will compare the sample spectrum to stored spectra of standard or known materials. The result of the search is a listing of probable candidates and the probability of a particular spectral match. Normally only matches greater than 95% should be considered an exact match without other corroborating evidence. The final identification may depend on comparing the printout of the sample spectrum with the database spectrum to confirm a match.

Databases are available for purchase, and some are extensive containing tens of thousands of compounds. More common are limited databases of specialized classes of compounds such as polymers, lubricants, solvents, inorganics, controlled drugs, and so on.

Infrared spectroscopy is used in the pharmaceutical or medical device industrial laboratory to identify or confirm a compound. Incoming raw materials are checked for identity by comparing their IR spectra to standard spectra. IR spectroscopy is also used for quantitative analysis. As in UV–VIS spectroscopy, the absorption intensity of a band is related to the concentration according to Beer's law. Because water is a strong IR absorber, it is usually necessary to dissolve materials in nonaqueous, organic solvents in order to perform quantitative analysis. IR spectra may be obtained on liquids, solids, and gases utilizing appropriate sampling techniques, some of which are described later.

6.6. BEERS LAW AND QUANTITATIVE ANALYSIS

6.6.1. Transmittance

When radiation interacts with matter, the incident radiation may either be absorbed by the sample, reflected from the surface of the sample, or scattered. Scattering may be due to the presence of particles in the solution or scratches on the cuvette. The analyst is mainly interested in how much radiation a compound absorbs. However, because there is no direct way of measuring the absorbance, this information is obtained indirectly by measuring the radiation that is transmitted from the sample. Radiation that is transmitted is incident radiation minus all the radiation that is absorbed, scattered, or reflected by the sample. Consequently, in order to obtain accurate absorbance information, all other kinds of processes, like reflectance and scattering, have to be minimized. During sample preparation, clarity of the solution is emphasized to eliminate scattering; scratching of cuvettes should be avoided either by air drying them or by using very soft materials to wipe them.

If 500 photons of light enter a UV cell and only 200 photons come out from the other side, the transmittance is 0.40 or 40%. Transmittance (T) is defined as the ratio of light transmitted (I) to incident light (I_0):

$$T = I/I_0 = 200/500 = 0.4 = 40\% \tag{6.7}$$

6.6.2. Effect of Concentration on Transmittance

If 100 photons of light enter a cell that is empty (i.e., the concentration of solute is zero), then 100 photons will come out from the other side of the cell. As we increase the concentration of the sample in the cell, the amount of radiation absorbed by the sample will increase; consequently the transmittance will decrease with increasing sample concentration. Table 6.5 shows the relationship between increasing sample concentration and decreasing transmittance.

Table 6.5. Effect of Concentration on %T

Concentration	%T
0	100
10	79.4
20	63.1
40	39.8
60	25.1

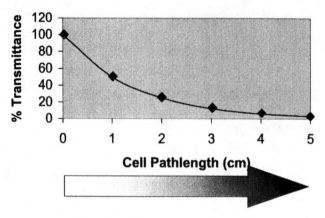

Figure 6.7. Effect of path length on %T.

If we were to plot % Transmittance versus concentration of the absorbing sample, we would note an exponential decay type of plot. The equation that expresses the relationship between % transmittance and concentration is called the Beer–Lambert law:

$$T = e^{-abc} \tag{6.8}$$

where a is the absorptivity of the sample at the given wavelength, b is the path length, c is the concentration of the sample, and e is the natural log base.

6.6.3. Effect of Path Length on Transmittance

If 100 photons of light enter a UV cell and only 50 photons come out from the other side, the percent transmission is 50%. If these 50 photons then pass through an identical sample cell with the same percent transmission, 25 photons will emerge, and so on. If the percent transmission following each cell is plotted against the number of cells, the curve would look like an exponential decay as shown in Figure 6.7.

The Beer–Bouguer law states that the absorbance (A) of a solution is equal to the product of absorptivity (a), path length (b), and sample concentration (c):

$$A = abc \tag{6.9}$$

The Beer–Lambert law in equation 6.8 provides an exponential relationship between transmittance and absorptivity, path length, and concentration. Mathematically, exponential relationships can be converted to linear relationships by applying a logarithmic transformation. In applying the logarithmic transformation to equation 6.8, we take logs of both sides of the equation and then we account for the conversion from natural log to base 10 log in a new constant, a, thus obtaining a linear relationship as in equation 6.9 (Beer's law) and arriving at the definition of absorbance, A, in equation 6.10:

$$A = -\log \%T = abc \qquad (6.10)$$

We can use these equations to determine the concentration of an unknown by using calibration curves.

6.7. INSTRUMENTATION

Next we will examine the general characteristics of instrumentation employed in the acquisition, display, and storage of spectra. This general discussion is applicable to ultraviolet (UV), visible (VIS), infrared (IR), Raman, and near-infrared (NIR) spectroscopy and will be followed by specific information for UV/VIS and for IR.

The components of a system necessary for obtaining absorption spectra are depicted as in Figure 6.8. They include a spectrophotometer, usually packaged in a singly housed unit. In a typical spectrophotometer, a sample absorbs a portion of the incident radiation and the nonabsorbed radiation is transmitted to a detector, which then displays and/or stores the absorption intensities as a function of wavelength, energy, or frequency, producing a spectrum.

A monochromator is the component of a spectrophotometer which separates light into its constituent wavelengths. Monochromators used in the UV-

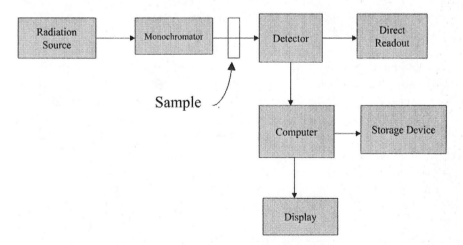

Figure 6.8. Components of a spectrophotometer.

VIS, IR, NIR, or Raman may consist of prisms, diffraction gratings, or filters. Although many modern instruments utilize computers to analyze, store, and display spectral data, a computer is not a necessary component except for Fourier transform (FT) techniques that are utilized mainly in infrared and NMR spectroscopy.

6.7.1. UV/VIS INSTRUMENTATION

The components that make up typical spectrophotometric instruments were sketched out in Figure 6.8. The specific components of a UV-VIS instrument—sources, monochromators, detectors, and readout devices—will be discussed in detail.

6.7.1.1. Sources. Measurement of a spectrum in the UV or visible region requires a light source emitting radiation in the wavelength region 200–800 nm. Ideally, such sources should emit radiation in a continuous manner over the entire spectral region. Usually, however, separate sources are used in the UV region and visible regions. Sources are expected to exhibit the following attributes:

- *Intensity.* To allow for meaningful signal to noise in absorption measurements.
- *Stability.* Source should not fluctuate in intensity from run to run or within a run.
- *Wide Spectral Range.* To permit examination of wide spectral regions when using a single source.
- *Uniform Output.* Ideally the output intensity of the lamp should not change with wavelength. This permits direct comparison of different spectral regions.
- *Ruggedness.* Lamp should not be delicate and be able to withstand a laboratory environment.
- *Optimum Size.* Because the lamp is usually contained in the spectrophotometer, it should be sized to minimize the size of the instrument housing.
- *Long Life.* Ideally the lamp should last over many runs. Minimal replacement of lamps reduces disturbance and need for recalibration of the instrument.

There are two major types of sources utilized in the UV and visible spectral regions:

- *Continuum Incandescent Source.* Emits radiation over a broad spectral region in a continuous fashion with no spectral gaps. This is usually achieved by glowing black bodies such as tungsten filaments, plasma arcs, and quartz halogen lamps. Hydrogen, deuterium, and xenon lamps, useful in the UV, show continuous output but have atomic emission lines superimposed on the continuum. It is possible to use these emission lines for wavelength calibration.

• *Line Source.* For certain types of spectroscopy, including atomic absorption and Raman, only single-wavelength output is required rather than continuum output. Elemental resonance lamps (e.g., Hg vapor) and lasers are useful for this purpose. For atomic absorption, separate lamps are required for each element of interest.

6.7.1.2. Monochromators. The white light coming from a radiation source contains superimposed wavelengths. In order to gain spectral information it is necessary to isolate individual wavelengths or wavelength regions and to measure the light absorption (or emission) at each wavelength separately. This can be accomplished using either prisms, gratings, or filters.

Prisms. White light incident on a prism is dispersed by the prism material into its constituent wavelengths. If a screen is placed at the exit of the prism, the separated wavelengths from red to violet may be observed spread out over the screen. In a spectrometer, each of these wavelengths may be separately focused on a detector by rotating the prism. The placement of mechanical slits further isolates the separated wavelengths. Slits are employed because the prism disperses and passes a wavelength band rather than a single wavelength. Narrow mechanical slits further isolate the wavelengths of interest. A schematic of a single-beam prism spectrophotometer is shown in Figure 6.9.

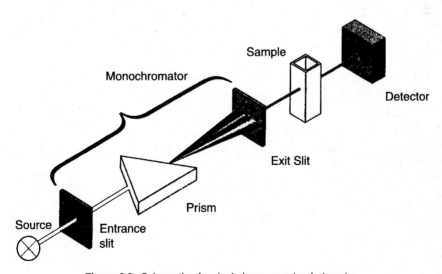

Figure 6.9. Schematic of a single beam spectrophotometer.

The degree to which a prism separates wavelengths is referred to as the dispersion or dispersive power. The greater the dispersion, the more separated the wavelengths from red to blue. A prism functions because the refractive index of light in a prism material differs for different wavelengths of light. The shorter wavelength violet light is bent to a greater degree than red light as it is transverses the prism. Typical prism materials useful in the visible region are constructed of glass, and quartz is used in the UV.

Diffraction Gratings. Another device used as a monochromater to disperse light is a diffraction grating. Unlike a prism, which separates light according to refractive index differences, a grating separates light according to constructive/ destructive interference of light waves in a process called diffraction.

A diffraction grating is an array of many parallel equidistant slits (lines) that are laid down onto a rigid substrate. A ruling engine using a diamond cutting tool etches lines onto a polished substrate. The collection of carefully ruled parallel lines has the property of separating incident light into its constituent wavelengths. For a given angle of the grating in relation to the incident light, only certain wavelengths will be in phase as they are reflected or transmitted from the ruled surface. By rotating the grating, it is then possible to separate the light into its constituent wavelengths. This optical effect can be appreciated by rotating a compact disc (CD) used in computers or audio/visual applications while shining a light on the surface. A spectrum of different colors is observed reflected from the CD, which contains many closely spaced concentric circles "etched" into its surface.

Originally etched gratings are called master gratings and are very expensive to produce because of the precise control necessary in cutting the grooves. Most gratings in use are replica gratings in which a film is allowed to form on the master. The peeled-off film contains the replica of lines from the original master grating. These replicas are then mounted on rigid supports. Laser technology is also used to prepare another type of grating referred to as a holographic grating.

The level of dispersion obtained with a grating depends upon the number of etched grooves. In the visible spectral region it is not unusual to use gratings having more than 3000 grooves/mm of grating surface.

Filters. In addition to dispersing devices such as gratings or prisms, the incoming light may be isolated into wavelength regions by optical filters. It is possible to isolate narrow spectral regions by appropriate choice of filters. There are many different types of filters including interference and absorption filters. Some operate by rejecting light outside narrow spectral regions, while other filters will only absorb particular wavelengths and transmit the rest. By combining filters, gratings, and prisms, it is possible to construct a monochromator having high wavelength selectivity.

Inexpensive instruments such as filter photometers may employ filters to isolate selected wavelengths, which are used for repetitive analyses at a given wavelength. Such instruments do not scan wavelengths, and so the actual spectrum is not obtained. Filter photometers have been used in clinical laboratories where the absorption at a particular wavelength is a measure of a metabolite or other moiety found in blood or biological specimens.

Generally speaking, a diffraction grating monochromator allows for better spectral dispersion within the size limitations of a benchtop instrument. In addition, the dispersion of light from a grating is linear in wavelength, whereas a prism exhibits nonlinear dispersion. Because we usually like to display spectra with equal wavelength spacing, the output of the grating allows for this direct linear display. For instruments controlled by computers the nonlinearity in wavelength may be adjusted by software before display.

6.7.1.3. Detectors.

6.7.1.3. Detectors. After the individual wavelengths of light are separated and allowed to pass through a sample, a detection system is required to measure the amount of energy that has been transmitted by the sample. By measuring the ratio of the intensity of the incoming light to the transmitted light the transmittance and absorbance may be calculated at any wavelength.

Typical detectors used in the UV-VIS region are transducers that convert light energy into a current flow, which is monitored by a readout device or stored in a computer. Examples of detectors include photomultipliers (PMT or multiplying phototubes) and photodiodes.

Photomultipliers. Vacuum tubes containing a photoemissive cathode that emits electrons when exposed to photons of light. The amount of emitted electrons is proportional to the number of incident photons. The photomultiplier employs a series of dynodes (electrodes) coated with materials emitting electrons. These dynodes are maintained at increasing voltages and give rise to an avalanche of current at the anode when even a single photon is incident on the initial cathode. Multiplication factors on the order of hundreds of millions of electrons for each incident light photon are readily achieved. (See Figure 6.10.)

Photomultipliers are utilized in many spectrophotometer detection systems because of their sensitivity but may only be used at fairly low power levels of incident light before they saturate or swamp the detector.

Photodiodes. These are semiconductor devices that generate electrical currents proportional to incident light intensities. They are useful in the visible and near infrared regions. The generated currents are orders of magnitude less than with photomultipliers, but the small size of diodes allows their placement into linear arrays. Photodiode Arrays (PDA) are positioned in a spectrophotometer with individual diodes at different angles with respect to the diffracted light from the grating. This permits the simultaneous detection of different wavelengths and the rapid recording of a spectrum. Some commercial UV-VIS instruments use separate photodiode arrays of over 200 diodes each to separately detect the UV and visible spectral regions. They are popular as liquid chromatography (LC) detectors and are further discussed in Chapter 9.

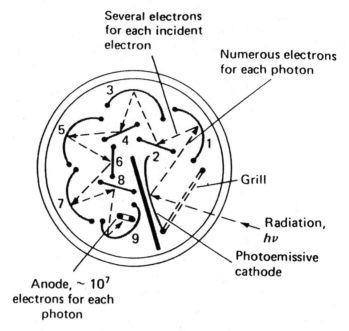

Figure 6.10. Schematic of a PMT.

6.7.1.4. Readout Devices. The spectrophotometer system takes the measured light intensities and permits visual display or storage of these data. Instruments of the nonscanning type may use a simple digital or analogue meter to display the light intensity. Scanning instruments may employ a strip chart recorder that plots a spectrum and provides paper hard copies. Most instruments utilize desktop computers to store and display the spectra. The software allows manipulation of data to enhance the display. These spectra may be stored on magnetic media or printed out as hard copy. More information about data systems can be found in Chapter 14.

6.7.1.5. Performance Characteristics. In choosing a spectrophotometer there are characteristics that relate to performance for which the manufacturer provides specifications. The choice of instrument may depend on the types of analyses intended by the user. Typical specifications will identify resolution, photometry precision, wavelength accuracy, stray light, and signal to noise.

6.7.1.6. Resolution. Resolution is defined as the ability of an instrument to separate adjacent spectral wavelengths. Usually for liquid samples the natural bandwidths are fairly broad (5–50 nm) and so most spectrophotometers would have little difficulty in observing such bands. However, in gases and solids, natural bandwidths are much less and it is important to assure the ability of the instrument to resolve these narrow spectral features.

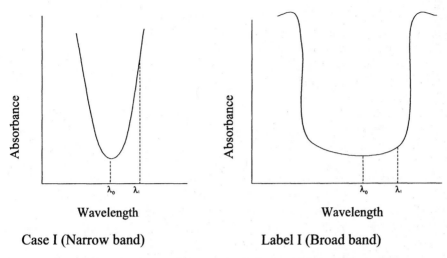

Figure 6.11. Effect of absorption bandwidth on the selection of wavelength.

6.7.1.6. Photometry Precision. Photometry precision is defined as the ability of an instrument to reproduce the transmittance and absorbance of samples over a broad spectral range. This is important in making quantitative measurements of species dissolved in solvents. The photometry precision must be demonstrated over a wide range of sample absorbance (concentration).

6.7.1.7. Wavelength Precision/Accuracy. It is important that an instrument be able to consistently return to the designated analytical wavelength for making repeatable quantitative measurements. This becomes important when using narrow bands for quantitation. For example, if one chooses the peak absorption to measure the concentration of an analyte such as in Figure 6.11, errors may occur.

In case I, an error in the wavelength reproducibility will cause a significant error in absorption because of the narrow spectral bandwidth and the steep change of absorption with wavelength. Case II, in which a broader band is used for quantitation, would show a much smaller error in absorption for the same error in wavelength reproducibility. Ideally an instrument should have wavelength reproducibility of better than 0.02 nm.

Wavelength accuracy may not be as important as precision but may be determined by using didymium glass or holmium oxide standards, which exhibit well-characterized and narrow absorption bands.

6.7.1.8. Stray Light. When an instrument is set at a given wavelength, it is expected that only light of that wavelength is received by the detector. Due to scattering of radiation in the instrument from dust, optical imperfections, and so on, light that hasn't gone through the sample may reach the detector. This detected light may include other than the expected wavelength and could give rise to anomalous absorption and deviations from Beer's law. Such errors will lead to inaccuracies in quantitative analysis.

Figures for stray light at given wavelengths are specified by instrument manufacturers and should be minimized.

6.7.1.9. Operating Parameters.
It is possible to affect a spectrophotometer's apparent resolution, wavelength precision, and photometry precision by changing operating variables such as scan speed, slit width, amplifier gain, and so on.

Scanning too quickly, especially over a narrow band, may cause distortion of the band shape. For weak signals one may open the slits to permit more energy. However, this will diminish the instrument's ability to resolve sharp peaks. An increase in amplifier gain will also enhance the signal but may introduce additional noise in the observed spectrum. Improper scanning parameters will also affect the photometric accuracy and lead to errors in quantitative analysis.

Figure 6.12 is an example of the effect of slit width on peak shape. At 20 nm, the observed spectral band has been distorted by the improper slit width setting.

The appropriate conditions for scanning a spectrum involve careful choice of

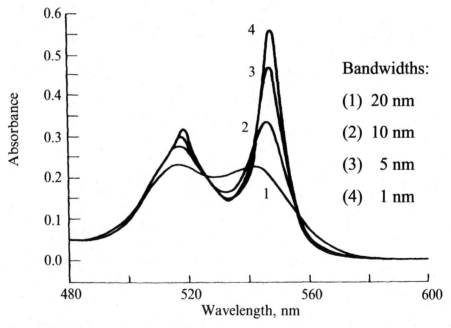

Figure 6.12. Effect of slit width on peak shape and resolution. (From *Instrumental Methods of Analysis*, 7th edition by H. H. Willard, L. L. Merritt, J. A. Dean, and F. A. Settle, Jr., Copyright 1988. Reprinted with permission of Brooks/Cole Publishing, a division of Thomson Learning, Fax 800-730-2215.)

instrument parameters to allow recording of undistorted spectra. Computer-controlled instruments allow for setting scan programs having optimized parameter settings.

Summary. The discussion of spectrophotometric systems as applied to the UV-VIS region is applicable to other types of absorption spectroscopy. The individual components are optimized for use in a specific spectral region by appropriate choice of materials that maximize the energy throughput.

6.7.2. IR Instrumentation

Dispersive. Typical IR dispersive spectrophotometers have the same components as UV-VIS spectrophotometers. Some of the materials employed are summarized in Table 6.6.

Because IR radiation is absorbed by glass, other materials are utilized for the components and for sampling cells. Prisms are constructed of crystalline materials such as NaCl, KBr, or CsBr. The use of such water-sensitive materials, however, requires appropriate desiccation of the instruments to minimize their degradation. Diffraction gratings have replaced prisms for the most part; these gratings, in turn, are being replaced by interferometers, which will be discussed below. The components of a typical IR spectrophotometer are depicted in Figure 6.13. The functioning of these components is similar to UV-VIS spec-

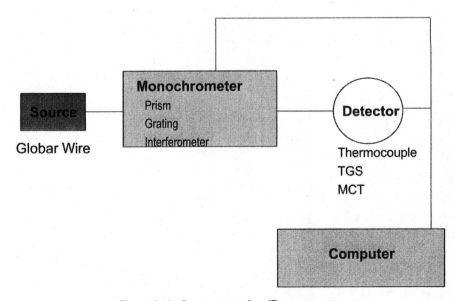

Figure 6.13. Components of an IR spectrometer.

Table 6.6. Components of IR Spectrophotometers

Spectral Region	Source	Absorption Cell	Dispersing Element	Detector
Mid-IR	Filaments, Nernst glower, globar	NaCl, KBr, sapphire, diamond	Grating, prism, interferometer	Thermocouple, bolometer, solid-state photoconductors, Golay cells
Visible	Tungsten filament, xenon arc	Glass	Prism, grating	Photomultiplier Photodiode array, photographic plate
Ultraviolet	Deuterium lamp, xenon arc	Quartz	Prism, grating	Photomultiplier Photodiode array, photographic plates

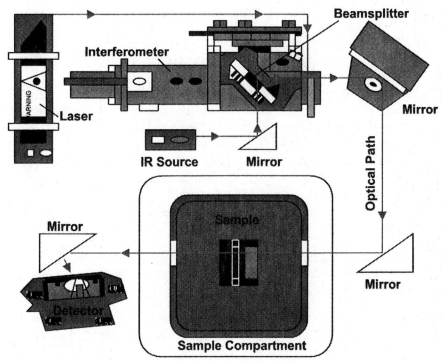

Figure 6.14. Diagram of an FTIR instrument. (Reprinted with permission from Nicolet, Inc.)

trophotometers and will not be discussed except for the interferometer that is shown in the Fourier transform infrared (FTIR) instrument in Figure 6.14.

6.7.2.1. Fourier Transform Infrared (FTIR).
The technique of collecting IR spectral information using a scanning Michelson interferometer is referred to as FTIR. This nondispersive technique for collecting absorption and wavelength information has many advantages compared to dispersive spectrophotometers. As the price of these instruments comes down, they are finding wide acceptance in quality assurance and analytical chemistry laboratories.

The heart of the FTIR instrument is a device called an interferometer, invented by A. A. Michelson, an American scientist. Unlike dispersive instruments in which the wavelengths of light are individually separated, an interferometer allows all wavelengths to pass through the instrument and sample to produce an interference pattern that may be analyzed by a computer to convert the data into an IR spectrum. As with a grating, an interferometer utilizes the wave nature of light and analyzes the constructive and destructive interference of superimposed light waves. The end result of this process when a sample is placed in the optical path is an IR spectrum similar to that produced in dispersive instruments.

The advantages of FTIR over dispersive IR include the following:

- *Speed.* All frequencies are measured simultaneously (Felgett's advantage).
- *Sensitivity.* There is high optical throughput because no slits are required (Jacquinot's advantage).
- *Enhanced Signal-to-Noise.* Fast scanning allows time for many scans, which can be added to improve the signal-to-noise ratio, S/N. Random noise will cancel when added, and as a result the S/N will increase. The improvement in S/N is proportional to the square root of the number of scans.
- *Frequency Accuracy.* An He–Ne laser is used as an internal wavelength calibration (Conne's advantage) and provides accuracy to 0.01 wavenumbers with no other calibration necessary.
- *Mechanical Simplicity.* The moving interferometric mirror is the only moving part.

Spectra, which could take hours to record using dispersive instruments, may now be acquired in seconds or minutes at comparable spectral resolution and photometric accuracy.

Using the Advantages of FTIR. Because the computer is an essential component of an FTIR spectrophotometer, it may be utilized to other advantage.

- *Background Subtraction.* An FTIR instrument normally scans as a single-beam instrument. It is therefore necessary to record a background spectrum to determine absorption by interfering water vapor and carbon dioxide in the sample compartment. The computer-stored spectrum or interferogram of the background may then be eliminated from the sample spectrum by mathematical subtraction.
- *Solvent Subtraction.* The ability to store solvent blank spectra permits the running of samples in solution, subtracting the solvent spectrum and then displaying the solute or analyte spectrum. These difference spectra are particularly useful for obtaining spectra of species in highly absorbing solvents such as water.
- *Kinetic Studies.* The ability to fast scan enables the recording of spectral changes that occur rapidly such as fast reaction kinetics, adsorption of materials onto substrate surfaces and catalysts.
- *Spectra of Mixtures and Multiphases.* The subtraction technique for obtaining difference spectra may be used to quantitatively determine levels of coatings on substrates, crystalline and amorphous contents in polymers, individual crystalline polymorphs in drug substances, and additives or excipients in materials.

6.7.2.2. Sampling of Solids. IR techniques are available for recording spectra of gases, liquids, and solids when the appropriate sampling cells and probes are used. Solids are normally placed in an IR transmitting matrix for spectral acquisition. A popular technique places small amounts of powdered solid into powdered KBr. A high-pressure device compresses the mixture and enables the KBr to flow and dissolve the analyte by forming a clear pellet or disc within a dye. Because KBr is transparent to IR radiation the recorded spectrum represents the IR spectrum of the dispersed analyte of interest. It takes practice and care in forming clear and dry pellets, but the spectra obtained may be of high quality.

Another useful technique for solids is to suspend a solid powder of an analyte into a grease-like mull. For this purpose, mineral oil (Nujol), chlorofluorocarbon greases (fluorolubes), and perfluorokerosene, amongst others, are employed as the sample carriers. A small amount of analyte is mixed with the oil by grinding in a mortar and pestle to obtain a uniform suspension of the analyte in the grease.

A thin film of the suspension is spread onto an IR-transmitting crystal such as KBr, AgCl, or Irtran, and the spectrum is recorded. Although the mulling agents are IR-transmitting, they may have absorption bands in certain regions which could interfere with the analyte spectrum. Judicious choices of a mulling agent or mixture of mulling agents will allow complete analyte spectra to be observed. It is also possible to subtract the spectrum of the mulling agent to obtain the difference spectrum.

IR spectra of materials such as polymers that form films may be obtained without suspension in a pellet or mull. It is necessary, however, to utilize thin films for this purpose. Such films may be prepared by dissolution of the sample in a volatile solvent and then allowing a few drops of the solution to evaporate from an IR transmitting substrate.

6.7.2.3. Sampling of Liquids. IR spectra of liquids may be obtained by spreading a few drops of sample between IR transmitting plates, pressing the plates together usually avoids sample evaporation. For quantitative analysis, specially constructed liquid cells that have known sample thickness are used. There are also techniques for confirming the cell thickness by observing and counting interference fringes from empty cells run in the IR spectrophotometer.

6.7.2.4. Internal Reflectance. The use of attenuated total reflectance (ATR) permits the recording of IR spectra of liquids and solids without the need for sample dilution or other treatment. A sample of analyte is placed in contact with a plate of high refractive index. If the IR radiation enters the plate at an angle of incidence greater than its critical angle, the light will be totally internally reflected such as in the prism shown in Figure 6.15.

A sample in contact with the prism at the point of total internal reflectance will absorb a small amount of the IR radiation. A spectrum obtained of the radiation exiting the prism will be the characteristic IR spectrum of the sample. This technique has been adapted to a variety of geometries and employs crystalline materials of differing refractive index. The prism is an example of a single internal reflection device.

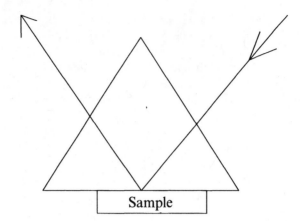

Figure 6.15. Internal reflectance cell for IR sampling using a prism.

The depth of penetration of the IR beam into a sample is related to the refractive index of the internal reflectance element. By choice of elements having a range of refractive indices, it is possible to obtain depth profile spectra. This technique is applied in examining laminates of films to obtain IR spectra of individual layers in the laminate. For contamination on a surface, depth profiling may identify the actual depth of the contaminant from the surface.

A more popular internal reflectance accessory is shown in Figure 6.16. The internal reflectance plate is constructed so that the entering IR beam internally reflects multiple times as it transverses the plate. At each reflection from the surface in contact with a sample, some of the radiation is absorbed by the sample. The multiple reflections increase the total energy absorbed by the sample, which enhances the observed spectrum. Normally the internally reflected beam only penetrates the sample slightly and the IR spectra so obtained are weak. Multiple internal reflection techniques provide for high-quality spectra.

Materials used for internal reflection spectroscopy include KRS5, Irtrans, ZnSe, AgCl, Si, germanium, diamond, sapphire, and so on. The convenience of the technique relies on simple physical contact of the sample and the internal

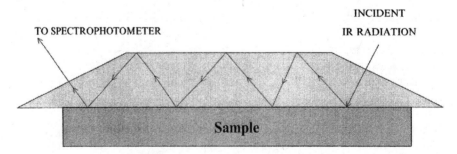

Figure 6.16. Multiple internal reflectance cell.

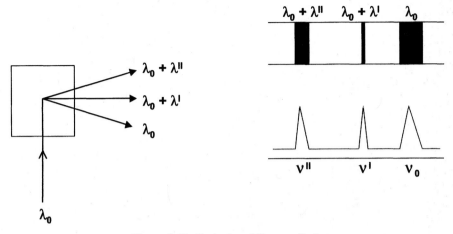

Figure 6.17. Illustration of Raman effect.

reflectance plate. Internal reflectance devices are available to record IR spectra by dipping a probe into liquids, solids, or gases. These probes have simplified IR sampling enabling faster throughput in the laboratory.

6.8. RAMAN SPECTROSCOPY

A form of spectroscopy that gives information similar to that of IR, was discovered by a Nobel Prize (1928) winner Sir C. V. Raman, an Indian physicist. He observed that when intense visible light of a single wavelength (λ_0) is incident on a material, the observed scattered light measured in a spectrophotometer contains, in addition to the incident wavelength, additional wavelengths as well. This phenomenon has become known as the Raman effect and is depicted in Figure 6.17. When the scattered light is sent through a spectrophotometer, an emission spectrum of the various scattered wavelengths is obtained as shown in Figure 6.18. The intense spectral line at the same wavelength as the incident wavelength (λ_0) is referred to as the Rayleigh line. The additional lines or bands, as they are usually called at higher and lower wavelengths compared to the incident wavelength, are referred to as Raman-shifted. The series of bands at higher wavelengths (lower frequency) is called the Stokes spectrum, and at lower wavelengths it is called the anti-Stokes spectrum.

Normally we deal only with the Stokes Raman spectrum. What makes this discovery intriguing is that when the observed wavelengths are converted to wave numbers the difference in wave number of each observed spectral line from the nonshifted incident line gives a set of frequencies that are in the IR region of the spectrum and include far and near IR frequencies as well.

A typical Raman spectrum of phenobarbital is shown in Figure 16.19. The spectrum was excited by a single laser wavelength at 488 nm in the visible

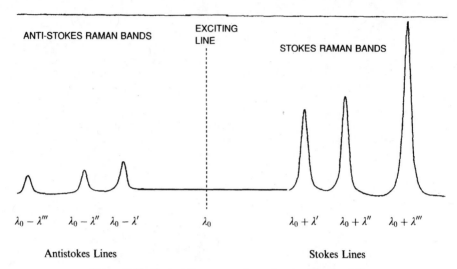

Figure 6.18. Typical Raman spectrum showing Raman shifts.

region. The spectral frequencies observed in Raman spectroscopy are related to the group frequencies discussed in the IR section of this chapter. The origin of Raman spectra depends upon polarizability changes in a molecule as the atoms vibrate. Unlike IR, no dipole moment change is necessary, but the molecular polarizability must change. As such the intensities of bands may not be the same as with IR spectra. This is a distinct advantage because symmetrical

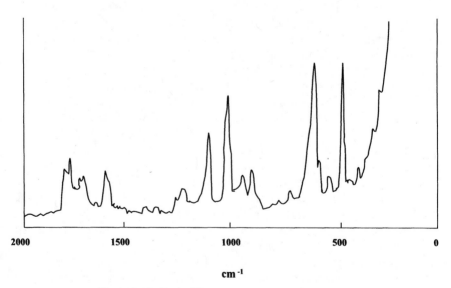

Figure 6.19. Typical Raman spectrum of phenobarbital.

vibrations with no dipole moment change are inactive in IR but may be fairly intense in Raman spectra. Examples of strong Raman bands are those originating in such functional groups as

$$-C=C-, \ -C\equiv C-, \ -C\equiv N, \ -C=S-, \ -C-C-, \ -S-S-, \ -N=N-, \ Br-Br, \ Cl-Cl$$

The great advantage of Raman spectroscopy is that a single wavelength source of light in the visible region of the spectrum is able to generate molecular information corresponding to the IR regions of the spectrum.

The use of visible light simplifies sample handling and instrumentation compared to IR. Because the Raman spectrum displayed is really a wave number difference spectrum, it is possible with a normal visible spectrophotometer to obtain spectra down to a few cm^{-1} in the far IR which contain information relating to the nature of crystalline or solid state phases of the material.

Another advantage of Raman spectroscopy is its ability to generate spectra in aqueous solvents. Unlike IR, water shows only weak Raman spectra and may be used as a solvent for obtaining Raman spectra. This advantage also simplifies the study of biological and cellular systems containing water.

There has been recent use of Raman spectroscopy to examine polymorphic content in drug substances. Although IR can do the same, the necessity of subjecting samples to the rigors of grinding, mulling or high-pressure KBr pelleting may change the polymorphic structures being examined.

Raman sampling is readily achieved by directing the focused incident laser radiation onto capillary tubes containing liquids or solids. (See Figure 6.20.) It is even possible to generate a Raman spectrum of a material in a jar (provided that the jar is transparent to visible light) or intact gel caps.

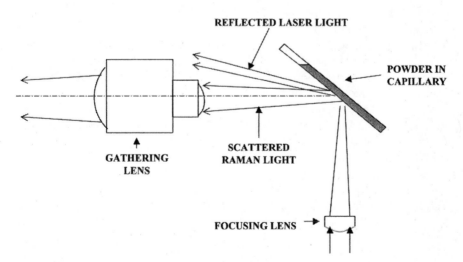

Figure 6.20. Capillary tube for solids.

Raman spectroscopy is a complement to IR, and both techniques are important for identifying and studying materials. Databases of fingerprint Raman spectra are available and lend themselves to search routines for identifying or confirming materials as in IR.

6.8.1. Raman Instrumentation

The components of a Raman spectrophotometer are similar to the UV-VIS instruments previously described. The main difference is the monochromatic source, which is usually a laser.

Fourier transform Raman spectroscopy using an interferometer is now beginning to replace dispersive monochromators by offering similar advantages of speed, energy throughput, and wavelength accuracy as in IR. When an interferometer is employed, the Raman exciting source is a laser with output wavelength in the near IR rather than in the visible region. The use of near-IR excitation minimizes sample fluorescence which could interfere with the Raman spectrum.

6.9. NEAR-IR (NIR) SPECTROSCOPY

The wave number region above 4000 cm^{-1} and extending to ~12,500 cm^{-1} is referred to as the near infrared. This region expressed in terms of wavelength extends from 2500 nm to 800 nm and borders on the visible region of the electromagnetic spectrum.

When molecules absorb radiation in the mid-infrared the spectrum observed is a collection of all the vibrating groups in the molecule. The vibrations, however, may interact with one another, giving rise to harmonics of the fundamental vibrations. These interactions led to complex spectra containing overtones and combinations of the fundamental vibrational frequencies. These harmonic spectra are observed in the NIR region of the spectrum and typically involve stretching vibrations of C–H, O–H, and N–H functional groups. For example, the OH overtone is observed at 7140 cm^{-1}, and the NH stretching overtone is observed at 6667 cm^{-1}.

NIR absorptions are much weaker than mid-IR, and so impurities play less of a role in the spectra. Absorption spectra in the NIR may be run on thick samples, and readily available quartz optics and sampling cells are used.

NIR spectroscopy has found extensive application in the quality assurance laboratory and is useful for identification of materials as well as quantitative analysis. Quantitative work requires careful correlation of observed absorption bands with concentration. It is important that the spectral correlation with content be well-validated over the range of expected conditions before NIR spectroscopy is routinely applied. Normally the method is validated against a different but acceptable method.

Wide use has also been made in the food industry, where NIR spectroscopy

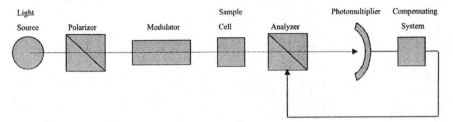

Figure 6.21. Schematic of single-beam polarimeter.

is employed to measure moisture content, protein, and fat content in foods and animal feeds. Reflectance probes are inserted into the bulk sample, and the incident and reflected radiation is transmitted and collected by fiber optics connected to the spectrophotometer. The simplicity of sampling makes this technique particularly inviting for a quality assurance (QA) environment. NIR spectroscopy is being used in the pharmaceutical QA lab as well for inspecting raw materials and for *in situ* monitoring of processes such as coating drying.

6.10. OTHER OPTICAL AND SPECTROSCOPIC TECHNIQUES

6.10.1. Polarimetry

Molecules, which are optically active, have the ability to rotate a plane of polarized light. The direction of rotation and the degree of rotation may be used to identify such optically active species in solution. An instrument to measure this rotation is a polarimeter and makes use of the components shown in Figure 6.21.

The analyzer that may be a polaroid sheet or Nicol prism is rotated with respect to the polarized incident light so that no light is passed through when no sample is present. An optically active sample placed between these crossed polars will rotate the plane of polarized light so that it is now transmitted by the analyzer. The amount of angular rotation of the analyzer necessary to cause no light to be transmitted is a measure of the optical rotation of the solution containing the sample. Typically, a 1-dm pathlength sample tube is used and the concentration prepared as 1 g/100 mL. The rotation observed under these conditions is called the specific rotation, α.

This technique is useful for measuring the concentration or purity of optically active isomers of drug substances. Instrumentation is simple and fully automated.

6.10.2. Inductively Coupled Plasma (ICP) and Atomic Absorption Spectroscopy (AAS)

These techniques are used to quantitate the presence of atomic species, especially cations and metals. ICP is an atomic emission technique that utilizes an argon plasma coupled to a high-frequency magnetic field to vaporize a sample and to bring it into an electronically excited state. The atom then emits visible radiation as it relaxes into its ground state. ICP has been successfully used to quantitate more than 70 elements present in a wide variety of matrices including chemicals, polymers, pharmaceuticals, drinking water, and so on.

The emitted radiation is sent into a visible spectrophotometer, and the emission is detected with photomultiplier or diode array detectors. This technique is sensitive for many atoms at concentrations in the low parts-per-million range.

To enhance the speed of the analysis, nonscanning spectrometers may be used. These direct readers are configured so that a series of slits and photomultipliers is positioned so that a wavelength identified with a particular atom is selectively monitored. When a collection of such readout detectors is used, it is possible to simultaneously monitor the concentration of many species. Computers may then be used to analyze and correct the data for display and storage.

Atomic absorption spectroscopy, referred to as AAS or sometimes as AA, is an optical absorption technique for measuring the concentration of elements present in a sample. The sample is vaporized or nebulized, and the radiation from a hollow-cathode lamp containing the same material as the analyte is allowed to pass through the vapor. The emitted resonance line from the lamp corresponds to the absorption wavelength of the analyte and is therefore absorbed. The amount of absorption is a function of the concentration of the analyte in the vapor and is expected to obey Beer's law. AAS requires that the analyte species be known so that the appropriate lamp may be chosen for the quantitative analysis. Separate lamps and separate spectral runs may be necessary to obtain quantitative data on more than one species, although some multielement lamps are available. The narrow emission lines from the hollow cathode lamp provide specificity for each element.

AAS instrumentation is much simpler than ICP and less expensive. ICP, however, offers the advantage of simultaneous determination of many elements, thereby increasing laboratory throughput. It is possible to use ICP to provide identification of all elements in a sample in a single run. AAS may then be used to quantitate the particular elements of interest.

6.10.3. Mass Spectroscopy (MS)

MS is a quantitative and qualitative technique involving the formation of gas-phase ions from a gas, liquid, or solid sample and separating and identifying these ions according to their mass to charge ratio. This is accomplished by precise tuning of electric and magnetic fields. Both negative and positive ion spectra may be recorded.

The technique is useful for determining molecular weights of species introduced into the instrument. This allows confirmation of the identity of these species. Databases of mass spectra are available for fingerprinting as in IR. The fingerprints, however, are the collection of mass numbers of ions and fragments rather than vibrational frequencies as in IR, NIR, or Raman spectroscopy. Large molecules with molecular weights greater than 1000 daltons are not typically determined by MS.

Mass spectra are not simple and contain contributions from many different ionic species capable of forming under the electron or chemical impact conditions of a mass spectrometer. The technique is extremely sensitive and may be used for detecting and quantifying trace impurities in drinking water, air, pharmaceuticals, and so on.

Instrumentation is generally large and requires very high vacuum for optimal performance. Mass spectrometers are extremely powerful when coupled with chromatographic separation techniques such as gas chromatography (GC-MS), liquid chromatography (LC-MS), and thermal techniques such as thermal gravimetric analysis (TGA-MS).

Benchtop mass spectrometers are available and used as mass detectors for the different chromatographic techniques. The ability to monitor the mass attributed to the elution of a species in a chromatograph allows for identification of the eluted peak according to its mass number. GC-MS and LC-MS are widely used in the pharmaceutical industry to monitor the purity of drug substances and formulations. It is also utilized in confirming structures of new drug molecules and chemical intermediates.

6.10.4. Nuclear Magnetic Resonance (NMR) Spectroscopy

NMR spectroscopy is an important technique for elucidating the structure of molecules including conformational and configurational information. An NMR spectrum is capable of being generated by atomic nuclei having nonzero nuclear spin. When such molecules are placed in a strong magnetic field the magnetic dipole of the atom is separated into different orientations. When a radiofrequency pulse is allowed to interact with these oriented nuclear quantum states, energy is absorbed. When the spin system is allowed to return to its equilibrium state a signal referred to as the free induction decay is generated and received by the detector.

The NMR signals are influenced by the chemical environment of surrounding and nearby atoms in a molecule. It is therefore possible to deduce the chemical structure of an unknown substance with a high degree of certainty. Typical nuclei examined in NMR spectroscopy include protons, fluorine, phosphorous. A useful form of NMR involves the spectra of naturally abundant carbon-13 found in all organic materials. Analysis of both proton and carbon-13 NMR spectra elucidates chemical structures and the composition of mixtures or purity of compounds.

Various sampling and instrumental techniques enable recording of solid-state spectra in addition to the usual solution spectra. These spectra yield structural and morphological information for polymers, proteins, and so on.

NMR spectrometers may employ Fourier transform techniques to compute the spectrum that is a signal intensity and frequency representation. There are databases of proton and carbon-13 NMR spectra available for thousands of compounds. Even without such comparison spectra, NMR spectroscopy quantitates the differing types of proton and carbon nuclei environments in a molecule by integration of the signal strengths arising from the different environments. These signal integrals enable reconstruction of the molecular structure. Integration of signals is also used to quantitate the composition of mixtures.

NMR is limited, however, to those nuclei having nonzero magnetic dipole moments; and since most NMR is recorded from solution, the solubility of the compound of interest may limit the acquisition of spectra.

6.11. SUMMARY

Molecular and atomic spectroscopy represent important analytical tools for identification and quantitation of chemical substances. The choice of technique is dependent on the information required as well as the availability of instrumentation. It is not necessary for a QA laboratory to have all the outlined techniques; however, the choices to be made may depend upon the particular product and material being evaluated.

Cost of instrumentation is certainly a factor, and maintenance of a full line of instrumentation may not be cost effective because there is a host of commercial laboratories providing extensive spectroscopic services.

The material in this chapter is not meant to be a complete and rigorous treatment of spectroscopy, and the following bibliography is provided for further information.

GENERAL REFERENCES

Anal. Chem. **1991**, 63, 91 (Spectroscopy Nomenclature).

G. J. Shugar and J. A. Dean, *The Chemists Ready Reference Handbook*, McGraw-Hill, New York, 1990.

Instrumental Methods of Analysis, 7th edition, H. H. Willard, L. L. Merritt, Jr., J. A. Dean and F. A. Settle, Jr., Wadsworth, Belmont, CA, 1988.

J. Michael Hollas, *Modern Spectroscopy*, Wiley New York, 1990.

John P. Sibilia (editor), *A Guide to Materials Characterization and Chemical Analysis*, 2nd edition, VCH Publishers, 1996.

William Kemp, *Organic Spectroscopy*, 3rd edition, W. H. Freeman, New York, 1991.

B. W. Cook and K. Jones, *A Programmed Introduction to Infrared Spectroscopy*, Heyden & Son, 1972.

7

CHROMATOGRAPHIC PRINCIPLES

James M. Miller

Of all the techniques used in the analytical laboratories of the pharmaceutical industry, chromatography is undoubtedly the most useful. In 1997, high-performance liquid chromatography (HPLC) was the instrumental technique most used by the pharmaceutical industry (35%), followed in second place by gas chromatography (GC) at 15% and gas chromatography/mass spectroscopy (GC/MS) in sixth place at 6% [1]. Another survey found that the pharmaceutical field was the largest user of gas chromatography [2].

This chapter presents the basic principles that govern all types of chromatography, and it is followed by separate chapters on GC and liquid chromatography (LC). Many of the concepts presented here are elaborated upon in the specific individual chapters; this chapter is intended as a general overview. It begins with some common, basic definitions, terms, and symbols.

7.1. DEFINITIONS, TERMS, AND SYMBOLS

7.1.1. Chromatography

Chromatography is a separation method in which the components of a sample partition between two phases: One phase is a stationary bed with a large surface area, and the other is a gas or liquid (mobile phase) that percolates through the stationary bed. The sample is carried by the mobile phase through the column.

Analytical Chemistry in a GMP Environment. Edited by J. M. Miller and J. B. Crowther
ISBN 0-471-31431-5 © 2000 John Wiley & Sons, Inc.

Samples partition (equilibrate) between the mobile phase and the stationary phase based on their solubilities or their relative tendencies to sorb on/in each of the respective phases. The components of the sample (called solutes or analytes) separate from one another based on their *relative* affinities for the two phases. This type of chromatographic process is called *elution*.

The "official" definition of the International Union of Pure and Applied Chemistry (IUPAC) is as follows: "Chromatography is a physical method of separation in which the components to be separated are distributed between two phases, one of which is stationary (stationary phase) while the other (the mobile phase) moves in a definite direction. Elution chromatography is a procedure in which the mobile phase is continuously passed through or along the chromatographic bed and the sample is fed into the system as a finite slug" [3].

Thus, a chromatographic system is made up of two phases, one stationary and one mobile. Because this is a general introduction, we will refer to the phases using these general terms and will use the symbol S for stationary and M for mobile.

In fact, chromatography is divided into two types according to the physical state of the mobile phase. Thus, if the mobile phase is a *gas*, the technique is called gas chromatography (GC), and if it is *a liquid*, it is called liquid chromatography (LC). A subclassification can be made according to the state of the stationary phase. If the stationary phase is a solid, the GC technique is called gas–solid chromatography (GSC), and if it is a liquid, it is called gas–liquid chromatography (GLC). The names used in liquid chromatography are more diverse and do not usually follow this simple pattern. A complete classification scheme is shown in Figure 7.1.

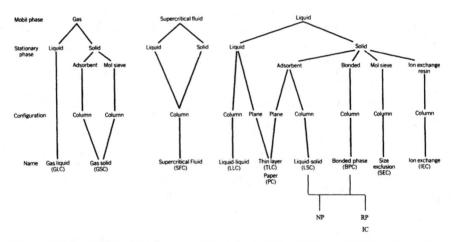

Figure 7.1. Classification of chromatographic methods. (Adapted from J. M. Miller, *Chromatography: Concepts and Contrasts*, Wiley, New York, 1988, p. 5. Reprinted by permission of John Wiley & Sons, Inc.)

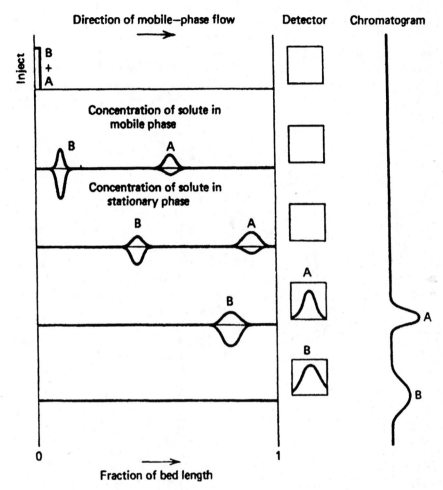

Figure 7.2. Schematic representation of the chromatographic process. (From H. M. McNair and J. M. Miller, *Basic Gas Chromatography*, Wiley, New York, 1998, p. 5. Reprinted by permission of John Wiley & Sons, Inc.)

7.1.2. The Chromatographic Process

Figure 7.2 is a schematic representation of the chromatographic process. The horizontal lines represent the column; each line is a snapshot of the chromatographic process at a different time (increasing time from top to bottom). In the first snapshot, the sample, composed of components A and B, is introduced onto the head of the column in a narrow zone. It is then carried through the column (from left to right) by the mobile phase.

Each component partitions between the two phases, as shown by the distributions or peaks above and below the line. Peaks above the line represent the concentration of solute in the mobile phase, and peaks below the line represent the concentration in the stationary phase. Component A has a greater affinity for the mobile phase; as a consequence, it is carried down the column faster than component B, which spends more of its time in the stationary phase. Thus, separation of A from B occurs as they travel through the column. Eventually the components leave the column and pass through the detector as shown. The output signal of the detector plotted versus time gives rise to a *chromatogram* shown at the right side of Figure 7.2.

Note that the figure shows how an individual chromatographic peak widens or broadens as it goes through the chromatographic process. The extent of this broadening, which results from the kinetic processes at work during chromatography, will be discussed later.

The tendency of a given component to be attracted to the stationary phase is expressed in chemical terms as an equilibrium constant called the distribution constant, K_c, sometimes also called the partition coefficient. The distribution constant is similar in principle to the partition coefficient that controls a liquid–liquid extraction. (See Chapter 5.) In chromatography the stationary-phase concentration is in the numerator, so the greater the value of the constant, the greater the attraction to the stationary phase, and the longer the retention time.

Alternatively, the attraction can be classified relative to the *type* of *sorption* by the solute. Sorption on the surface of the stationary phase is called *ad*sorption, and sorption into the bulk of a stationary liquid phase is called *ab*sorption. These terms are depicted in comical fashion in Figure 7.3. However, most chro-

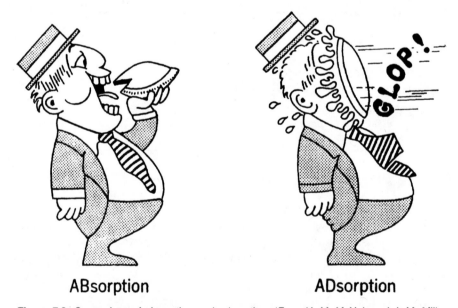

ABsorption   **ADsorption**

Figure 7.3. Comparison of absorption and adsorption. (From H. M. McNair and J. M. Miller, *Basic Gas Chromatography*, Wiley, New York, 1998, p. 6. Reprinted by permission of John Wiley & Sons, Inc.)

matographers and the IUPAC use the term *partition* to describe the absorption process. Thus they speak about adsorption on the surface of the stationary phase and speak about partitioning as passing into the bulk of the stationary phase. Usually one of these processes predominates for a given column, but both can be present.

The distribution constant provides a numerical value for the sorption by a solute *on* or *in* the stationary phase. As such, it expresses the extent of interaction and regulates the movement of solutes through the chromatographic bed. In summary, differences in distribution constants (parameters controlled by thermodynamics) effect a chromatographic separation.

7.1.3. Some Chromatographic Terms and Symbols

The IUPAC has attempted to codify chromatographic terms, symbols, and definitions for all forms of chromatography [3], and their recommendations are used in this book. Unfortunately, some of them differ from those given in the 24th *United States Pharmacopeia* (USP) [4]. Indeed, until the IUPAC publication in 1993, uniform nomenclature did not exist at all in the field, and some confusion may result from reading older publications. Table 7.1 compares some older conventions with the new IUPAC recommendations, and Table 7.2 compares the IUPAC with the USP.

The distribution constant, K_c, has just been discussed as the controlling factor in the partitioning between a solute and the stationary phase. It is defined as the concentration of the solute A in the stationary phase divided by its concentration in the mobile phase:

Table 7.1. Chromatographic Terms and Symbols

Symbol and Name Recommended by the IUPAC [3]		Other Symbols and Names in Use	
K_c	Distribution constant	K_p	Partition coefficient, distribution coefficient
k	Retention factor	k'	Capacity factor, capacity ratio, partition ratio
N	Plate number	n	Theoretical plate number, No. of theoretical plates
H	Plate height	HETP	Height equivalent to one theoretical plate
R	Retardation factor (in columns)	R_R	Retention ratio
R_s	Peak resolution	R	Resolution
α	Separation factor	—	Selectivity, solvent efficiency
t_R	Total retention time	t	Elution time,
V_R	Total retention volume	—	elution volume
V_M	Hold-up volume	V_0	Volume of the mobile phase, void volume, dead volume

Table 7.2. Comparison of USP and IUPAC Terms and Symbols

USP		IUPAC	
α	Relative retention	α	Separation factor
c	Concentration	C	Concentration
C_A, C_S	Concentration ratios	—	
f	Distance (in tailed peak)	f	Relative detector response factor
k'	Capacity factor	k	Retention factor
N	Number of theoretical plates	N	Plate number
q	Total quantity	—	
Q_A	Quantity ratio	Q_A	Practical specific capacity
Q_S	Quantity ratio	—	
R_r	Relative retention	r	Relative retention
r_S, r_U	Peak response	—	
R	Resolution	R_s	Peak resolution
—		R	Retardation factor in column chromatography
R_f	Retardation factor (in TLC)	R_F	Retard. factor in plane chromatography
R_S, R_U	Peak response ratios	—	
$S_R(\%)$	RSD	—	
T	Tailing factor	T	Temperature (in general)
t	Retention time	t_R	Total retention time
t_a	Retention time of nonretarded comp.	t_M	Mobile-phase hold-up time
W	Width of peak at baseline	w_b	Peak width at base
$W_{h/2}$	Width of peak at half-height	w_h	Peak width at half-height
$W_{0.05}$	Width of peak at 5% height	—	

$$K_c = \frac{[A]_S}{[A]_M} \tag{7.1}$$

This constant is a true thermodynamic value and is temperature-dependent; it expresses the relative tendency of a solute to distribute itself between the two phases. Differences in distribution constants result in differential migration rates of solutes through a column.

Figure 7.4 shows a typical chromatogram for a single solute, A, with an additional small peak early in the chromatogram. Solutes like A are retained by the column and are characterized by their *retention volumes*, V_R; the retention volume for solute A is depicted in the figure as the distance from the point of injection to the peak maximum. It is the volume of mobile phase necessary to elute solute A. This characteristic of a solute could also be specified by the retention time, t_R, if the column flow rate, F_c, were constant:

$$V_R = t_R \times F_c \tag{7.2}$$

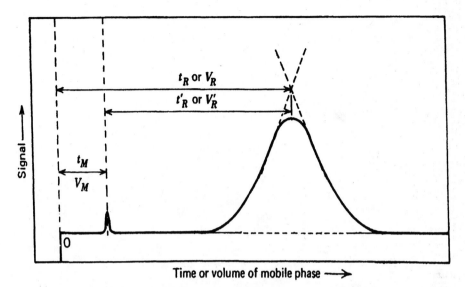

Figure 7.4. A typical chromatogram. (From H. M. McNair and J. M. Miller, *Basic Gas Chromatography*, Wiley, New York, 1998, p. 8. Reprinted courtesy of John Wiley & Sons, Inc.)

Unless specified otherwise, a constant flow rate is assumed and retention time is proportional to retention volume, and either can be used to represent the same concept.

The small early peak represents a solute that does not sorb in the stationary phase—it passes straight through the column without partitioning. In GC, this behavior is often shown by air or methane, and the peak is often called an "air peak." In LC, it may be difficult to find a suitable nonretained solute, but even changes in refractive index can cause a deflection in the baseline (for an ultraviolet detector) corresponding to the position of a nonretained solute. The symbol V_M, sometimes called the hold-up volume or void volume, serves to measure the interstitial or interparticle volume of the column. Other IUPAC approved symbols include V_0, representing the volume of the mobile phase in the column. The term dead volume, while not recommended, is also widely used.

Another retention parameter is the *adjusted* retention volume, V_R', or adjusted retention time, t_R'. From Figure 7.4, it can be seen that this parameter represents the retention time (or volume) measured from the nonretained peak to the solute peak. Alternatively,

$$V_R' = V_R - V_M \tag{7.3}$$

It is a useful definition because the adjusted retention volume measures the differing retention behavior (solubility or adsorption) for various solutes, regardless of the column volume, V_M.

Equation 7.4, one of the fundamental chromatographic equations,* can be used to relate the chromatographic *retention volume* to the distribution constant:

$$V_R = V_M + K_C V_S \qquad (7.4)$$

And substituting equation 7.3 into equation 7.4 yields

$$V_R - V_M = V_R' = K_C V_S \qquad (7.5)$$

where V represents a volume and the subscripts R, M, and S stand for retention, mobile, and stationary, respectively. V_M and V_S represent the volumes of mobile phase and stationary phase in the column respectively, and the total retention volume, V_R can be described by reference to Figure 7.4. Thus, equation 7.5 provides us with a surprisingly simple relationship between the experimental adjusted retention volume (V_R') and the theoretical distribution constant K_C.

An understanding of the chromatographic process can be deduced by a further examination of equation 7.4. The total volume of mobile phase that flows during the elution of a solute can be seen to be composed of two parts: One part is that which fills the column or, alternatively, the volume through which the solute must pass in its journey through the column, and it is represented by V_M. The second part is the volume of mobile phase that flows while the solute is not moving but is stationary on, or in, the column bed. The latter is determined by the distribution constant (the solute's tendency to sorb) and the amount of stationary phase in the column, V_S. Thus, there are only two things a solute can do: Move with the flow of mobile phase when it is in the mobile phase, or sorb into the stationary phase and remain immobile. The sum of these two effects is the total retention volume, V_R.

7.1.4. The Normal Distribution

Individual solute molecules act independently of one another during the chromatographic process; as a result, they produce a randomized aggregation of retention times after repeated sorptions and desorptions. The result for a given solute is a distribution, or peak, whose shape can be approximated as being *normal* or *Gaussian*, the peak shape that represents the ideal. Symmetrical, Gaussian peaks are shown in all figures in the book except for those real chromatograms whose peaks are not ideal.

The characteristics of a Gaussian shape are well known; Figure 7.5 shows an ideal chromatographic peak. The inflection points occur at 0.607 of the peak height, and tangents to these points produce a triangle with a base width, w_b,

---

* For a derivation of this equation, see B. L. Karger, L. R. Snyder, and C. Horvath, *An Introduction to Separation Science*, Wiley, New York, 1973, pp. 131 and 166.

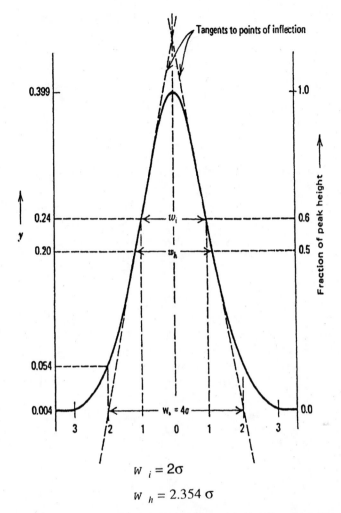

Figure 7.5. A normal distribution. (From H. M. McNair and J. M. Miller, *Basic Gas Chromatography*, Wiley, New York, 1998, p. 37. Reprinted by permission of John Wiley & Sons, Inc.)

equal to four standard deviations, 4σ. Sigma (σ) is used to express the peak width similar to its use to express the width of a standard distribution in Chapter 4. The width of the peak is 2σ at the inflection point, w_i (60.7% of the height) and 2.354σ at half-height, w_h. These characteristics are used in the definitions of some parameters, including the plate number.

7.1.5. Asymmetry and Tailing Factor

Nonsymmetrical peaks usually indicate that some undesirable interaction has taken place during the chromatographic process. Figure 7.6 shows some shapes

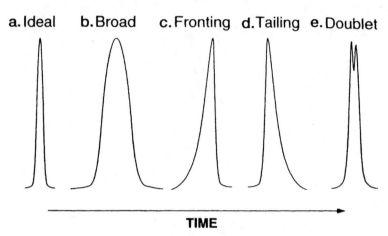

Figure 7.6. Peak shapes: (*a*) Ideal, (*b*) broad, (*c*) fronting, (*d*) tailing, (*e*) doublet. (From H. M. McNair and J. M. Miller, *Basic Gas Chromatography*, Wiley, New York, 1998, p. 35. Reprinted courtesy of John Wiley & Sons, Inc.)

that can occur in actual samples compared to the ideal shape shown in (a). Broad peaks like (b) in Figure 7.6 are undesirable and usually indicate that the kinetics of mass transfer are too slow (see discussion on the Rate Theory in this chapter). However, it is the chromatographer's goal to make the peaks as narrow as possible in order to achieve the best separations.

Asymmetric peaks can be classified as tailing (d) or fronting (c) depending on the location of the asymmetry. One definition of the extent of asymmetry is the tailing factor, TF (Fig. 7.7):

$$\text{TF} = \frac{b}{a} \tag{7.6}$$

Both *a* and *b* are measured at 10% of the peak height as shown in Figure 7.7. As can be seen from the equation, a tailing peak will have a TF greater than 1. The opposite symmetry, fronting, or leading will yield a TF of less than 1.

While the definition was designed to provide a measure of the extent of tailing and is so named, it also measures fronting. A tailing factor of less than 1 indicates fronting. Hence the limits on the tailing factor for meeting system suitability ought to include both upper and lower limits. For example, one's method might say that to meet specifications, the TF should be less than 1.2 and greater than 0.8, depending of course on the requirements of the specific method.

Unfortunately, the IUPAC did not give a computational definition of tailing factor in its nomenclature report [3], and the USP does not use the one just described. The USP definition [4] can be seen by referring to Figure 7.8.

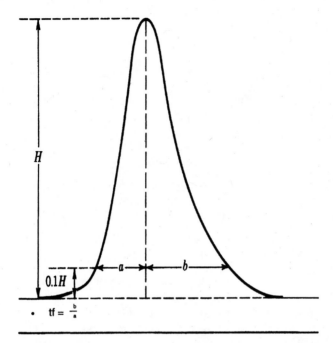

Figure 7.7. Figure defining asymmetric ratio or tailing factor. (From H. M. McNair and J. M. Miller, *Basic Gas Chromatography*, Wiley, New York, 1998, p. 36. Reprinted courtesy of John Wiley & Sons, Inc.)

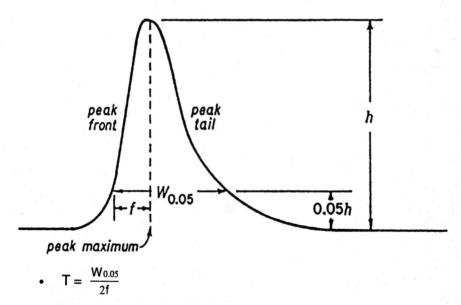

Figure 7.8. Figure defining USP tailing factor. (Copied with permission from the United States Pharmacopeia 24 – National Formularly 19. All rights reserved. Copyright 1999, The United States Pharmacopeial Convention, Inc.).

$$T = \frac{W_{0.05}}{2f} \qquad (7.7)$$

Note that both the definition and the height at which the measurements are made differ from the earlier definition. However, the relative values are similar: a value greater than 1 indicates tailing, and a value less than 1 indicates fronting. Many data systems give the user the opportunity to select the fraction of the peak height at which these measurements are made, so one must be careful to select the correct one in setting up a method. The higher on the peak the measurements are made, the smaller will be the calculated tailing factor. To meet USP requirements, however, the measurements must be made at 5%.

At the present time there is no definition for fronting factor, and thus it is wise to include upper and lower limits on the tailing factor as mentioned above.

7.1.6. Plate Number

To describe the efficiency of a chromatographic column, we need a measure of the peak width, but one that is relative to the retention time of the peak because width increases with retention time as we have noted before. Figure 7.9 illustrates this broadening phenomenon, which is a natural consequence of the chromatographic process.

The most common measure of the efficiency of a chromatographic system is the plate number, N:

$$N = \left(\frac{t_R}{\sigma}\right)^2 = 16\left(\frac{t_R}{w_b}\right)^2 = 5.54\left(\frac{t_R}{w_h}\right)^2 = 4\left(\frac{t_R}{w_i}\right)^2 \qquad (7.8)$$

Figure 7.10 shows the measurements needed to make this calculation. Different

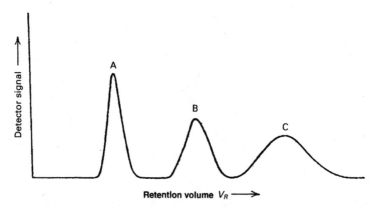

Figure 7.9. Typical elution chromatogram.

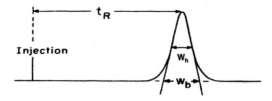

$$N = \left(\frac{t_R}{\sigma}\right)^2 = 16\left(\frac{t_R}{W_b}\right)^2 = 5.54\left(\frac{t_R}{W_h}\right)^2 \quad H = \frac{L}{N}$$

Figure 7.10. Figure used to define plate number, N. (From H. M. McNair and J. M. Miller, *Basic Gas Chromatography*, Wiley, New York, 1998, p. 38. Reprinted courtesy of John Wiley & Sons, Inc.)

terms arise because the measurement of σ can be made at different heights on the peak. At the base of the peak, w_b is 4σ, so the numerical constant is 4^2 or 16. At half-height, w_h is 2.354σ and the constant becomes 5.54 (refer to Figure 7.5). The USP allows measurements to be made at the base and at half-height.

Independent of the symbols used, both the numerator and the denominator must be given in the same units, and therefore N is unitless. Typically, both the retention time and the peak width are measured as distances on the chromatographic chart. Alternatively, both could be in either volume units or time units. No matter which calculation is made, a large value for N indicates an efficient column and is highly desirable.

For a chromatogram containing many peaks, the values of N for individual peaks may vary (they should increase slightly with retention time) depending on the accuracy with which the measurements are made. It is common practice, however, to assign a value to a particular column based on only one measurement even though an average value would be better.

A related parameter that expresses the efficiency of a column is the plate height, H:

$$H = \frac{L}{N} \tag{7.9}$$

where L is the column length. H has the units of length and is better than N for comparing efficiencies of columns of differing length. It is called the *height equivalent to one theoretical plate* (HETP), a term that is carried over from distillation terminology. Further discussion of H can be found later in this chapter. A good column will have a large N and a small H.

7.2. COMPARISON OF GC AND LC

As indicated earlier, the two types of chromatography, GC and LC, are named for their respective mobile phases, and a complete classification was shown in Figure 7.1. Let us take a look at some additional comparisons.

The gaseous mobile phase in GC is compressible under the pressure needed to force it through the column—unlike liquids (used in LC), which are virtually incompressible. Consequently, the volumetric flow rate through a GC column varies with the position in the column—getting faster as the gas moves toward the exit. The flow rate is usually measured at the column exit, so a correction factor is required to get the average column flow rate. It is called the compressibility factor, j, and when it is multiplied by the outlet mobile gas–flow rate, the average F_c, $\bar{F}_c$, is obtained:

$$jF_c = \bar{F}_c \tag{7.10}$$

Multiplication of the compressibility factor times the retention time produces a "corrected" retention volume, V_R^0:

$$jV_R = V_R^0 = j(F_c t_R) \tag{7.11}$$

So, for GC, it is possible to have a retention time that is both "adjusted" and "corrected"; it is called the "net" retention volume, V_N.

$$jV_R' = V_N = V_R^0 - V_M^0 \tag{7.12}$$

This correction is not usually necessary in LC, so the total and corrected retention volumes are identical as are the adjusted and net retention volumes.

Because the mobile phase is a gas in GC, it is also necessary to use a column to contain the stationary phase and prevent the escape of the gas. In LC, the stationary phase can be contained in a column too, but it can also be spread out in a thin layer on a plane surface. The two common plane surface techniques (see Figure 7.1) are paper chromatography (PC) and thin-layer chromatography (TLC).

The column techniques in LC can be alternatively subdivided according to their efficiencies and their capacities. Thus we have low-pressure (or no pressure, namely gravity feed) LC (often called "column chrom"), in comparison with high-performance LC (so-called HPLC). In the latter case, the high performance also requires higher pressures, but the acronym HPLC is best used to denote high performance, not high pressure. A third alternative is preparative LC, which uses large stationary phase beds and thus large columns so that large quantities of material can be separated. (See Chapter 9.)

Figure 7.1 also includes a third type of chromatography in addition to GC and LC; it is supercritical fluid chromatography (SFC). Supercritical fluids have properties intermediate between gases and liquids, giving SFC some unique properties. However, the field has not developed to the extent of GC and LC; and few, if any, applications are routine in the pharmaceutical industry. Consequently, this type of chromatography is not included in this book.

The instruments required for GC and LC are, of course, different and so are the columns that contain the stationary phases. Therefore, the instrumental requirements will be discussed in the respective chapters. However, the data from the detectors are similar, so data systems are covered for both techniques in one chapter, Chapter 14.

7.3. TWO IMPORTANT FUNDAMENTALS

A closer examination of the chromatographic process reveals two questions that need further explanation. The first question relates to the cause of the differential migration (selective sorption or retention) that causes solutes to travel at different speeds and thus become separated from each other. The second relates to the cause of the broadening of the solute zones during the procedure. This broadening works against the separation and should be minimized. The answer to the first question can be found in thermodynamics, and the answer to second one can be found in kinetics. Each of these topics will be discussed briefly.

7.3.1. Thermodynamics of Chromatography

For a simple treatment of thermodynamics of the chromatographic process, we can begin with the definition of the distribution constant, K_c, which was introduced earlier:

$$K_c = \frac{[A]_S}{[A]_M} \tag{7.13}$$

Although the distribution constant is a classic equilibrium constant, it can be used to describe the nonequilibrium situation that exists in chromatography because the chromatographic system, if operated ideally, will not be far from equilibrium even though it is a dynamic process. Like all thermodynamic constants, it is temperature-dependent.

The use of brackets (in equation 7.13) to represent solute concentration is a simplification because solute activities should be used, but the simpler version is sufficient for this discussion. Because both numerator and denominator are concentrations of the same species (and it is assumed that there is only one species for each solute in each of the two phases), equation 7.13 can also be shown with the units (mass)/(volume) in addition to the conventional (moles)/(volume).

The larger the value of K_c, the greater the fraction of the solute in the stationary phase, and the longer the retention time. So, separations occur when the solutes in a sample have differing distribution constants. If we knew what in-

Table 7.3. Intermolecular Forces and Polarity

Name	Type of Force
Dispersion	Induced dipole–induced dipole
Induction	Dipole–induced dipole
Orientation	Dipole–dipole
Hydrogen bonding	Hydrogen bond

teractions caused solutes to behave differently, we could predict retention times, and we would know how to select phases for a given separation.

Unfortunately, intermolecular interactions are quite complex, and our knowledge is so limited that prediction is not usually accurate. We do classify intermolecular interactions according to those general forces listed in Table 7.3, but efforts to attach numerical values to specific cases have not been successful. For an introductory discussion such as this one, it is sufficient to note that chemists rely on the old adage that "like dissolves like."

The interactions under consideration are those between:

- The solute and the stationary phase
- The solute and the mobile phase
- (For GC) the solute–solute interactions that can be measured by the solute's vapor pressure

Actually, in GC, the mobile phase is an inert gas and serves only to push the solute through the column, and normally there are no solute–mobile-phase interactions. So, the important forces in GC are as follows:

1. Between solute and stationary phase
2. The vapor pressure of the solute

The former is controlled by one's selection of the stationary phase (we say, choosing the best column), and the latter is controlled by the temperature of operation.

For LC, there are three important parameters:

1. Solute–stationary-phase interactions
2. Temperature
3. Solute-mobile phase interactions

The effect of temperature on the distribution constant is often not exploited in LC, but the solute–mobile-phase interactions provide LC with considerably more flexibility than is available in GC and extends the range of potential applications. The wide range of types of LC was shown by the numerous subdivisions of LC in Figure 7.1.

Distribution constants are generally not known, although some attempts have been made in LC to make use of the partition coefficients for the (equilibrium) distribution of solutes between 1-octanol and water [5]. The current practice for selecting the phases for HPLC, going beyond the "like-dissolves-like" guideline, is covered in Chapter 9. The discussions in Chapter 8 (GC) and Chapter 9 (HPLC) should be consulted for more information. This discussion continues with a search for alternative parameters that can be used instead of the distribution constant to evaluate the thermodynamic processes at work in a given separation.

The distribution constant can be factored into two terms:

$$K = \frac{(Wt_A)_S / V_S}{(Wt_A)_M / V_M} \tag{7.14}$$

$$= k \times \beta \tag{7.15}$$

where k is the retention factor,

$$k = \frac{(Wt_A)_S}{(Wt_A)_M} \tag{7.16}$$

and β is the phase volume ratio,

$$\beta = \frac{V_M}{V_S} \tag{7.17}$$

Because the retention factor can be calculated by equation 7.18,

$$k = \frac{V_R - V_M}{V_M} = \frac{V_R'}{V_M} = \frac{t_R'}{t_M} \tag{7.18}$$

it can easily be obtained from a chromatogram as long as the void volume is known, as was shown in Figure 7.4. The void volume is obtained by measuring the retention time for a nonretained solute, t_M. Because the phase volume ratio (β) is fixed for a given column, the retention factors can be used to represent relative distribution constants, and they frequently are used in equations and calculations.

The ease of calculation (or even estimation) of the retention factor can be seen with reference to Figure 7.11, where t_M is 0.66 min. For the last peak, which has a retention time of 7.50 min, the k value can be calculated as

$$k = \frac{7.50 - 0.66}{0.66} = \frac{6.84}{0.66} = 10.4 \tag{7.19}$$

which is just a little greater than 10. Even without a calculator, the retention

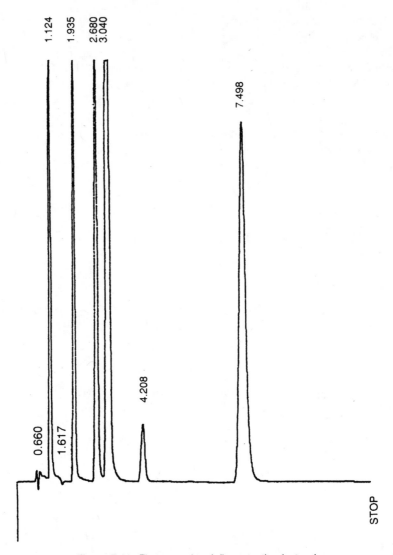

Figure 7.11. Figure used to define retention factor, k.

factor can easily be estimated by visual observation. As another example, the retention factor for peak number 3 can be estimated as about 3. The actual calculation is

$$k = \frac{2.02}{0.66} = 3.06 \qquad (7.20)$$

Alternatively, we can determine the *ratios* of distribution constants by measuring the separation factor, α,

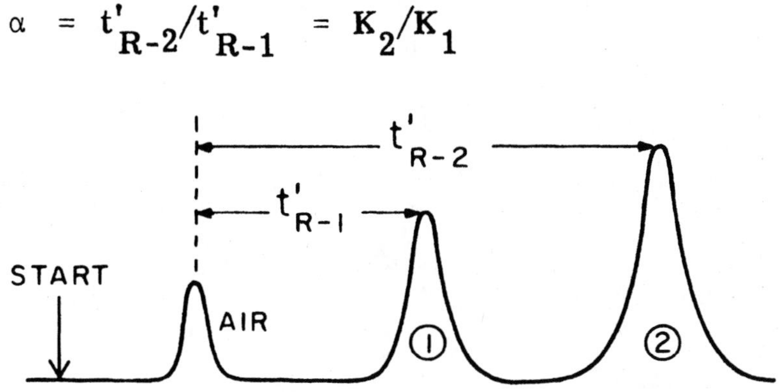

$$\alpha = t'_{R-2} / t'_{R-1} = K_2 / K_1$$

Figure 7.12. Figure used to define separation factor, α.

$$\alpha = \frac{(V'_R)_B}{(V'_R)_A} = \frac{(K_C)_B}{(K_C)_A} = \frac{k_B}{k_A} \tag{7.21}$$

as illustrated in Figure 7.12. In effect, the separation factor measures (a) the relative magnitudes of the distribution constants or (b) the relative forces acting on the solutes and producing their separation.

In summary, chromatographers use the retention factor, k, and the separation factor, α, to describe the thermodynamic effects operating in a particular chromatographic system. Both are easy to measure from a chromatogram, and both find use in chromatographic calculations and method development.

7.3.2. Kinetics

Kinetics has been used to explain the peak broadening that is a natural consequence of the chromatographic process. The earliest attempts to explain chromatographic band broadening were applied to GC and were based on an equilibrium model that came to be known as the Plate Theory. While it was of some value, it did not deal with the nonequilibrium conditions that actually exist in the column and did not address the causes of band broadening. However, an alternative approach describing the kinetic factors was soon presented; it became known as the Rate Theory.

7.3.2.1. The Original van Deemter Equation or Rate Equation. The most influential paper using the kinetic approach was published by van Deemter, Zuiderweg, and Klinkenberg in 1956 [6]. It identified three effects that contribute to band broadening in packed GC columns: eddy diffusion (the A term), longitudinal molecular diffusion (the B term), and mass transfer in the station-

ary liquid phase (the C term). The band broadening was expressed in terms of the plate height, H, as a function of the average linear gas velocity, u. In its simple form, the van Deemter equation is

$$H = A + \frac{B}{u} + Cu \qquad (7.22)$$

Because plate height is inversely proportional to plate number, a small value indicates a narrow peak—the desirable condition. Thus, each of the three constants, A, B, and C, should be minimized in order to maximize column efficiency.

In equation 7.22, the speed of the mobile phase was expressed as its linear velocity, u, rather than its flow rate, F. The two are proportional for a given column, of course, because

$$F = u(A_c) \qquad (7.23)$$

where A_C is the cross-sectional area of the column.* Chromatographers often use the two interchangeably. The linear velocity can be measured chromatographically from the length of the column, L, and the retention time for a nonretained solute, t_M:

$$u = \frac{L}{t_M} \qquad (7.24)$$

Typical units for the velocity are cm/sec, and those for flow rate are ml/min.

Returning to the rate equation, let us examine the three terms as proposed by van Deemter et al. The A term is called the eddy diffusion or Multipath term, the B term is caused by molecular diffusion, and the C term concerns mass transfer of the solute.

Eddy Diffusion. As originally proposed by van Deemter et al., the A term dealt with eddy diffusion in packed columns as shown in Figure 7.13. The diffusion paths of three molecules are shown in the figure. All three start at the same initial position, but they find differing paths through the packed bed and arrive at the end of the column having traveled different distances. Because the flow rate of mobile phase is constant, they arrive at different times and are separated from each other. Thus, for a large number of molecules, the eddy diffusion process or the multipath effect results in band broadening as shown.

---

* For an open tubular column, $A_c = \pi r^2$, where r is the radius of the column; for a packed column, the cross-sectional area exposed to mobile phase is less due to the packing.

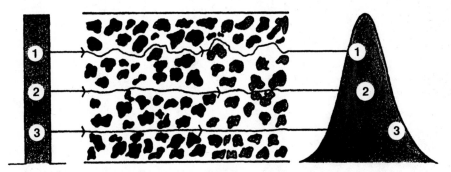

Figure 7.13. Illustration of eddy diffusion or multipath effect. (From H. M. McNair and J. M. Miller, *Basic Gas Chromatography*, Wiley, New York, 1998, p. 46. Reprinted courtesy of John Wiley & Sons, Inc.)

The A term in the van Deemter equation is

$$A = 2\lambda d_p \qquad (7.25)$$

where d_p is the diameter of the particles packed in the column and λ is a packing factor. To minimize A, small particles should be used and they should be tightly packed. In practice, the lower limit on the particle size is determined by the pressure drop across the column and the ability to pack very small particles uniformly. Small ranges in mesh size also promote better packing (minimal λ).

Molecular Diffusion. The B term of equation 7.22 accounts for the well-known molecular diffusion. The equation governing molecular diffusion is

$$B = 2\gamma D_M \qquad (7.26)$$

where D_M is the diffusion coefficient for the solute in the mobile phase and γ is called a tortuosity factor, which allows for the nature of packed beds. Open tubes (as in capillary GC) are not packed, and in that case the B term would not include a tortuosity factor.

Figure 7.14 illustrates how a zone of molecules diffuses from the region of high concentration to that of lower concentration with time. The equation tells us that a small value for the diffusion coefficient is desirable so that diffusion is minimized, yielding a small value for B and for H. In general, a low diffusion coefficient can be achieved in GC by using carrier gases with larger molecular weights such as nitrogen or argon; in HPLC, viscous liquids would promote low diffusion. However, in both cases these choices are not desirable for other reasons, so the choice of mobile phase is seldom based on the B term.

In the van Deemter equation, this term is divided by the linear velocity, so a large velocity or flow rate will also minimize the contribution of the B term to the overall peak broadening. That is, a high velocity will decrease the time a

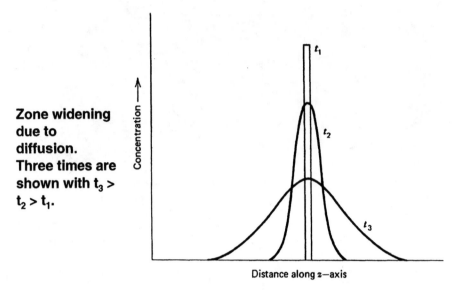

Figure 7.14. Illustration of molecular diffusion. (From H. M. McNair and J. M. Miller, *Basic Gas Chromatography*, Wiley, New York, 1998, p. 42. Reprinted courtesy of John Wiley & Sons, Inc.)

solute spends in the column and thus decrease the time available for molecular diffusion. Because high velocities are desirable to minimize the time of an analysis, the *B* term is usually not an important contributor to the overall band spreading (*H*), and it is of little interest in selecting chromatographic parameters.

Mass Transfer. The *C* term in the van Deemter equation concerns the transfer of solute into and out of the stationary phase—that is, sorption and desorption. This mass transfer can be described by referring to Figure 7.15. In both parts of the figure, the upper peak represents the distribution of a solute in the mobile phase while the lower peak represents the distribution in the stationary phase. A distribution constant of 2 is used in this example so the lower peak has twice the area of the upper one. At equilibrium the solute achieves relative distributions like those shown in part (a), but an instant later the mobile phase moves the upper curve downstream, giving rise to the situation shown in (b). The solute molecules in the stationary phase are stationary; the solute molecules in the mobile phase have moved ahead of those in the stationary phase, thus broadening the overall zone of molecules. The solute molecules that have moved ahead must now partition into the stationary phase, and those that were in the stationary phase must now equilibrate with the mobile phase, as shown by the arrows. The faster they can make this transfer, the less will be the band broadening.

The *C* term in the van Deemter equation is

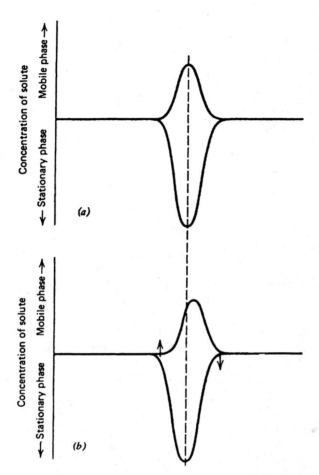

Figure 7.15. Illustration of mass transfer in the stationary phase. (From H. M. McNair and J. M. Miller, *Basic Gas Chromatography*, Wiley, New York, 1998, p. 44. Reprinted courtesy of John Wiley & Sons, Inc.)

$$C = \frac{8kd_f^2}{\pi^2(1+k)^2 D_S} \tag{7.27}$$

where d_f is the average film thickness of the liquid stationary phase and D_S is the diffusion coefficient of the solute in the stationary phase. To minimize the contribution of this term, the film thickness should be small and the diffusion coefficient should be large. Rapid diffusion through thin films allows the solute molecules to stay closer together. Thin films can be achieved by coating small amounts of liquid on the stationary support or the column walls, but diffusion coefficients cannot usually be controlled except by selecting low-viscosity stationary liquids.

Minimization of the C term results when mass transfer into and out of the stationary liquid is as fast as possible. An analogy would be to consider a person jumping into and out of a swimming pool; if the solvent is water and shallow, the process can be done quickly; if the solvent is motor oil (high viscosity) and deep, the process is slow.

If the stationary phase is a solid, modifications in the C term are necessary to relate it to the appropriate adsorption–desorption kinetics. Again, the faster the kinetics, the closer the process is to equilibrium, and the less is the band broadening.

The other part of the C term is the ratio $k/(1 + k)^2$. This ratio is minimized at large values of k, but very little decrease occurs beyond a k value of about 20. Because large values of retention factor result in long analysis times, little advantage is gained by k values larger than 20.

7.3.2.2. Other Rate Equations.

The C term in the van Deemter equation concerned mass transfer in only the stationary phase, because mass transfer in the mobile phase was not thought to be significant since the equation was derived for GC and mass transfer is fast in gases. However, when the van Deemter equation was applied to LC and thin-film capillary GC columns, it became necessary to include C terms for mass transfer in both the stationary and the mobile phases. Ideally, fast solute sorption and desorption will keep the solute molecules close together and keep the band broadening to a minimum.

The so-called *extended* van Deemter equation includes two C terms, namely, a C_S term for mass transfer in the stationary phase and a C_M term for mass transfer in the mobile phase. The mass transfer term included in the van Deemter equation was for mass transfer in the stationary phase.

A simplified version of the extended equation is

$$H = A + \frac{B}{u} + C_S u + C_M u \qquad (7.28)$$

This equation has found general acceptance although some modifications have been proposed and are discussed in the next section. First, we should describe mass transfer in the mobile phase.

Mass transfer in the mobile phase can be visualized by referring to Figure 7.16, which shows the profile of a solute zone resulting from nonturbulent flow through a tube. Inadequate mixing (slow kinetics) in the mobile phase can result in band broadening because the solute molecules in the center of the column move ahead of those at the wall. Small-diameter columns minimize this broadening because the mass transfer distances are relatively small. The C_M term used for open tubular columns in the so-called Golay equation is

$$C_M = \frac{(1 + 6k + 11k^2)r_c^2}{24(1 + k)^2 D_M} \qquad (7.29)$$

where r_c is the radius of the column.

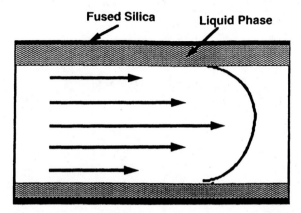

Figure 7.16. Illustration of mass transfer in the mobile phase. (From H. M. McNair and J. M. Miller, *Basic Gas Chromatography*, Wiley, New York, 1998, p. 45. Reprinted courtesy of John Wiley & Sons, Inc.)

Additional modifications to the original van Deemter equation have been proposed by other workers. For example, one can argue that eddy diffusion (the A term) is part of mobile-phase mass transfer (the C_M term) or is coupled with it. Giddings [7] has thoroughly discussed mass transfer and prefers a coupled term combining eddy diffusion and mass transfer to produce a new equation.

Others have defined rate equations that would serve both GC and LC, and an interesting discussion summarizing much of this work has been published by Hawkes [8]. His summary equation is in the same form as Golay's, but it is less specific.

7.3.2.3. Van Deemter Plots. When the rate equation is plotted (H versus u), the so-called van Deemter plot takes the shape of a nonsymmetrical hyperbola, shown in Figure 7.17. As one would expect from an equation in which one term is multiplied by velocity while another is divided by it, there is a minimum in the curve—an optimum velocity that provides the highest efficiency and smallest plate height.

It is logical to assume that chromatography would be carried out at the optimum velocity represented by the minimum in the curve because it yields the least peak broadening. However, if the velocity can be increased, the analysis time will be decreased. Consequently, chromatographers have devoted their time to manipulating the van Deemter equation to get the best performance in the shortest analysis time. By examining the relative importance of the individual terms to the overall equation in Figure 7.17, one sees that the upward slope as velocity is increased comes about from the increasing contribution of the C terms. Therefore, most attention has been focused on minimizing them, a topic that will be covered shortly.

There is no single van Deemter plot for all chromatographic systems and Figure 7.17 most closely resembles a plot for a GC column. Figure 7.18 is

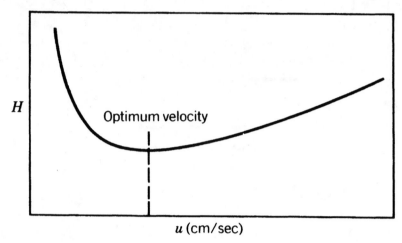

Figure 7.17. Typical van Deemter plot for GC. (From J. M. Miller, *Chromatography: Concepts and Contrasts*, Wiley, New York, 1988, p. 28. Reprinted courtesy of John Wiley & Sons, Inc.)

slightly different and is more typical of plots for HPLC columns. In it, Huber [9] has shown the various contributions of the four terms from the extended van Deemter equation.

While the Rate Theory is a theoretical concept, it is a useful one in practice. It is common to obtain a van Deemter plot for one's column in order to evaluate it and the operating conditions. A solute is chosen and run at a variety of flow rates, being sure to allow sufficient time for flow equilibration after each change. The plate number is calculated from each chromatogram using equation 7.8 and then used to calculate the plate height (equation 7.9). The plate height values are plotted versus linear velocity (obtained by equation 7.24). From this plot, one can compare the optimum velocity (or flow) with the operating velocity being used in practice to determine if a change in velocity could be made to produce higher efficiencies.

7.3.2.4. A Summary of the Rate Equations.

Let us conclude this discussion by considering only three rate equations—one for open tubular GC columns, one for packed GC columns, and one for packed HPLC columns. The first is represented by the Golay equation [10]:

$$H = \frac{2D_M}{u} + \frac{2kd_f^2 u}{3(1+k)^2 D_S} + \frac{(1 + 6k + 11k^2)r_c^2 u}{24(1+k)^2 D_M} \qquad (7.30)$$

and the second by the extended van Deemter equation;

$$H = 2\lambda d_p + \frac{2\gamma D_M}{u} + \frac{8kd_f^2 u}{\pi^2(1+k)^2 D_S} + \frac{\omega d_p^2 u}{D_M} \qquad (7.31)$$

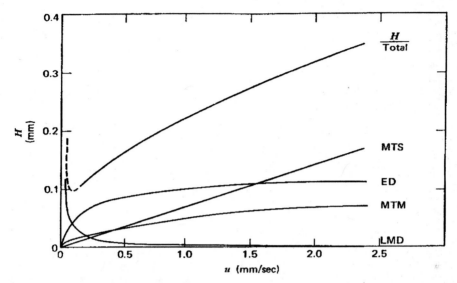

Figure 7.18. Typical van Deemter plot for LC. (Reproduced from J. F. K. Huber, *J. Chromatogr. Sci.* **1969**, *7*, 85 by permission of Preston Publications, A division of Preston Industries Inc.)

The extended van Deemter equation could also be used for HPLC, but it has become common to use an equation described by Knox [11]. He and his co-workers have written an equation in so-called reduced parameters, namely, the reduced plate height, h, and the reduced velocity, v. The use of reduced parameters facilitates comparisons between columns, especially between GC and LC columns. The definitions of the reduced parameters are

$$h = \frac{H}{d_p} \tag{7.32}$$

and

$$v = u\frac{d_p}{D_M} \tag{7.33}$$

The Knox equation (in reduced parameters) is

$$h = Av^{0.33} + \frac{B}{v} + Cv \tag{7.34}$$

For many HPLC columns, A equals 1, B equals 2, and C equals 0.1. Figure 7.19 shows a typical plot of the Knox equation; note that the indices are log–log, not linear as in the other plots. Further discussion of the Knox equation and reduced parameters can be found in Chapter 10.

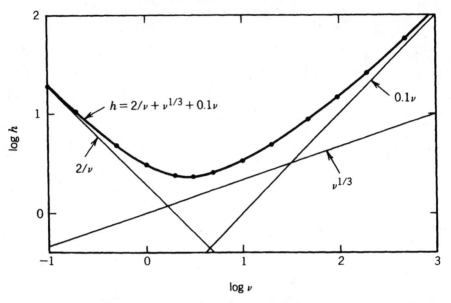

Figure 7.19. Plot of rate equation for LC using reduced parameters (Knox plot). (Reprinted with permission from P. A. Bristow and J. H. Knox, *Chromatographia* **1977**, *10*, 279.)

7.4. SOME ADDITIONAL TERMS

7.4.1. Resolution

Because the objective of chromatography is to separate the components of a mixture, a measure of the extent of that separation is needed. The common term used is peak resolution, R_s, and its definition is

$$R_s = \frac{(t_R)_B - (t_R)_A}{\dfrac{(w_b)_B + (w_b)_A}{2}} = \frac{2d}{(w_b)_B + (w_b)_A} \tag{7.35}$$

where the distance of separation of two peaks, d, is divided by their average peak width at base. Technically, for this equation to be valid, the peaks need to be of the same height as shown in Figure 7.20.

Because the two peaks being separated are adjacent, their base widths should be approximately equal; that is, $(w_b)_A = (w_b)_B$. Making this substitution in equation 7.36 yields the approximation

$$R_s = \frac{d}{(w_b)_B} \tag{7.36}$$

and when d equals w_b, the value of the peak resolution is 1.0. In this case, as shown in Figure 7.20, the peaks are not resolved to the baseline but the peak

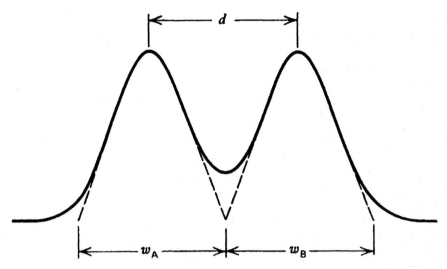

Figure 7.20. Illustration of definition of resolution, R_s. (From H. M. McNair and J. M. Miller, *Basic Gas Chromatography*, Wiley, New York, 1998, p. 40. Reprinted courtesy of John Wiley & Sons, Inc.)

tangents are just touching. A resolution of 1.5 is obtained for baseline resolution, and any value greater than 1.5 designates complete peak separation.

Recently the USP has presented an alternative computional formula using the width at half-height, w_h, rather than the width at base, w_b. Thus, Equation 7.35 becomes

$$R_s = \frac{2d}{1.70[(w_h)_B + (w_h)_A]} \tag{7.37}$$

7.4.2. Retardation Factor

The retardation factor is another parameter used to designate the chromatographic behavior of a given solute. Slightly different definitions and symbols are used in columnar and planar chromatography, but the concept of retention factor is the same. The retardation factor measures the extent of retardation of a solute as it passes through the chromatographic system compared with the speed of the mobile phase. For columns the definition is

$$R = \frac{v}{u} \tag{7.38}$$

where v is the velocity of the solute and u is the velocity of the mobile phase. R will always be less than 1 for retained solutes, and it expresses the fractional rate at which a solute is moving. It also represents the fraction of molecules of a given solute in the mobile phase at any given time and, alternatively, the fraction of time an average solute molecule spends in the mobile phase as it travels through the column.

Each of the velocities in equation 7.38 can be measured experimentally. Equation 7.24, given earlier, can be used for the mobile-phase velocity, and a similar equation (7.39) can be used for the solute velocity:

$$v = \frac{L}{t_R} \qquad (7.39)$$

Substituting these two equations into equation 7.39 gives a working definition of the retardation factor:

$$R = \frac{t_M}{t_R} = \frac{V_M}{V_R} \qquad (7.40)$$

In thin-layer chromatography (TLC), the velocities cannot easily be measured, so a slightly different definition and symbol are used.

$$R_F = \frac{\text{distance solute moved}}{\text{distance solvent front moved}} \qquad (7.41)$$

These two definitions can be compared by referring to Figure 7.21. In this figure, the ratio a/b is used to emphasize the similarity between the two definitions. Thus, if one wanted to compare some HPLC column data with some TLC data in which the same stationary and mobile phases were used, the preferred parameter would be the retardation factors, R and R_F.

A comparison of equations 7.18 and 7.39 clearly shows that the retardation factor, R, and the retention factor, k, are inversely related. The exact relation-

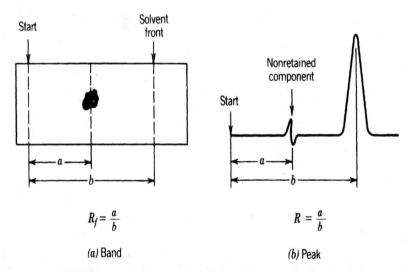

$$R_f = \frac{a}{b}$$

$$R = \frac{a}{b}$$

(a) Band

(b) Peak

Figure 7.21. Comparison of equations for calculating retardation factor in (a) TLC and (b) HPLC. (From J. M. Miller, *Chromatography: Concepts and Contrasts*, Wiley, New York, 1988, p. 71. Reprinted courtesy of John Wiley & Sons, Inc.)

ship is given in equation 7.42:

$$R = \frac{1}{1 + k} \tag{7.42}$$

7.4.3. System Suitability

In the chromatography part ⟨621⟩ of USP, there is a section on system suitability [4, p. 776]. It states: "System suitability tests are an integral part of gas and liquid chromatographic methods. They are used to verify that the resolution and reproducibility of the chromatographic system are adequate for the analysis to be done. The tests are based on the concept that the equipment, electronics, analytical operations, and samples to be analyzed constitute an integral system that can be evaluated as such."

Two of the parameters just defined—resolution and tailing—are to be a part of this process. As an alternative to resolution, plate number (N) can also be specified, especially if there is only one peak of interest. These measurements are normally to be made on the analyte peak, and appropriate limits need to be specified. (See Chapter 15 for further discussion.)

The other system suitability requirement is the relative standard deviation (RSD), determined on five or six replicate injections; it was discussed in Chapter 4.

7.5. SUMMARY

The plate height was discussed at several places in this chapter, and it now seems appropriate to ask, "What is the real meaning of plate height?" The terminology is a holdover from distillation theory, but the concept is that it is a measure of the dispersivity of a solute in the column or of the rate of band broadening in the column. In this context, the appropriate definition is

$$H = \frac{\sigma^2}{L} \tag{7.43}$$

where σ^2 is the variance or square of the standard deviation representing the width of a peak, and L refers to the length or distance of movement of a solute. In columnar chromatography, the real meaning of L is better represented by the retention time:

$$H = \frac{\sigma^2}{t_R} \tag{7.44}$$

When equation 7.43 is solved for σ, we find that it is proportional to the square root of the retention time. It is this fact that explains why chromatographic separations are possible: The solutes separate from each other on a linear basis, but their peaks get wider only on a square root basis. Consequently, the solutes are separated faster than they are broadened, and thus they get resolved.

Another useful equation describes the dependency of resolution on three variables:

$$R_s = \left(\frac{\sqrt{N}}{4}\right)\left(\frac{\alpha - 1}{\alpha}\right)\left(\frac{k}{k + 1}\right) \qquad (7.45)$$

Poor separations can be improved by adjusting each of three variables. The application of this equation to HPLC can be found in Chapter 12.

REFERENCES

1. K. C. Associates. *Analytical Instrument End User Study*, Wilmington, DE, 1997.
2. R. E. Majors, *LC-GC* **1990**, *8*, 442.
3. L. S. Ettre, *Pure Appl. Chem.* **1993**, *65*, 819–872. See also L. S. Ettre, *LC-GC* **1993**, *11*, 502.
4. United States Pharmacopeial Convention, *U.S. Pharmacopeia*, 24th edition/*National Formulary*, 19th edition, Rockville, MD, 2000, ⟨621⟩, Chromatography, pp. 1914–1926.
5. I. Benito, J. Saz, and M. L. Marina, *J. Liq. Chromatogr.* **1998**, *21*, 331.
6. J. J. van Deemter, F. J. Zuiderweg, and A. Klinkenberg, *Chem. Eng. Sci.* **1956**, *5*, 271.
7. J. C. Giddings, *Dynamics of Chromatography*, Part 1: *Principles and Theory*, Marcel Dekker, New York, 1965, p. 4.
8. S. J. Hawkes, *J. Chem. Educ.* **1983**, *60*, 393–398.
9. J. F. K. Huber, *J. Chromatogr. Sci.* **1969**, *7*, 85.
10. M. J. E. Golay, *Gas Chromatography 1958*, D. H. Desty (editor), Butterworths, London, 1958, p. 36.
11. J. H. Knox and M. Saleem, *J. Chromatogr. Sci.* **1969**, *7*, 279.

General References

A. Braithwaite and E. J. Smith, *Chromatographic Methods*, 5th edition, Kluwer Academic Press, Hingham, MA, 1996.

E. Heftmann (Editor), *Chromatography*, 5th edition, Parts A and B, Elsevier, New York, 1992.

J. M. Miller, *Chromatography: Concepts and Contrasts*, John Wiley, New York, 1987.

J. P. Parcher and T. Chester (editors), *Unified Chromatography*, ACS Symposium Series, American Chemical Society, Washington, D.C., 1999.

C. F. Poole and S. K. Poole, *Chromatography Today*, Elsevier, New York, 1991.

Chiral Methods

T. E. Beesley and R. P. W. Scott, *Chiral Chromatography*, John Wiley, New York, 1998.

8

GAS CHROMATOGRAPHY

James M. Miller and Harold McNair

Gas chromatography (GC) is not as widely used in the pharmaceutical industry as liquid chromatography (LC) because GC analysis requires samples that are volatile, and many drugs are not. However, GC is usually the method of choice for the analysis of residual solvents in drug products and for the quality control of those raw materials and contaminants in air or water that are volatile. Probably there are other applications that could be performed routinely by GC, but the prevalence of, and familiarity with, high-performance liquid chromatography (HPLC) equipment results in its being chosen for many analyses for which both techniques are applicable. Some United States Pharmacopeia (USP) tests and assays using GC are summarized at the end of the chapter.

8.1. SOME HISTORICAL NOTES

Although the original work in chromatography by Tswett [1] was in LC, the entire field took a large step forward when GC was invented. Credit for suggesting that the mobile phase could be a gas goes to Martin and Synge [2], which resulted in the first GC publication in 1952 [3]. It was so easy to construct and use a gas chromatograph that the field of GC grew quickly in the 1950s and 1960s. Commercial instruments were rapidly produced, thereby stimulating new applications. Theories of chromatography were proposed, tested, and found to predict accurately chromatographic performance.

It was the application to LC of the newly devised GC theories that led to the

Analytical Chemistry in a GMP Environment. Edited by J. M. Miller and J. B. Crowther
ISBN 0-471-31431-5 © 2000 John Wiley & Sons, Inc.

subsequent development of modern HPLC. So, GC preceded HPLC in popularity, and it has turned out that the two techniques are complementary. However, the pharmaceutical industry has found more applications for LC than for GC.

The first GC instruments used columns packed with the stationary phase (a solid or a liquid coated on a solid). Many of the older GC methods (some of which are still in use) are based on packed column technology. When Golay suggested the use of open tubes for columns in 1958 [4], the instrumentation had to be modified to utilize these capillary-sized columns. This gave rise to two types of GC instruments, and normally only one type of column can be accommodated in each type. The capillary GCs did not achieve their current level of popularity until the original patent on them expired about 20 years ago. This resulted in capillary promotion by most equipment manufacturers. Today the more popular column is the open tubular (OT) or capillary type, but many packed columns are still in use, so this chapter discusses both types and points out the differences in them and the reasons for choosing one over the other.

8.2. ADVANTAGES AND DISADVANTAGES

Gas chromatographs can provide high resolution, separating hundreds of components in complex matrices such as coffee aroma, petroleum, or natural products like the essential oils. Sensitive detectors like the flame ionization detector can quantitate 50 ppb of organic compounds with a relative standard deviation of 2–5%. Automated systems can handle more than 100 samples per day with minimum downtime, and all of this can be accomplished with an investment of less than $20,000 [5].

The important advantages of GC are summarized in Table 8.1. Chromatographers have always been interested in fast analyses, and GC has been the fastest chromatographic technique. Its high efficiency (expressed in *plate numbers*) is typically several hundred thousand, making possible remarkable separations. Because GC is excellent for quantitative analysis, it has found wide use for many different applications. Sensitive, quantitative detectors provide fast, accurate analyses, and at a relatively low cost.

Table 8.1. Advantages and Disadvantages of GC

Advantages	Disadvantages
Fast analysis time	Limited to volatile samples
Efficient; good resolution	Limited to thermally stable compounds
Good for quantitative analysis	Not good for qualitative analysis
Sensitive; low detection limits	Not good for preparative-size samples
Reliable, simple, inexpensive	
Nondestructive	

GC has replaced distillation as the preferred method for separating volatile materials. In both techniques, temperature is a major variable, but GC separations are also dependent upon the chemical nature (polarity) of the stationary phase. This additional variable makes GC more powerful. In addition, the fact that solute concentrations are very dilute in GC eliminates the possibility of azeotropes, which often plagued distillation separations.

Both GC and distillation are limited to volatile samples. A practical upper temperature limit for GC operation is about 380°C, so samples need to have an appreciable vapor pressure (60 torr or greater) at that temperature. Solutes usually do not exceed boiling points of 500°C and molecular weights of 1000 daltons. This, a major limitation of GC is listed in Table 8.1 along with other disadvantages of GC.

In summary: For the separation of volatile materials, GC is usually the method of choice due to its speed, high resolution capability, reproducibility, good quantitative results, and ease of use.

8.3. CLASSIFICATION OF GC

Figure 8.1 shows the various divisions of GC. The two major subdivisions, as mentioned earlier, are packed and OT. A further division is made according to the state of the stationary phase (SP): If it is a solid, the technique is called gas–solid chromatography (GSC), and if it is a liquid, the technique is called gas–liquid chromatography (GLC). Both types are included in this chapter, but the most common type is GLC. The other acronyms used in the figure refer to commonly used designations of capillary column types: WCOT = wall-coated open tubular (classed as GLC); SCOT = support-coated open tubular (classed as GLC); and PLOT = porous-layer open tubular (classed as GSC).

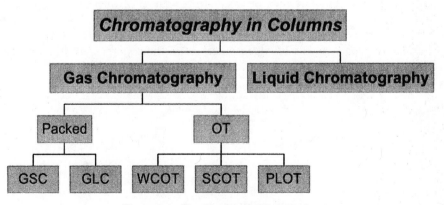

Figure 8.1. Classification of GC methods.

8.4. COLUMNS

Chromatographers are fond of saying, "the heart of the chromatograph is the column," so columns will be discussed first, followed by the other parts of the instrument. As we have seen, there are two types of column, namely, packed and OT. They are packed with, or coated with, the SP. If the SP is a liquid, it must be immobilized on an inert support or on the wall of the column.

8.4.1. Stationary Phases

Virtually every nonvolatile liquid found in a common chemical laboratory has been tested as a possible stationary phase. As a consequence, there is an over-abundance of liquid phases in many commercial suppliers' catalogs (typically about 200 of them). Forty-four phases are listed for GC in the USP [6]; see Appendix IV. The problem is to restrict this long list of phases to a few that will solve most analytical problems.

Liquid phases for capillary columns are very similar to those used for packed columns. In both cases the liquid phase must show high selectivity for the compounds of interest. In addition, they should be capable of operation at high temperatures with minimal column bleed.

Table 8.2 lists the most commonly used liquid phases for both packed and capillary columns. Basically, there are two types of liquid phases in use today. One is a series of siloxane polymers, of which OV-1, SE-30, DB-1 (100% methyl polysiloxane) and OV-17, OV-275, DB-1701, DB-710 (mixtures of methyl, phenyl, and cyano) polysiloxanes are the most popular. The other common liquid phase is a polyethylene glycol (Carbowax 20M®, Superox®, and DB-Wax®).

Schematic structures of both a dimethyl polysiloxane and a polyethylene glycol liquid phase are given in Figure 8.2. There is, however, one difference between packed-column and capillary-column liquid phases: Capillary-column phases are extensively cross-linked. By heating the freshly prepared capillary column at high temperatures or irradiating with ultraviolet (UV) light, the methyl groups form free radicals which readily cross-link to form a more stable, higher-molecular-weight gum phase. There is even some chemical bonding with the silanol groups on the fused silica surface of the column wall. These cross-

Table 8.2. Most Commonly Used Stationary Phases

Type	Capillary Column Example	Packed Column Example
Silicone polymer		
Dimethyl silcone	DB-1	SE-30,OV-1, OV-101
5% Phenyl	DB-5	SE-54
50% Phenyl	DB-17	OV-17
Trifluoropropyl	—	OV-210
Dicyanoallyl	DB-23	OV-275
Polyglycol	DB-WAX	Carbowax 20M

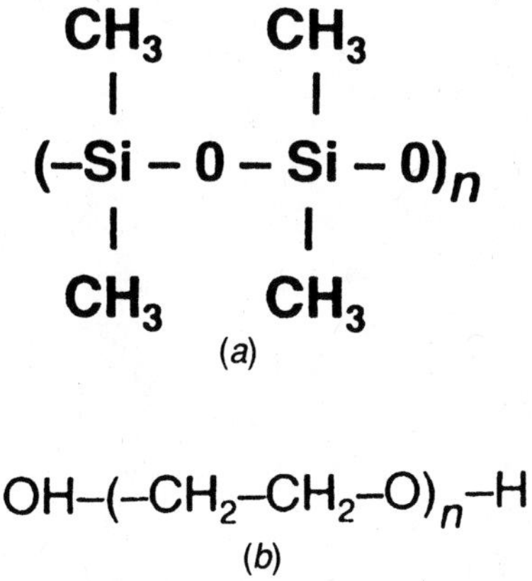

$$
\begin{array}{ccc}
CH_3 & & CH_3 \\
| & & | \\
(-Si & - O - Si & - O)_n \\
| & & | \\
CH_3 & & CH_3
\end{array}
$$

(a)

$$OH-(-CH_2-CH_2-O)_n-H$$

(b)

Figure 8.2. Chemical structures of two polymer types used as stationary phases: (a) Dimethylpolysiloxane, (b) polyethyleneglycol.

linked and chemically bonded phases are more temperature-stable, last longer, and can be more easily cleaned by rinsing with solvents (when cold). Most commercial capillary columns are cross-linked.

8.4.2. Column Materials

Packed columns are typically made of stainless steel and have outside diameters of 1/4 or 1/8 in. and lengths of 2–10 ft. For applications requiring greater inertness, alternative materials have been used for columns including glass, nickel, fluorocarbon polymers (Teflon), and steel that is lined with glass or Teflon. Copper and aluminum are conveniently soft for easy bending, but are not recommended due to their reactivity resulting in columns with low efficiencies.

OT columns are made of fused silica with the stationary liquid phase coated on the inside surface of the capillary wall; hence, they are called wall-coated open tubular (WCOT) columns. The tube can also be made of glass or stainless steel, but almost all capillary columns are now made of fused silica. The two common types of column are shown in Figure 8.3.

Two other types of capillary columns are also shown in Figure 8.3: the SCOT and the PLOT. SCOT columns contain an adsorbed layer of very small solid support (such as Celite) coated with a liquid phase. They can hold more liquid phase and have a higher sample capacity than the thin films common to the early WCOT columns. However, with the introduction of cross-linking techniques, stable thick films are possible for WCOT columns, and the need for SCOT columns has disappeared. A few SCOT columns are still commercially available, but only in stainless steel tubing. PLOT columns contain a porous layer of a solid adsorbent such as alumina, molecular sieve, or Porapak. PLOT columns are well-suited for the analysis of light fixed gases and other volatile compounds.

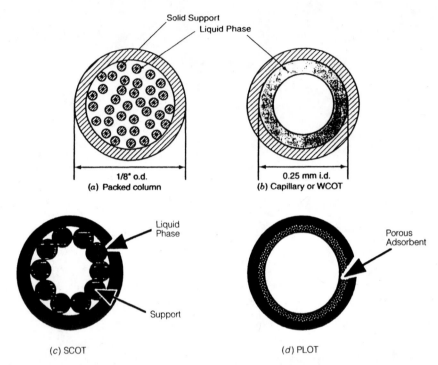

Figure 8.3. GC column types: (*a*) Packed, (*b*) WCOT, (*c*) SCOT, (*d*) PLOT. (From H. M. McNair and J. M. Miller, *Basic Gas Chromatography*, Wiley, New York, 1998, pp. 13, 88. Reprinted courtesy of John Wiley & Sons, Inc.)

8.4.3. Comparison of Column Types

Table 8.3 presents a comparison of the typical operating parameters for two common columns: a WCOT capillary column and a 1/8-inch packed column. In general, WCOT columns are longer, faster, more efficient (large N and small

Table 8.3. Comparison of Packed and Capillary (OT) Columns

	Capillary (WCOT)	Packed (1/8 inch)
Typical ID	0.10–0.32 mm	2.2 mm
Typical OD	0.40 mm	3.1 mm
Typical length	12–60 m	1–3 m
SP film thickness	0.10–1.0 μm	5 μm
Phase volume ratio	250	15–30
Nominal linear Vel.	30 cm/sec	5 cm/sec
Nominal flow rate	1 mL/min	20 mL/min
Maximum plate number	300 K	4 K
Minimum plate height	0.2 mm	0.5 mm

H), and capable of handling higher boiling samples. Packed columns offer the advantage of greater variety because they can be prepared from a wide range of stationary phases and even mixed stationary phases. Hence they may provide larger separation factors (α values). See, for example, the USP phase G-44 in Appendix IV; it is a special mixture that, like many others, is suitable only for packed column use.

Also, many *solid* stationary phases (in GSC) are not commercially available (as SCOT columns), so these packed columns are unique. (See the section on GSC which follows).

Often packed columns are specified as the column of choice in older methods, because the method was developed before OT columns became popular. As methods are revised, the newer OT columns are often substituted for packed columns. However, packed columns do have some other advantages. The amount of stationary phase in a packed column is usually much greater than for OT columns. This means that they can be used with larger samples, even preparative-sized samples. Heavily loaded packed columns also have a greater retentive power for volatile compounds, so packed columns find use in gas analysis and volatile solvent analysis (see Section 8.9.1).

As indicated earlier, instruments are usually designed to use only one type of column, either packed or capillary. However, an intermediate type of column has become popular, and it can fit in both types (with some minor modifications). It is a large-diameter capillary column [usually 530-μm inside diameter (ID)], commonly referred to as a "wide-bore" or a "megabore" column. Its properties are intermediate between small ID WCOT columns and packed columns, and it meets a special need.

8.4.4. Solid Supports

For packed columns, the stationary liquid phase is coated on a *solid support* that is chosen for its high surface area and inertness. Many materials have been used, but those made from diatomaceous earth (Chromosorb® is one registered trademark for diatomaceous earth) have been found to be best. The surfaces of the diatomaceous earth supports are often too active for polar GC samples. They contain free hydroxyl groups that can form undesirable hydrogen bonds and thus cause tailing peaks. Even the most inert support material (white Chromosorb W) needs to be acid washed (designated AW) and silanized to make it still more inert [7]. Some typical silanizing reagents are dimethyldichlorosilane (DMDCS) and hexamethyldisilazane (HMDS). The deactivated white supports are known by names such as Supelcoport®, Chromosorb® W-HP, Gas Chrom® Q II, and Anachrom® Q. One disadvantage of deactivation is that these supports become hydrophobic, and coating them with a polar stationary liquid can be difficult.

The USP includes four "siliceous earth" types in its list of supports for GC [8]: S1A, S1AB, S1C, and S1NS. The entire list is given in Appendix IV; the other 11 supports are discussed in the next section.

Table 8.4. Some Common GC Adsorbents

Adsorbent	Commercial Trade Names
Silica gel	Davidson Grade 12, Chromasil, Porasil
Activated alumina	Alumina F-1, Unibeads-A
Zeolite molecular sieves	MS 5A, MS 13X
Carbon molecular sieves	Carbopack, Carbotrap, Cargograph, Graphpac
Porous polymers	Porapak, HayeSep, Chromasorb Century Series
Tenax polymer	Tenax TA, Tenax GR

Narrow ranges of small particles produce more efficient columns. Particle size is usually given according to mesh range, determined by the pore sizes of the sieves used for screening. Common choices for GC are 80/100 or 100/120 mesh. The amount of liquid phase coated on the solid support varies with the support and can range from 3% to 25%. Low loadings are better for high-efficiency and high-boiling compounds, and the high loadings are better for large samples or volatile solutes—gases, for example. The coated material, even with as much as 25% liquid stationary phase (on Chromosorb® P), will appear *dry* and will pack easily into a column.

8.4.5. Solid Stationary Phases (GSC)

Common adsorbent solids like silica gel and alumina are used in GSC, but most of the solids used as stationary phases have been developed for specific applications in GSC. Table 8.4 lists some in common use, and those specified by USP are listed in Appendix IV.

It is easy to separate oxygen and nitrogen using solids known as molecular sieves, naturally occurring Zeolites, and synthetic materials like alkali metal aluminosilicates which have specific pore sizes and effect separations based on size. The classic separation on a synthetic molecular sieve is shown in Figure 8.4.

Carbosieves® are typical of solids that have been made for GC, in this case by pyrolysis of a polymeric precursor that yields pure carbon containing small pores and serving as a molecular sieve. The Carbosieves® will separate oxygen and nitrogen and can be substituted for the molecular sieves just described. They also find use for the separation of low-molecular-weight hydrocarbons and formaldehyde, methanol, and water. Other trade names include Ambersorb® and Carboxen®.

Another class of carbon adsorbents are graphitized carbon blacks that are nonporous and nonspecific and separate organic molecules according to geometric structure and polarity. Often they are lightly coated with a liquid phase to enhance their performance and minimize tailing.

In 1966 Hollis [9] prepared and patented a porous polymer (polystrene cross-

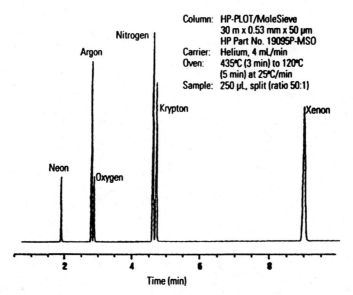

Figure 8.4. Separation of fixed gases on a molecular sieve PLOT column. (Copyright 1994 Hewlett-Packard Company. Reproduced with permission from p. 180 of the 1998–99 Chemical Analysis Catalog.)

linked with 8% divinyl benzene) that has been marketed under the trade name Porapak. It provided a good solution to the problem of separating and analyzing water in polar solvents. Because of its strong tendency to hydrogen-bond, water usually tails badly on most stationary phases, but Porapak solves that problem as shown in Figure 8.5.

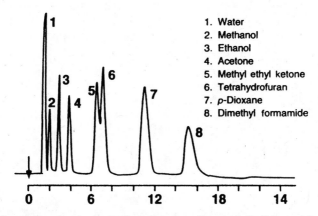

Figure 8.5. Separation of water and polar solvents on Porapak® Q. Column 6 ft. ×1/4 in. OD, 150/200 mesh Porapak® Q, 220°C, flow rate 37 mL/min He; TCD. (From H. M. McNair and J. M. Miller, *Basic Gas Chromatography*, Wiley, New York, 1998, p. 78. Reprinted courtesy of John Wiley & Sons, Inc.)

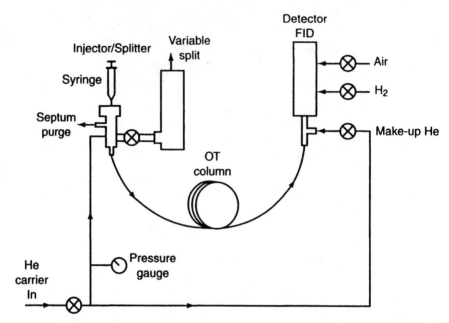

Figure 8.6. Schematic of typical GC instrument. (From H. M. McNair and J. M. Miller, *Basic Gas Chromatography*, Wiley, New York, 1998, p. 15. Reprinted courtesy of John Wiley & Sons, Inc.)

8.5. OTHER INSTRUMENT COMPONENTS

Figure 8.6 shows schematically a GC system. The components that will be discussed include: (1) carrier gas, (2) flow control, (3) sample inlet and sampling devices, (4) controlled temperature zones (ovens), and (5) detectors.

A gas chromatograph functions as follows: An inert carrier gas (like helium) flows continuously from a large gas cylinder through the injection port, the column, and the detector. The flow rate of the carrier gas is carefully controlled to ensure reproducible retention times and to minimize detector drift and noise. The sample is injected (usually with a microsyringe) into the heated injection port where it is vaporized and carried onto the column, typically a capillary column 15–30 m long, coated on the inside with a thin (0.2 µm) film of high-boiling liquid (the stationary phase). The sample partitions between the mobile and stationary phases and is separated into individual components based on their relative solubility in the liquid phase and their relative vapor pressures.

After the column, the carrier gas and sample pass through a detector. This device measures the quantity of sample, and it generates a proportionate electrical signal. This signal goes to a data system/integrator that generates a chromatogram (the written record of analysis). In most cases the data-handling system automatically integrates the peak area, performs calculations, and prints out a report with quantitative results and retention times.

8.5.1. Carrier Gas

The main purpose of the carrier gas is to carry the sample through the column. It is the *mobile phase*; it is inert and does not interact chemically with the sample. A secondary purpose is to provide a suitable matrix for the detector to measure the sample components. High purity is a necessity. The most commonly used gases are helium, nitrogen, and hydrogen.

8.5.1.1. Purity. It is important that the carrier gas be of high purity because impurities such as oxygen and water can chemically attack the liquid phase in the column and destroy it. Polyester, polyglycol, and polyamide columns are particularly susceptible to hydrolysis. Trace amounts of water can also desorb other column contaminants and produce a high detector background or even "ghost peaks." Trace hydrocarbons in the carrier gas cause a high background with most ionization detectors and thus limit their detectability.

Highly purified gases can be purchased, but most GC systems use a gas of moderately high purity and include some scrubbers in the gas line. As a minimum, a molecular sieve trap should be used to remove the water and methane from the carrier gas. Other traps can be used to remove hydrocarbons and oxygen. The sequence for multiple traps (starting at the gas cylinder) should be as follows: water, then hydrocarbon, then oxygen. All traps should be installed in a vertical position, and they should be replaced or reactivated according to the manufacturers' instructions. The frequency and responsibility for this maintenance should be incorporated into the laboratory's standard operating procedures (SOPs).

8.5.1.2. Efficiency. Van Deemter plots were shown in Chapter 7, and they illustrate the effect of column flow rate on band broadening, H. There is an optimal flow rate for a minimum of band broadening. With packed columns, and also with thick film megabore columns, nitrogen is the carrier gas of choice because the van Deemter B term (longitudinal diffusion in the gas phase) dominates. Nitrogen being heavier than helium minimizes diffusion in the mobile phase (the B term) and produces more efficiency.

In capillary columns, however, particularly those with thin films, hydrogen and helium are the best carrier gases (refer to Figure 8.7). With capillary columns the efficiency (N) is usually more than sufficient and the emphasis is on speed. Thus, capillary columns are usually run at faster-than-optimal flow rates where the C_M term (mass transfer in the mobile phase) dominates. Hydrogen provides a much faster analysis with a minimal loss in efficiency because it allows faster diffusion in the mobile phase and minimizes the C_M term in the Golay equation. However, helium is nearly as good as hydrogen and it is not as hazardous, so it is usually the carrier gas of choice.

8.5.1.3. Detector Requirements. Shown in Table 8.5 are the carrier gases preferred for various detectors. For the thermal conductivity detector, helium is

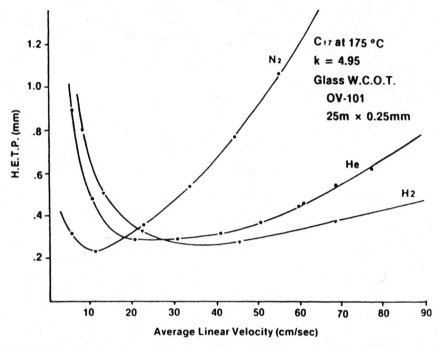

Figure 8.7. Effect of carrier gas on van Deemter curve. (Reprinted from R. R. Freeman (editor), *High Resolution Gas Chromatography*, 2nd edition, Hewlett-Packard Company, Wilmington, DE, 1981. Copyright Hewlett-Packard Company. Reproduced with permission.)

Table 8.5. Carrier Gas Requirements for Some Detectors

Detector	Suitable Carrier Gases	Also Requires
FID	Helium, nitrogen	Hydrogen and air
TCD	Helium (or hydrogen)	—
ECD	Nitrogen, very dry or argon with 5% methane	—
NPF	Helium, nitrogen	Hydrogen and air

the most popular. While hydrogen is commonly used in some parts of the world (where helium is very expensive), it is not recommended because of the potential for fire and explosions.

With the flame ionization detector and the nitrogen–phosphorus detector, either nitrogen or helium may be used. Nitrogen provides slightly more sensitivity, but a slower analysis, than helium. For the electron capture detector, very dry, oxygen-free nitrogen, or a mixture of argon with 5% methane is recommended.

8.5.2. Flow Control and Measurement

The measurement and control of carrier gas flow is essential for both column efficiency and for qualitative analysis. Column efficiency depends on the proper linear gas velocity that can be easily determined by changing the flow rate until the maximum plate number is achieved. Typical optimum values are: 75–90 mL/min for 1/4-inch-outside-diameter (OD) packed columns; 25 mL/min for 1/8-inch-OD packed columns; and 0.75 mL/min for a 0.25-μm-ID open tubular column. These values are merely guidelines; the optimum value for a given column should be determined experimentally. For qualitative analysis it is essential to have a constant and reproducible flow rate so that retention times can be reproduced.

The first flow control in any system is a two-stage regulator connected to the carrier gas cylinder to reduce the tank pressure of 2500 psig down to a usable level of 20–60 psig. For isothermal operation, constant pressure is sufficient to provide a constant flow rate assuming that the column has a constant pressure drop. In temperature programming, even when the inlet pressure is constant, the flow rate will decrease as the column temperature increases. This decrease is due to the increased viscosity of the carrier gas at higher temperatures. In all temperature-programmed instruments, a differential flow controller is used to ensure a constant mass flow rate.

Sometimes, however, it is not desirable to control the flow rate with such a controller. For example, split and splitless sample injection on OT columns both depend on a constant *pressure* for correct functioning. Constant pressure maintains the same flow rate through the column, independent of the opening and closing of the purge valve. Under these conditions, the carrier gas pressure can be increased electronically during a programmed run in order to maintain a constant flow. An electronic sensor is used to detect the (decreasing) flow rate and increase the pressure to the column, thus providing a constant flow rate by electronic pressure control (EPC).

8.5.3. Sample Inlets and Sampling Devices

8.5.3.1. Liquid and Solid Sampling. Because liquids expand considerably when they vaporize, only small sample sizes are desirable, typically microliters. Syringes are almost the universal method for injection of liquids. The most commonly used syringe sizes for liquids are 1, 5, and 10 μL. In those situations where the liquid samples are heated (as in all types of vaporizing injectors) to allow rapid vaporization before passage into the column, care must be taken to avoid thermal decomposition of the sample.

Solids are best handled by dissolving them in an appropriate solvent, and by using a syringe to inject the solution.

Syringes. Figure 8.8 shows a 10-μL liquid syringe typically used for injecting 1–5 μL of liquids or solutions. The stainless steel plunger fits tightly inside a

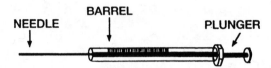

Figure 8.8. Typical 10 µL microsyringe. (From H. M. McNair and J. M. Miller, *Basic Gas Chromatography*, Wiley, New York, 1998, p. 22. Reprinted courtesy of John Wiley & Sons, Inc.)

precision barrel made of borosilicate glass. The needle, also stainless steel, is epoxyed into the barrel. Other models have a removable needle that screws onto the end of the barrel. For smaller volumes, a 1-µL syringe is also available. A useful suggestion is to always use a syringe whose total sample volume is at least two times larger than the volume to be injected.

It is very difficult to inject exactly a particular sample size with a syringe. Syringe injections can be precise (reproducible), but are rarely accurate due to additional vaporization of sample from the heated needle. Consequently, syringe injection does not provide good accuracy for some methods of quantitative analysis. In such cases, the internal standard method of analysis should be used (see Chapter 13).

Autosamplers. Samples can be injected automatically with mechanical devices that are often placed on top of gas chromatographs. These autosamplers mimic the human injection process just described using syringes. After flushing with solvent, they draw up the required sample several times from a sealed vial and then inject a fixed volume into the standard GC inlet. Autosamplers consist of a tray that holds a large number of samples, standards, and wash solvents, all of which are rotated into position under the syringe as needed. They can run unattended and thus allow many samples to be run overnight. Autosamplers provide better precision than manual injection—typically 0.2% relative standard deviation (RSD).

Septa. Syringe injection is accomplished through a self-sealing septum, a polymeric silicone with high temperature stability. Many types of septa are commercially available; some are composed of layers and some have a film of Teflon on the column side. In selecting one, the properties that should be considered are as follows: high temperature stability, amount of septum "bleed" (decomposition), size, lifetime, and cost. With repeated injection, the septa develop leaks and must be replaced regularly. Ideally, the laboratory SOP should state the frequency of replacement; often it is at the start of every new run of samples.

8.5.3.2. Gas Sampling. Gas sampling methods require that the entire sample be in the gas phase under the conditions in use. Mixtures of gases and liquids pose special problems. If possible, mixtures should be heated to convert all components to gases, or they should be pressurized to convert all components to liquids. Unfortunately, this is not always possible.

Gas-tight syringes and gas sampling valves are the most commonly used methods for gas sampling. The syringe is more flexible, is less expensive, and is the most frequently used device. A gas sampling valve on the hand gives better repeatability, requires less skill, and can be more easily automated. These valves are similar to the ones used in HPLC, and they are described in Chapter 9.

8.5.3.3. Injection Ports. Ideally, the sample is injected instantaneously onto the column, but in practice this is impossible and a more realistic goal is to introduce it as a sharp symmetrical band or plug. For the best peak shape and the maximum resolution, the smallest possible sample size should be used.

Flash Vaporization Versus On-Column Injection. There are two different methods of injecting the sample into a GC inlet. In the first method the sample is vaporized and, in the process, expands in volume by a factor of about 500–1000. To make this process efficient, the injection port needs to be hot and have a large volume. Unfortunately, for thermally labile samples, the high temperature may cause decomposition and the large volume may cause an excessively wide sample band.

An alternative is to inject the sample directly on the column. Each of these methods differs somewhat depending on the type of column being used—packed or capillary (OT). Therefore, each will be described separately.

Packed Column Inlets. For on-column operation, the column is positioned as shown in the Figure 8.9, with the column packing beginning at a position just reached by the syringe needle. The column is lined up collinearly with the syringe needle. When the syringe is pushed as far as it will go into the port, its contents will be delivered into the first part of the column packing—ideally on a small glass wool plug used to hold the packing in the column. There the ana-

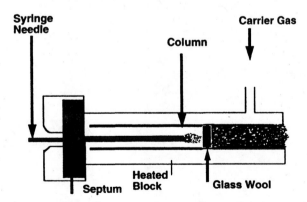

Figure 8.9. Injection port for on-column injections on packed columns. (From H. M. McNair and J. M. Miller, *Basic Gas Chromatography*, Wiley, New York, 1998, p. 82. Reprinted courtesy of John Wiley & Sons, Inc.)

lytes will be sorbed onto the column or evaporated, depending on their relative distribution constants. For most samples, the majority of the sample will go into the stationary phase; hence the name *on-column injection*. When purchasing a commercial column for on-column injection, it is necessary to specify the length of column that should be left empty in accordance with the geometric requirements just discussed. An alternative is to use a precolumn liner that can be replaced or cleaned when it gets dirty.

In the second configuration, the column is placed so that its front end (and its packing) barely extends into the injection port and is beyond the reach of the syringe needle. Efficient sampling for this configuration requires that the sample evaporate quickly (flash vaporization) when injected into the heated injection port.

Capillary Column Inlets. Capillary columns can accommodate only very small sample sizes (nanograms), too small to be injected with a syringe. Hence, the injection port is constructed to split the vaporized sample and admit only a small fraction of it to the column.

Split injection is the oldest, simplest, and easiest injection technique to use. The procedure involves injecting about 1 µl of the sample by a standard syringe into a heated injection port that contains a deactivated glass liner. The sample is rapidly vaporized, and only a fraction, usually 1–2%, of the vapor enters the column (see Figure 8.10). The rest of the vaporized sample and a large flow of carrier gas passes out through a split or purge valve.

There are several advantages to split injections. The technique is simple

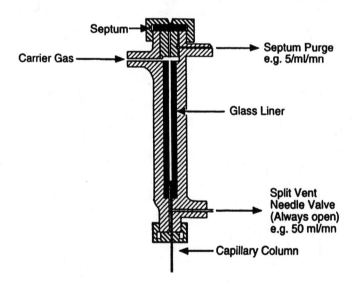

Figure 8.10. Split injector configuration for capillary columns. (From H. M. McNair and J. M. Miller, *Basic Gas Chromatography*, Wiley, New York, 1998, p. 98. Reprinted courtesy of John Wiley & Sons, Inc.)

because the operator has only to control the split ratio by opening or closing the split (purge) valve. The sample amount introduced to the column is very small (and easily controlled), and the flow rate up to the split point is fast (the sum of both column and vent flow rates). The result is high-resolution separations. Another advantage is that "neat" samples can be introduced, usually by using a larger split ratio, so there is no need to dilute the sample. A final advantage is that "dirty" samples can be introduced by putting a plug of deactivated glass wool in the inlet liner to trap nonvolatile compounds.

One disadvantage is that trace analysis is limited because only a fraction of the sample enters the column. Consequently, splitless or on-column injection techniques are recommended for trace analysis. A second disadvantage is that the splitting process sometimes discriminates against high-molecular-weight solutes in the sample so that the sample entering the column is not representative of the sample injected. For these reasons, another vaporization mode, splitless injection, is sometimes used.

Splitless injection uses the same hardware as split injection (Figure 8.11), but the split valve is initially closed. The sample is diluted in a volatile solvent (like hexane or methanol), and 1–5 µL is injected in the heated injection port. The sample is vaporized and slowly (flow rate of about 1 mL/min) carried onto a cold column where both sample and solvent are condensed. After 45 sec, the split valve is opened (flow rate of about 50 mL/min), and any residual vapors left in the injection port are rapidly swept out of the system. Septum purge is essential with splitless injections.

The column is now temperature-programmed, and initially only the volatile

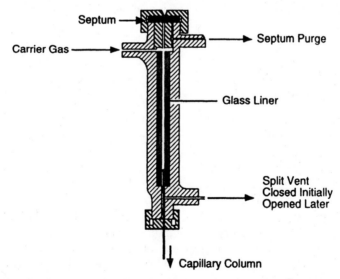

Figure 8.11. Splitless injector configuration for capillary columns. (From H. M. McNair and J. M. Miller, *Basic Gas Chromatography*, Wiley, New York, 1998, p. 99. Reprinted courtesy of John Wiley & Sons, Inc.)

solvent is vaporized and carried through the column. While this is happening, the sample analytes are being refocused into a narrow band in the residual solvent. At some later time, these analytes are vaporized by the hot column and chromatographed. High resolution of these higher boiling analytes is observed.

The big advantage of splitless injection is the improved sensitivity over split. Typically, 20- to 50-fold more sample enters the column, and the result is improved trace analysis for environmental, pharmaceutical, or biomedical samples.

Splitless has several disadvantages. It is time-consuming; you must start with a cold column, and you must temperature program. You must also dilute the sample with a volatile solvent, and you must optimize both the initial column temperature and the time of opening the split valve. Finally, splitless injection is not well-suited for volatile compounds. For good chromatography the first peaks of interest must have boiling points 30°C higher than the solvent.

For more information on split/splitless injections, see reference 10.

Three other types of capillary inlets are "direct injection," "on-column," and "cold on-column." *On-column* means inserting the precisely aligned needle into the capillary column, usually a 0.53-mm-ID megabore, and making injections inside the column. Direct injection involves injecting a small sample (usually 1 μL or smaller) into a glass liner where the vapors are carried directly to the column. Both of these techniques require thick film capillaries and wide-diameter columns with faster-than-normal flow rates (~10 mL/min). Even with these precautions, the resolution is not as good as with split or splitless injection. The advantages can be better trace analysis and good quantitation.

Both high resolution and good quantitation result from *cold on-column* injections. A liquid sample is injected into either a cold inlet liner or a cold column. The cold injector is rapidly heated, and the sample is vaporized and carried through the column. Minimal sample decomposition is observed. For thermolabile compounds, cold on-column is the best injection technique.

8.5.4. Detectors

A detector senses the effluents from the column and provides a record of the chromatography in the form of a chromatogram. The detector signals are proportionate to the quantity of each solute (analyte) making possible quantitative analysis.

The most common detector is the flame ionization detector (FID). It has the desirable characteristics of high sensitivity and wide range of linearity, and yet it is relatively simple and inexpensive. Other popular detectors are the thermal conductivity cell (TCD), the electron capture detector (ECD), and the nitrogen–phosphorus detector (NPD). Only these four will be discussed in this chapter, but over 60 GC detectors have been described and about a dozen are commercially available and in common use [11], including the established identification methods, FTIR and MS. Table 8.5 lists the carrier gases needed for those detectors.

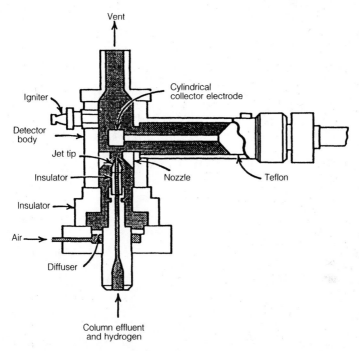

Vent

Igniter

Cylindrical
collector electrode

Detector
body

Jet tip

Insulator

Nozzle

Teflon

Insulator

Air

Diffuser

Column effluent
and hydrogen

Figure 8.12. Schematic of a typical FID. (Courtesy of Perkin-Elmer. From H. M. McNair and J. M. Miller, *Basic Gas Chromatography*, Wiley, New York, 1998, p. 115. Reprinted courtesy of John Wiley & Sons, Inc.)

8.5.4.1. Flame Ionization Detector (FID).

The FID is the most widely used GC detector, and it is an example of the ionization detectors invented specifically for GC. The column effluent is burned in a small oxygen–hydrogen flame, producing ions in the process. These ions are collected and form a small current that, when amplified, becomes the signal that is proportional to the number of carbon atoms in the sample. When no sample is being burned, there should be little ionization, with the small current (10^{-14} A) arising from impurities in the hydrogen and air supplies. Thus, the FID is a specific property-type detector with characteristic high sensitivity for organic molecules.

A typical FID design is shown in Figure 8.12. The column effluent is mixed with hydrogen and led to a small burner tip that is surrounded by a high flow of air to support combustion. An igniter is provided for remote lighting of the flame. The collector electrode is biased at +300 V relative to the flame tip, and the collected current is amplified by a high-impedance circuit. Because water is produced in the combustion process, the detector must be heated to at least 125°C to prevent condensation of water and high-boiling samples. Most FIDs are run at 250°C or hotter.

For efficient operation, the gases (hydrogen and air) must be pure and free of organic material that would increase the background ionization. Their flow

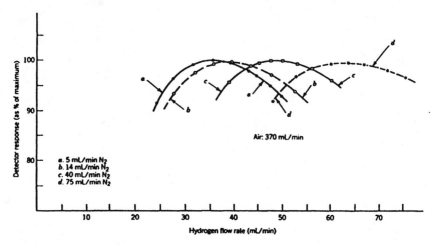

Figure 8.13. Effect of hydrogen flow Rate on FID response. (Courtesy of Perkin-Elmer. From H. M. McNair and J. M. Miller, *Basic Gas Chromatography*, Wiley, New York, 1998, p. 115. Reprinted courtesy of John Wiley & Sons, Inc.)

rates need to be optimized for the particular detector design (and to a lesser extent, the particular analyte). As shown in Figure 8.13, the flow rate of hydrogen goes through a maximum sensitivity for each carrier gas flow rate, with the optimum occurring at about 30–40 mL/min. For open tubular columns that have flows around 1 mL/min, make-up gas is added to the carrier gas to bring the total up to about 40 mL/min.

Hydrogen can be used as the carrier gas, but changes in gas flows (a separate source of hydrogen is still required) and detector designs are required [12] in addition to the safety precautions that must be taken.

The flow rate of air is much less critical, and a value of 300–400 mL/min is sufficient for most detectors as shown in Figure 8.14.

Compounds not containing organic carbon do not burn and are not detected. The most important ones are listed in Table 8.6. Most significant among those listed is water, a compound that often produces badly tailed peaks. The absence of a peak for water permits the FID to be used for analysis of samples that contain water because it does not interfere in the chromatogram. A typical application is the common analysis of residual solvents in drug products discussed later.

Table 8.7 summarizes the characteristics of the FID. Its advantages are good sensitivity, a wide range of linearity, simplicity, ruggedness, and adaptability to all sizes of columns.

8.5.4.2. Thermal Conductivity Detector (TCD). Nearly all of the early GC instruments were equipped with thermal conductivity detectors. They have remained popular, particularly for packed columns and for inorganic analytes such as air, H_2O, CO, CO_2, and H_2, which cannot be detected by the FID.

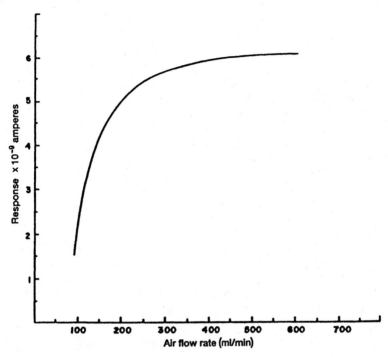

Figure 8.14. Effect of air flow rate on FID response. (Courtesy of Perkin-Elmer. From H. M. McNair and J. M. Miller, *Basic Gas Chromatography*, Wiley, New York, 1998, p. 115. Reprinted courtesy of John Wiley & Sons, Inc.)

Table 8.6. Compounds Not Detected by the FID

He	O_2	NO	CS_2
AR	N_2	NO_2	COS
Kr	CO	N_2O	$SiCl_4$
Ne	CO_2	HN_3	$SiHCl_3$
Xe	H_2O	SO_2	SiF_4

Table 8.7. Summary of FID Characteristics

1. LOD of about 10^{-11} g or 50 ppb
2. Nearly universal but not for fixed gases (see Table 8.6)
3. Excellent linear range of 10^6
4. Temperature limit of about 400°C
5. Highly stable and easy to operate
6. Requires conventional amplifier

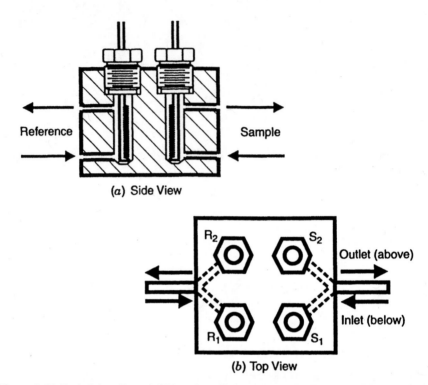

Figure 8.15. Typical four-filament TCD cell. (*a*) Side view, (*b*) top view. (Provided by: GOW-MAC Instrument Co., Bethlehem, PA. GOW-MAC is a registered trademark of GOW-MAC Instrument Co. From H. M. McNair and J. M. Miller, *Basic Gas Chromatography*, Wiley, New York, 1998, p. 117. Reprinted courtesy of John Wiley & Sons, Inc.)

The TCD is a differential detector that measures the thermal conductivity of the analyte in carrier gas, compared to the thermal conductivity of pure carrier gas. In a conventional detector at least two cell cavities are required, although a cell with four cavities is more common. The cavities are drilled into a metal block (usually stainless steel), and each contains a resistance wire or filament (known as *hot wires*). The filaments are either mounted on holders, as shown in Figure 8.15, or held concentrically in the cylindrical cavity, a design that permits the cell volume to be minimized. They are made of tungsten or a tungsten–rhenium alloy (so-called WX filaments) of high resistance.

The filaments are incorporated into a Wheatstone Bridge circuit, the classic method for measuring resistance (Figure 8.16). A DC current is passed through them to heat them above the temperature of the cell block, creating a temperature differential. With pure carrier gas passing over all four elements, the bridge circuit is balanced with a "zero" control. When an analyte elutes, the thermal conductivity of the gas mixture in the two sample cavities is decreased, their filament temperatures increase slightly, causing the resistance of the fila-

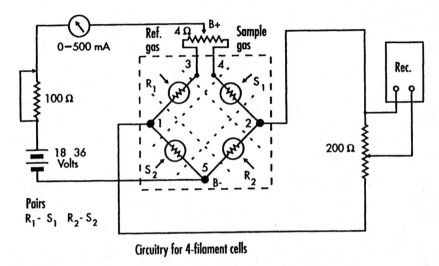

Circuitry for 4-filament cells

Figure 8.16. Wheatstone bridge circuit for four-filament TCD. (Provided by: GOW-MAC Instrument Co., Bethlehem, PA. GOW-MAC is a registered trademark of GOW-MAC Instrument Co. From H. M. McNair and J. M. Miller, *Basic Gas Chromatography*, Wiley, New York, 1998, p. 117. Reprinted courtesy of John Wiley & Sons, Inc.)

ments to increase greatly, and the bridge becomes unbalanced; that is, a voltage develops across opposite corners of the bridge. That voltage is dropped across a voltage divider (the so-called *attenuator*), and then all or part of it is fed to a recorder, integrator, or other data system. After the analyte is fully eluted, the thermal conductivity in the sample cavities returns to its former value and the bridge returns to balance.

The larger the heating current applied to the filaments, the greater the temperature differential and the greater the sensitivity. However, high filament temperatures also result in shorter filament life because small impurities of oxygen readily oxidize the tungsten wires, ultimately causing them to burn out. For this reason, the GC system must be free from leaks and operated with oxygen-free carrier gas.

The Wheatstone bridge can be operated at constant voltage or constant current, but a more elaborate circuit can be used to maintain constant filament *temperature*. Thus, the detector controls may specify setting a current, a voltage, a temperature, or a temperature difference (ΔT), depending on the particular type of control. Controlling the filament temperature to keep it constant will null the bridge, unlike the simpler circuit that directly measures the bridge unbalance. Nulling provides a larger linear range, greater amplification, lower detection limits, and less noise [13].

A summary of TCD characteristics is given in Table 8.8. In summary, it is a rugged, universal detector with moderate sensitivity widely used in analysis of gases and water.

Table 8.8. Summary of TCD Characteristics

1.	LOD of about 10^{-9} g or 10 ppm
2.	Universal
3.	Linear range of 10^4
4.	Temperature limit of about 400°C
5.	Requires good temperature control; otherwise very stable and easy to operate
6.	Needs no amplification
7.	Requires helium carrier gas for optimum performance

8.5.4.3. Electron Capture Detector (ECD). The invention of the ECD (for GC) is generally attributed to Lovelock, based on his publication in 1961 [14]. It is a selective detector that provides very high sensitivity for those compounds that "capture electrons." These compounds include halogenated materials such as pesticides, and, consequently, one of its primary uses is in pesticide residue analysis. It is an ionization-type detector, but unlike most detectors of this class, samples are detected by causing a *decrease* in the level of ionization. A radioactive source (typically ^{63}Ni or ^{3}H) causes the ionization of the carrier gas, resulting in a high standing current. When an electronegative analyte is eluted from the column and enters the detector, it captures some of the free electrons and the standing current is decreased. The mathematical relationship for this process is similar to Beer's law (used to describe the absorption process for electromagnetic radiation). Thus, the extent of the absorption or capture is proportional to the concentration of the analyte.

The necessity to use a radioactive source that may require a license or at least regular "wipe tests" for radioactivity is one drawback of the ECD. A new innovation is an ECD operated with a pulsed discharge so that it does not require a radioactive source [15]. This detector is commercially available and can also be operated as a helium ionization detector under different conditions.

The ECD is one of the most easily contaminated detectors and is adversely affected by traces of oxygen and water. Ultrapure, dry gases, freedom from leaks, and clean samples are necessary. Evidence of contamination is usually a noisy baseline or peaks that have small negative dips before and after each peak. Cleaning can sometimes be accomplished by operation with hydrogen carrier gas at a high temperature to burn off impurities, but dismantling and cleaning is often required.

8.5.4.4. Nitrogen–Phosphorus Detector (NPD). Like many other ionization detectors, the NPD was invented for use in GC. It was originally known as the alkali flame ionization detector (AFID) because it consisted of an FID to which a bead of alkali metal salt was added [16]. The presence of the alkali salt gave it selective sensitivity for nitrogen, phosphorus, and some halogen compounds. The current version, the NPD, uses a rubidium silicate bead heated by a platinum resistance wire and is operated with a hydrogen flow so low that combustion is not supported; that is, there is no longer a flame [17]. It is selective only for nitrogen and phosphorous-containing compounds and is the most sensitive GC detector available for these elements.

In the pharmaceutical laboratory, its major use has been for drugs containing nitrogen. A typical example is the analysis of benzodiazepines, many of which can be detected with the ECD [18]. Figure 8.17 compares chromatograms of some representative drugs using the NPD and ECD in tandem. The method was developed for screening plasma samples after solid-phase extraction, and these selective detectors are required to provide the necessary sensitivity without the interference of matrix peaks that would be present with an FID.

8.6. TEMPERATURE CONSIDERATIONS

Column temperature is one of the two important parameters in GC, the other being the nature of the stationary phase. If the analysis is run at constant temperature, the process is called isothermal GC. Programmed temperature gas chromatography (PTGC) is the process of increasing the column temperature during a GC run. It is a very effective method for optimizing an analysis and is often used for screening new samples. Before describing it in detail, let us consider the broader topic of temperature in general.

8.6.1. Temperature Zones

The column is thermostated so that a good separation will occur in a reasonable amount of time. It is often necessary to maintain the column at a wide variety of temperatures, from ambient to 360°C. The control of column temperature is one of the easiest and most effective ways to influence the separation. The column is fixed between a heated injection port and a heated detector, so it seems appropriate to discuss the temperature levels at which these components are operated.

8.6.1.1. Injection-Port Temperature. The injection port should be hot enough to vaporize the sample rapidly so that no loss in efficiency results from the injection technique. On the other hand, the injection-port temperature must be low enough so that thermal decomposition or chemical rearrangement is avoided.

For flash vaporization injection, a general rule is to have the injection temperature about 50°C hotter than the boiling point of the sample. A practical test is to raise the temperature of the injection port. If the column efficiency or peak shape improves, the injection-port temperature was too low. If the retention time, the peak area, or the shape changes drastically, the temperature may be too high and decomposition or rearrangement may have occurred. For on-column injection, the inlet temperature can be lower.

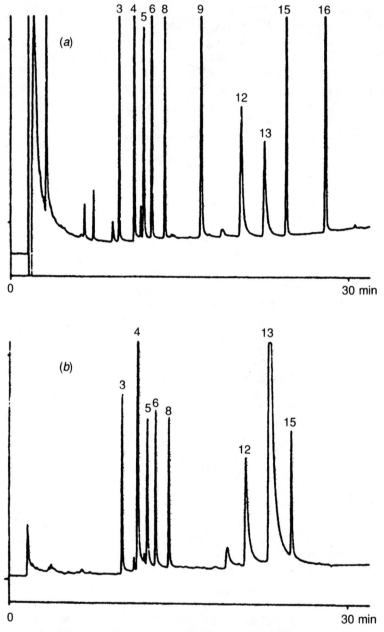

Figure 8.17. Dual detector gas chromatogram of a typical drug sample. (a) NPD, (b) ECD. Peak identifications: 3, diazepam; 4, clotiazepam; 5, clobazam; 6, midazolam; 8, prazepam, 9, zolpidem; 12, alprazolam; 13, triazolam; 15, aldipem; 16, buspirone. (Reprinted from Y. Gaillard, J-P Gay-Montchamp, and M. Ollagnier, *J. Chromatogr.* **1993**, *622*, 197–208, with permission from Elsevier Science.)

8.6.1.2. Column Temperature. The column temperature should be high enough so that sample components pass through it at a reasonable speed. It need not be higher than the boiling point of the sample; in fact it is usually preferable if the column temperature is considerably below the boiling point. If that seems illogical, remember that the column operates at a temperature where the sample is in the *vapor* state—it need not be in the *gas* state. In GC, the column temperature must be kept above the "dew point" of the sample, but not above its boiling point. Recall that fused silica OT columns that are coated with polyimide cannot be used above 380°C without degrading the coating.

8.6.1.3. Detector Temperature. The detector temperature depends on the type of detector employed. As a general rule, however, the detector and its connections from the column exit must be hot enough to prevent condensation of the sample and/or liquid phase. A good rule is to have the detector about 20°C hotter than the maximum column temperature. If the temperature is too low and condensation occurs, peak broadening and even the total loss of peaks is possible.

The thermal conductivity detector temperature must be controlled to ± 0.1°C or better for baseline stability and maximum detectivity. Ionization detectors do not have this strict a requirement; their temperature must be maintained high enough to avoid condensation of the samples and also of the water or by-products formed in the ionization process. A reasonable minimum temperature for the flame ionization detector is 125°C.

8.6.2. Programmed Temperature GC (PTGC)

Retention times and retention factors decrease as temperature increases because the distribution constants are temperature-dependent in accordance with the Clausius–Clapeyron equation,

$$\log p^0 = -\frac{\Delta\mathscr{H}}{2.3\mathscr{R}T} + \text{constant} \tag{8.1}$$

where $\Delta\mathscr{H}$ is the enthalpy of vaporization at absolute temperature, T, $\mathscr{R}$ is the gas constant, and p^0 is the compound's vapor pressure at this temperature. The equation indicates that as the (absolute) temperature decreases, the vapor pressure of the solute decreases logarithmically. A decrease in vapor pressure results in a decrease in the relative amount of solute in the mobile phase—that is, an increase in the retention factor, k, and an increase in retention time.

Figure 8.18, a plot of the log of net retention volume versus $1/T$ for a few typical solutes, illustrates this relationship. Straight lines are obtained over a limited temperature range in accordance with our prediction based on equation 8.1. The slope of each line is proportional to that solute's enthalpy of vaporization and can be assumed to be constant over the temperature range shown.

To a first approximation, the lines in Figure 8.18 are parallel, indicating that

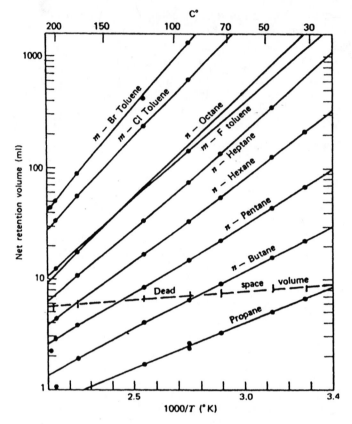

Figure 8.18. Temperature dependence of retention volume. (Reprinted with permission from reference 20. From H. M. McNair and J. M. Miller, *Basic Gas Chromatography*, Wiley, New York, 1998, p. 143. Reprinted courtesy of John Wiley & Sons, Inc.)

the enthalpies of vaporization for these compounds are nearly the same. A closer inspection reveals that many pairs of lines diverge slightly at low temperatures. From this observation we can draw the useful generalization that *GC separations are usually better at lower temperatures.* But look at the two solutes, *n*-octane and *m*-fluorotoluene; their lines cross at about 140°C. At 140°C, they cannot be separated; at a lower temperature the toluene elutes first, but at a higher temperature the *reverse* is true. While it is not common for elution orders to reverse, it can happen, resulting in misidentification of peaks! See, for example, the work of Hinshaw [19] with chlorinated pesticides.

The consequences of increasing the temperature for a GC analysis include: decreased retention time, slight increase in *N*, and a change (usually a decrease) in α. The effect of temperature on efficiency is quite complex [20] and does not always increase. Usually it is a minor effect and less important than the effect on column thermodynamics (selectivity). Overall, however, temperature effects are very significant and PTGC is very powerful.

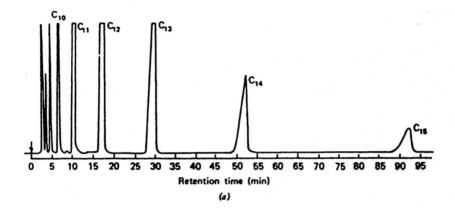

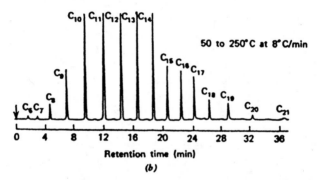

Figure 8.19. Comparison of (a) isothermal and (b) programmed temperature separations of n-paraffins. (From H. M. McNair and J. M. Miller, *Basic Gas Chromatography*, Wiley, New York, 1998, p. 145. Reprinted courtesy of John Wiley & Sons, Inc.)

8.6.2.1. Advantages and Disadvantages of PTGC. If a sample being analyzed by GC contains components whose vapor pressures (boiling points) extend over a wide range, it is often impossible to select one temperature that will be suitable for an isothermal run. As an example, consider the separation of a wide range of homologs like the kerosene sample shown in Figure 8.19a. An isothermal run at 150°C prevents the lighter components ($< C_8$) from being totally separated and still takes over 90 min to elute the C_{15} paraffin, which looks like the last one. Even so, this is probably the best isothermal temperature for this separation.

The separation can be significantly improved using programmed temperature. Figure 8.19b shows one such programmed run in which the temperature starts at 50°C, less than the isothermal temperature used in Figure 8.18a, and is programmed at 8 degrees per minute up to 250°C, a temperature higher than the isothermal temperature. Increasing the temperature during the run decreases

Table 8.9. Adavantages and Disadvantages of PTGC

Advantages	Disadvantages
Optimizes the separation of a sample with a wide boiling point range	Special instrumentation required for reproducibility
Peak widths remain narrow, resulting in better separations and higher peaks and lower LODs	Restricted to stationary phases capable of of high-temperature operation
Good for sample screening	May be noisier at high temperatures
Good for column cleaning	Baseline may drift
Usually faster analysis time	Cooling required between samples

the partition coefficients of the analytes still on the column, so they move faster through the column, yielding decreased retention times.

Some major differences between the two runs illustrate the advantages of PTGC. For a homologous series, the retention times are logarithmic under isothermal conditions, but they are linear when programmed. The programmed run facilitated the separation of the low-boiling paraffins, easily resolving several peaks before the C_8 peak while increasing the number of paraffins detected. The C_{15} peak elutes much faster (in about 21 min), and it turns out that it is not the last peak: Six more hydrocarbons are observed by PTGC. All of the peak widths are about equal in PTGC; in the isothermal run, some fronting is evidenced in the higher boilers. Because the peak widths do not increase in PTGC, the heights of the late-eluting analytes are increased (peak areas are constant), providing better detectivity. Table 8.9 summarizes the advantages and disadvantages of PTGC.

Programmed temperature operation is good for screening new samples. A maximum amount of information about the sample composition is obtained in minimum time. Usually one can tell when the entire sample has been eluted, often a difficult judgment to make with isothermal operation. Modern programmers typically provide as many as five temperature ramps.

8.6.2.2. Requirements of PTGC. PTGC requires a more versatile instrument than does isothermal GC. Most important is the ability to control the programmed temperature increase in the column oven while keeping the detector and injection port at constant temperatures. An electronic temperature programmer is needed along with an oven design that has a low mass, a high volume fan, and a vent to outside air, also controlled by the programmer.

Some means is usually provided to control the carrier gas flow. In a packed column chromatograph, this is usually accomplished with a differential pneumatic flow control valve placed in the gas line upstream of the injection port.

In a capillary column chromatograph, constant pressure regulation is required for split/splitless sampling, and a flow control valve cannot be used. As a consequence, the flow rate of carrier gas decreases during the programmed temperature run due to the increase in gas viscosity. Because the pressure drop across an OT column is relatively low, the change in flow rate is less severe than in packed columns. One solution is to set the initial flow rate above the optimum value and closer to the flow expected about 70% of the way through the program. This will ensure an adequate flow at the higher temperatures. However, electronic pressure control (EPC) is available on some instruments, and it can be used to maintain a constant flow by increasing the pressure during the run [21].

8.6.2.3. Theory of PTGC.
The theory of PTGC has been thoroughly treated by Harris and Habgood [22] and by Mikkelsen [23]. The following summary and guidelines have been taken from a simple but adequate treatment by Giddings [24].

The operation of PTGC can be envisioned as follows: The sample is injected onto the end of the cool column, and its components remain sorbed there; as the temperature increases, the analytes vaporize and move down the column at increasing rates until they elute. It is for these reasons that the injection technique is not critical in PTGC and that all peaks have about the same peak widths—they spend about the same amount of time actively partitioning down the column.

A useful rule of thumb is the generalization that a doubling of the retention time for a given compound will result from a decrease in temperature of 30°C, and vice versa. This value will vary for different samples, of course, but it makes a useful approximation that helps when one wants to adjust the time of analysis.

Three important variables are the programming rate, the flow rate, and the column length. In general, one does not vary the length but uses a short column (and lower temperatures) and relatively high flow rates. The programming rate is often chosen to be fast enough to save time but slow enough to get adequate separations, somewhere between 4°C/min and 10°C/min. However, for OT columns, one group of workers concluded that slow rates (around 2.5°C/min) and high flow rates (about 1 mL/min) are preferable [25]. Another study by Hinshaw [19] of a chlorinated pesticide mixture found that 8°C/min was preferable to either slower (down to 1.5°C/min) or faster (up to 30°C/min) rates.

For a variety of reasons, isothermal operation is often preferred in the workplace. If an initial screening is done by PTGC, one might wish to know which isothermal temperature would be the best one to use. Giddings has called this isothermal temperature the *significant temperature*, T'. Using reasoning based on the 30°C value, he has found that

$$T' = T_f - 45 \tag{8.2}$$

where T_f is the final temperature, the temperature at which the analyte(s) of interest eluted in the PTGC run. Thus, for example, a solute eluting at a temperature of 225°C on a PTGC run would be best run isothermally at 180°C [24].

Table 8.10. Recommended Columns for Scouting Runs

	Column[a]	
	Capillary	Packed
1. Stationary phase	DB-1	OV-101
2. Loading	0.25 μm	3% (w/w)
3. Column length	10 m	2 ft
4. Column inner diameter	0.25 mm	2 mm
5. Temperature program range (Hold for 5 min at maximum)	60–320°C	100–300°C

[a] Packed column is glass; capillary column is fused silica.

8.7. OPTIMIZATION AND METHOD DEVELOPMENT

8.7.1. Column Selection

As was stated in the previous section, choosing the best column (liquid phase) for a given sample is one of the two important parameters in setting up a GC method. It is a complex process and is more difficult than choosing the temperature. As a starting point, chemists usually rely on the maxim "like dissolves like." For GC, this means that the stationary phase is usually chosen to be "like" the sample in polarity and functional groups. For capillary columns, the efficiency is so great that the column polarity is of less importance.

Table 8.10 lists two possible columns that could be used as starting points for screening a new sample. However, it is very likely that the literature contains reports of similar separations, and it should be searched for information. A useful reference providing nearly 200 examples of actual separations on packed columns has been made available recently by Supelco [26]. Other suppliers also provide application information.

8.7.2. Optimization According to Basic Principles

The rate equation correctly predicts that the best columns have thin films of stationary phase (low percentage of stationary phase on packed columns) and small internal diameters. In addition, for packed columns the solid support must be very inert and have small, uniform diameters (e.g., 100–120 mesh). A light carrier gas such as helium will give the best results in the least amount of time.

Sample sizes should be as small as practical. Programmed temperature will improve the separation if the sample has a wide boiling range. Usually small program rates (2–5°/min) will give better separations.

Equation 8.3 indicates the three variables that can be adjusted to maximize the resolution of a given separation:

$$R_s = \frac{1}{4}\sqrt{N}\left(\frac{\alpha - 1}{\alpha}\right)\left(\frac{k}{k+1}\right) \tag{8.3}$$

The most powerful parameter is the separation factor, α. The most common way to change the separation factor is to choose a new stationary phase. However, with OT columns, the very large plate number, N, is often sufficient to provide a separation even though the improvement in resolution is only by the square root of N. A change in k is most often achieved by a change in temperature. A decrease in temperature will cause an increase in k and in resolution, but little advantage is gained above a k value of about 20, especially because increases in k result in increases in analysis times.

8.8. SOME SPECIAL TOPICS

There are many additional topics that could be included in this chapter, but space will permit only a few, and they can be introduced only briefly. For additional information see the general GC books listed at the end of the chapter.

8.8.1. Gas Chromatography/Mass Spectrometry (GC/MS)

Earlier in the chapter, it was noted that GC is not capable of providing qualitative identifications for samples that are totally unknown. A retention factor (k) can be helpful in identifying a particular analyte from among a limited small number of possibilities, but more powerful spectroscopic methods such as infrared (IR and FTIR) and mass spectrometry (MS) are required for positive identifications in many instances. Both of these spectroscopic methods have been attached to the end of a GC instrument to aid in identification, and they are somewhat complementary. However, the one combination that has proven of most use is GC/MS.

The mass spectrometer used in this application is usally a quadrupole type, one configuration of which is known as the ion trap. It has the advantages of being inexpensive, small, fast-scanning, and computerizable. Usually the GC column is led directly into the MS analyzer, which is under high vacuum; capillary columns are most advantageous because of their low flows and the flexibility of the fused silica of which they are made. The combination instrument is small and is often referred to as a "benchtop" GC/MS.

Data can be acquired in at least two ways: as the total ion (TIC) that provides a conventional chromatogram, or as only a few individual ions [selected ion monitoring (SIM)] that provide peaks for only one mass fragment in a scan (chromatogram). The former contains all the data necessary for identification which can be accomplished by comparing the mass spectrum in each peak with a reference spectrum contained in the computer. The SIM data can pinpoint a particular compound whose mass spectrum has been previously determined; and because only one ion is being measured, more data points can be taken and lower detection limits are possible for quantitative analysis.

For further information consult the books listed in the GC/MS section of the references.

8.8.2. Derivatization

Many of the drug products analyzed in the pharmaceutical laboratory are not sufficiently volatile to permit analysis by GC, even using open tubular columns. But GC does have the advantage over HPLC of being fast, efficient, and more simple, so there is some incentive for attempting analysis by GC. One alternative is to make volatile derivatives and run them by GC.

The functional groups that can be made more volatile by derivatization include carboxylic acids, alcohols, amines, amids, amino acids, catecholamines, carbohydrates, and sugars. The common reaction types are silylation, alkylation, and acylation. See the references listed for more information.

Incorporatiion of one or more derivatization steps into an analysis method may give rise to additional errors and will require extra method validation. However, the use of an internal standard (see Chapter 13) may improve quantitation.

8.8.3. Headspace Sampling

If the sample to be analyzed contains analytes that are relatively volatile but contains other matrix ingredients that are not, sampling of the headspace may facilitate analysis. The headspace is the vapor in a closed container above a solution. Heating is usually desirable, and sufficient time must be allowed for the liquid–vapor equilibrium to be established. Samples can simply be taken from the headspace with a syringe and injected into the gas chromatograph. The nonvolatile matrix components do not get injected into the GC, avoiding the contamination of the column.

The procedure just described is a static one; dynamic headspace sampling is also used. The dynamic process is often also called "purge-and-trap", a name which describes the process. Gas is purged through the sample and trapped in a solid adsorbent that can be later transferred to a GC inlet and desorbed.

A brief introduction to headspace sampling was published in 1990 [27], and several more extensive works are also listed in the references. It is being used in the analysis of residual solvents in pharmaceuticals and is further discussed in Section 8.9.1.

The newest type of headspace sampling is one version of the technique called solid-phase microextraction (SPME). This topic was introducted in Chapter 5. Briefly, it consists of a coated fiber that is exposed to the headspace vapors, sorbing them onto it. The fiber is then transferred to the heated injection port of a GC and desorbed. SPME is being evaluated as a replacement for many static headspace methods because it offers greater selectivity and perhaps greater detectivity. Some recent books are included in the headspace references.

8.8.4. USP

Gas chromatography is covered in the USP in the chromatography section of the General Chapters, ⟨621⟩ on pages 1919 and 1920. Included are directions for conditioning and testing of columns. Other parts of section ⟨621⟩ are also relevant, including the Interpretation of Chromatograms, System Suitability, Glossary of Symbols, and Chromatographic Reagents, most of which is discussed in Chapter 7.

The Chemical Tests Section includes GC methods for dimethylaniline (limit test 223, p. 1858), antimicrobial agents (341, p. 1864), barbiturates (assay 361, p. 1866), and organic volatile impurities (467, p. 1877). The latter includes four methods, three for direct injection and one headspace plus a method specifically for methylene chloride in coated tablets.

8.9. APPLICATIONS

Several applications of GC have already been mentioned, including those in the USP. Some are as follows: (1) the determination of water for which the Karl Fischer titration (see Chapter 5) has been the classic method, but near-infrared (NIR) spectroscopy is now gaining in popularity (see Chapter 6). Recently [28], GC has been proposed as a reference method for moisture detemination by NIR, and this paper contains several relevant references on water analysis by GC; (2) organic solvent extractables (OSEs) of petrolatum [29]; and (3) gas analysis, including the separation of oxygen and nitrogen by molecular sieves as was shown in Figure 8.4. Of these, only the analysis of residual solvents will be described further in this chapter.

8.9.1. Analysis of Residual Solvents

The manufacture and purification of drugs requires the use of solvents, traces of which can remain in the final product. Clearly undesirable, they need to be found and analyzed at low levels. The method of choice is usually GC. The ICH guidelines for residual solvents analysis were described in Chapter 2. Solvent classification schemes and methods were given in Table 3.5.

The problems associated with the GC analysis are as follows. The solvent needs to dissolve or at least disintegrate solid drug product to release the volatile impurities that must be soluble in the solvent, but it must also be separated from the impurities in the chromatogram and never be present in any sample. Some solvents that have been used are water, benzyl alcohol [30, 31], DMF [32], and methylcyclohexane [33].

A unique, new approach is to use a relatively high-boiling solvent such as N-methylpyrrolidone which elutes well after the volatiles [34]. It can be removed from the column by fast programming and need not be recorded or used in the calculations. Figure 8.20 shows a typical separation; n-butanol is used as the internal standard.

Another problem is the difficulty in retaining the volatiles sufficiently long to effect their separation. As mentioned earlier in this chapter, packed columns are

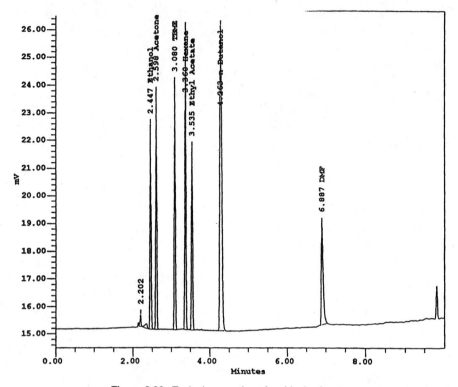

Figure 8.20. Typical separation of residual solvents.

preferred for this type of analysis; but because OT columns are gaining in popularity, they are often used. Consequently, OT columns should be long, heavily loaded, and of large diameter [34].

If water is present as one of the volatiles, it may tail badly and obscure other peaks. This problem can be eliminated by using an FID detector, but then water must be determined by an alternative method.

And finally, the problems associated with matrix effects can be addressed by using a headspace method. Both dynamic [35] and static [36–39] methods have been reported.

REFERENCES

1. M. Tswett, *Ber. d. deut. botan. Ges.* **1906**, *24*, 316, 384.

2. A. J. P. Martin and R. L. M. Synge, *Biochem. J.* **1941**, *35*, 1358.

3. A. T. James and A. J. P. Martin, *Biochem. J.* **1952**, *50*, 679.

4. M. J. E. Golay, in *Gas Chromatography 1958 (Amsterdam Symposium)*, D. H. Desty (editor), Butterworths, London, 1958, pp. 36–55 and 62–68.

5. H. M. McNair and J. M. Miller, *Basic Gas Chromatography*, Wiley, New York, 1998, p. 1.

6. United States Pharmacopeia Convention, *United States Pharmacopeia* 24 (USP 24/ NF 19), ⟨621⟩ Rockville, MD, 2000, p. 1925.

7. D. M. Ottenstein, *J. Chromatogr. Sci.* **1973**, *11*, 136.

8. United States Pharmacopeia Convention, *United States Pharmacopeia* 24 (USP 24/ NF 19), Rockville, MD, 2000, p. 1926.

9. O. L. Hollis, *Anal. Chem.* **1966**, *38*, 309.

10. K. Grob, *Classical Split and Splitless Injection in Capillary Gas Chromatography*, 3rd edition, Heuthig, Heidelberg, 1993.

11. H. H. Hill and D. G. McMinn (editors), *Detectors for Capillary Chromatography*, Wiley, New York, 1992.

12. R. K. Simon, Jr., *J. Chromatogr. Sci.* **1985**, *23*, 313.

13. R. T. Wittebrood, *Chromatographia* **1972**, *5*, 454.

14. J. E. Lovelock, *Anal. Chem.* **1961**, *33*, 162.

15. J. Mudabushi, H. Cai, S. Stearns, and W. Wentworth, *Am. Lab.* **1995**, *27*(15), 21– 30. H. Cai, W. E. Wentworth, and S. D. Stearns, *Anal. Chem.* **1996**, *68*, 1233.

16. A. Karmen and L. Giuffrida, *Nature* **1964**, *201*, 1204.

17. B. Kolb and J. Bischoff, *J. Chromatogr. Sci.* **1974**, *12*, 625.

18. Y. Gaillard, J. P. Gay-Montchamp, and M. Ollegnier, *J. Chromatogr. B* **1993**, *622*, 197–208.

19. J. V. Hinshaw, *LC-GC* **1991**, *9*, 470.

20. W. E. Harris and H. W. Habgood, *Talanta* **1964**, *11*, 115.

21. S. S. Stafford (editor), *Electronic Pressure Control in Gas Chromatography*, second printing, Hewlett-Packard Co., Wilmington, DE, 1994. Hewlett-Packard also has available EPC software called "HP Pressure/Flow Calculator."

22. W. E. Harris and H. W. Habgood, *Programmed Temperature Gas Chromatography*, Wiley, New York, 1966.

23. L. Mikkelsen, *Adv. Chromatogr. N.Y.* **1966**, *2*, 337.

24. J. C. Giddings, *J. Chem. Educ.* **1962**, *39*, 569.

25. L. A. Jones, S. L. Kirby, C. L. Garganta, T. M. Gerig, and J. D. Mulik, *Anal. Chem.* **1983**, *55*, 1354.

26. Anonymous, *Packed Column Application Guide*, Bulletin 890, Supelco, Bellefonte, PA., **1995**.

27. J. V. Hinshaw, *LC-GC* **1990**, *8*, 362.

28. X. Zhou, P. A. Hines, K. C. White, and M. W. Borer, *Anal. Chem.* **1998**, *70*, 390–394.

29. M. D. Spangler and M. B. Sidhom, *J Pharm. Biomed Anal.* **1996**, *15*, 139–143.

30. D. W. Foust and M. S. Bergen, *J. Chromatogr.* **1989**, *469*, 161–173.

31. B. S. Kerstern, *J. Chromatogr. Sci.* **1992**, *30*, 115–119.

32. A. P. Micheel, C. Y. Ko, and C. R. Evans, *J. Pharm. Biomed. Anal.* **1993**, *11*, 1233– 1238.

33. I. D. Smith and D. G. Waters, *Analyst* **1991**, *116*, 1327–1331.

34. Document # 30.03.028A, *Determination of Residual Solvents in Thymopentin Bulk Drug Substance*, IRI, Division J&J. 1995.

35. T. P. Wampler, W. A. Bowe, and E. J. Levy, *J. Chromatogr. Sci.* **1985**, *23*, 64–67.

36. A. Tinto, *Pharmaeuropa* **1992**, *4*, 150–154.

37. M. V. Russo, *Chromatographia* **1994**, *39*, 645–648.

38. R. B. George and P. D. Wright, *Anal. Chem.* **1997**, *69*, 2221.

39. P. Klaffenbach, C. Bruse, C. Coors, D. Kronenfeld, and H. Schulz, *LC-GC* **1997**, *15*, 1052.

General References

GC

R. Grob (editor), *Modern Practice of Gas Chromatography*, 3rd edition, Wiley, New York, 1995.

W. Jennings, E. Mittlefehldt and P. Stremple, *Analytical Gas Chromatography*, 2nd edition, Academic Press, 1997.

H. M. McNair and J. M. Miller, *Basic Gas Chromatography*, Wiley, New York, 1998.

GC/MS

F. G. Kitson, B. S. Larsen, and C. N. McEwen, *Gas Chromatography and Mass Spectrometry: A Practical Guide*, Academic Press, San Diego, 1996.

M. McMaster and C. McMaster, *GC/MS: A Practical User's Guide*, Wiley, NY, 1998.

G. M. Message, *Practical Aspects of Gas Chromatography/Mass Spectrometry*, Wiley, New York, 1984.

M. Oehme, *Practical Introduction to GC-MS with Quadrupoles*, Wiley, New York, 1999.

Derivatization

K. Blau and J. M. Halket, *Handbook of Derivatives for Chromatography*, Wiley, New York, 1993.

Handbook of Derivatization, Pierce Chemical Co., Rockford, IL.

G. Simchen and J. Heberle, *Silylating Agents*, Fluka Chemical Corp., Ronkonkoma, NY, 1995.

Headspace Sampling

K. Bruno and L. S. Ettre, *Static Headspace Gas Chromatography: Theory and Practice*, Wiley, New York, 1997.

J. V. Hinshaw, *LC-GC* **1989**, *7*, 904.

J. Pawliszyn, *Solid Phase Microextraction: Theory and Practice*, Wiley, New York, 1997.

Z. Zhang, M. J. Yang, and J. Pawliszyn, *Anal. Chem.* **1994**, *66*, 844A.

9

LIQUID CHROMATOGRAPHY: BASIC OVERVIEW

Lee N. Polite

9.1. INTRODUCTION

9.1.1. Importance of HPLC in the Pharmaceutical Industry

We have established that high-performance liquid chromatography (HPLC) is the most popular instrumental technique in the pharmaceutical industry. But why is HPLC so much more popular than gas chromatography (GC) for pharmaceutical analyses? The answer comes from the underlying operational requirements of the two techniques. GC requires all analytes to be volatile, while HPLC only requires the solubility of the analytes in the mobile phase (MP).

Volatility is inversely related to both the molecular weight and the polarity of the analytes. In the pharmaceutical industry, many analytes tend to be highly polar, water-soluble molecules with relatively high molecular weights. The highly polar nature of pharmaceutical molecules is not a coincidence, but due to the polar, water-based matrix of the intended recipient (the human body). Therefore, an HPLC system with a polar mobile phase (water, acetonitrile, and/or methanol) is ideally suited for pharmaceutical analysis.

Figure 9.1 shows a classification of chromotographic methods. The two

Analytical Chemistry in a GMP Environment. Edited by J. M. Miller and J. B. Crowther
ISBN 0-471-31431-5 © 2000 John Wiley & Sons, Inc.

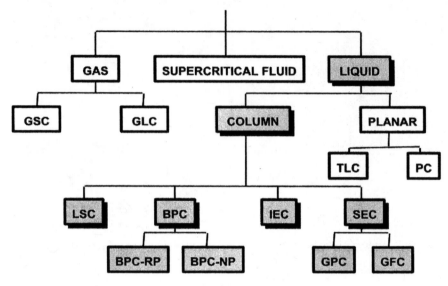

Figure 9.1. Classification of chromatographic techniques.

main types under liquid chromatography (LC) are column and planar. Most of the classifications are denoted by acronyms, and they will be defined and described in this section.

9.1.2. Column Versus Planar Liquid Chromatography

Most of the discussion in this text focuses on the use of column liquid chromatography (especially HPLC). However, there is another approach to LC—*planar chromatography*. Planar chromatography includes the techniques of paper chromatography (PC) and thin-layer chromatography (TLC). These two techniques are generally portrayed as old-fashioned approaches. While PC is rarely used today, TLC provides useful information in the high-tech pharmaceutical laboratory. Both PC and TLC rely on the same separation mechanism as HPLC. They are techniques that easily can be used in such applications as monitoring organic reactions for starting material, intermediates, and final products. They also can be used as a quick screening techniques to search for known impurities. A complete library of the most current HPLC methods for analyzing most pharmaceuticals in use today has recently been published in both print and CD-ROM versions [1].

9.1.3. Low-Pressure Versus High-Pressure Liquid Chromatography

Low-pressure, open-column chromatography was the original approach developed by Mikhail Tswett in 1906. This approach is still used today, primarily in the biological laboratory for the purification of samples that require only

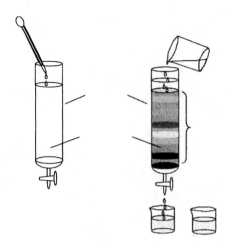

Figure 9.2. Low-pressure LC schematic.

modest resolution. The technique usually employs a large-diameter column (>1 cm ID) packed at ambient pressures with relatively large particles (>37 μm) and little other instrumentation. (See Figure 9.2.) The result of this approach is a loosely packed (low-resolution) column with low pressure requirements. The advantages and disadvantages are summarized in Table 9.1 and 9.2, respectively.

HPLC has the advantages of the low-pressure techniques but avoids their limitations by mechanizing the process. HPLC is designed around a column that has been tightly packed (>6000 psi) with small particles (<10 μm). This column design results in high-resolution separations but adds demanding instrumental requirements to the system. The tightly packed column does not allow for the reliance on gravity for mobile-phase flow. This results in the need for a high-pressure pump and, accordingly, a specialized injection device that isolates the sample (ambient pressure) from the high-pressure flow (up to 5000 psi). After the column, an on-line detector is introduced, allowing for the continuous monitoring of the separation. Finally, the components are connected with low-volume tubing and connectors in order to minimize the dispersion of

Table 9.1. Advantages of Low-Pressure Liquid Chromatography

Advantage	Explanation
Low cost	No expensive instruments to buy or maintain.
Simple to use	No complicated components to assemble.
High mass throughput	Large amount of stationary phase handles a high quantity of sample.

Table 9.2. Disadvantages of Low-Pressure Liquid Chromatography

Disadvantages	Explanation
Slow analysis	Flow through the column is limited to the force of gravity.
Low resolution	Large particles and low-density packing process results in low efficiency.
Difficult quantitation	Lack of on-line detector ⇒ must collect and analyze each fraction.

the sample as it travels through the system. A schematic of a complete instrument is shown in Figure 9.3.

9.1.4. Advantages and Disadvantages of HPLC

Clearly, HPLC has important advantages over the low-pressure method, including the possibility of performing an analysis with a gradient. These advantages are summarized in Table 9.3, and Figures 9.4 to 9.6 clearly illustrate the high speed, resolution and sensitivity, respectively. The main disadvantage at the present time is the lack of a detector which is universal and sensitive as listed in Table 9.4.

9.1.5. Isocratic Versus Gradient Elution

Isocratic elution refers to the technique of using constant solvent composition throughout the chromatographic analysis. During gradient elution, the MP is changed from a "weak" to a "strong" solvent during the analysis. Gradients

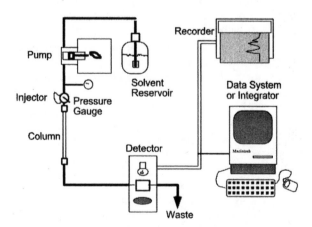

Figure 9.3. HPLC schematic.

Table 9.3. Advantages of HPLC

Advantage	Explanation
High speed	Analysis times measured in minutes or seconds.
High resolution	Columns tightly packed with small, uniform particles.
High sensitivity	Parts-per-million (ppm) to sub-parts-per-billion (ppb) detection limits.
High accuracy	High-precision sampling devices and good standards yield accurate numbers.
Automated systems	Unattended operation, from sample preparation to report generation.

are generally chosen for samples with large numbers of components or for those in a dirty or unknown matrix.

If a sample contains analytes that have widely divergent affinities for the column, a gradient is useful in shortening the analysis time and improving the shape of the peaks. Figure 9.7 illustrates a typical isocratic analysis of a complex mixture. The first few peaks elute too close to the void volume of the column. This suggests that the mobile phase is too strong. Also, the last peaks are short and broad with very long retention times. This suggests that the mobile phase is too weak. The solution to these problems is to begin with a weaker solvent and gradually strengthen the solvent throughout the course of the analysis. This is the definition of a gradient. Figure 9.8 shows that with the gradient, the resolution of the early eluting peaks is improved and the widths of the later peaks have been decreased while their heights have increased. The overall gradient separation yields more consistent peak widths, improved sensitivity, and shorter analysis times.

The components of the gradient for the MP can be binary or tertiary. The starting composition should be weak, which means that it should be very different in polarity from stationary phase (SP). In reversed-phase HPLC (to be

Table 9.4. Disadvantages of HPLC

Disadvantage	Explanation
Expensive instrumentation	Typical HPLC systems $30K to $50K.
Experience required	Complex three-way chemical interactions (analyte $\leftrightarrow$ SP; SP $\leftrightarrow$ MP; MP $\leftrightarrow$ analyte) and complex instrumentation.
No universal and sensitive detector	Maybe someday LC/MS will fill this void.
Expensive supplies	Columns, fittings, and consumables are expensive.
Requires spectroscopy for confirmation	Retention time characteristic of compound, not unique to that compound $\Rightarrow$ Need extra information (spectroscopic) for identity confirmation.

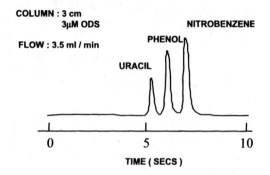

Figure 9.4. Example of high speed HPLC (small particles, short column, fast flow rate).

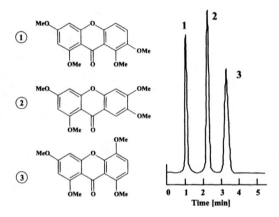

Figure 9.5. Example of high resolution HPLC.

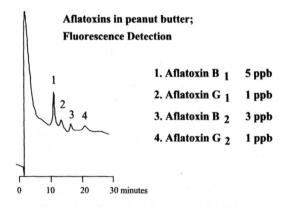

Figure 9.6. Example of high sensitivity HPLC.

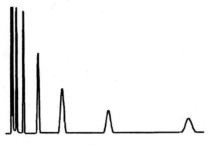

Figure 9.7. Typical isocratic separation.

discussed shortly), the SP is nonpolar, so the starting MP should be very polar. During the gradient rum, the MP is made "stronger" by increasing the proportion of the less polar component (usually acetonitrile, methanol, or THF). The opposite is required for a normal phase gradient; that is, the proportion of the polar solvent is increased during the run.

9.2. COLUMN METHODS

9.2.1. Normal Phase

Normal phase, defined as a polar stationary phase with a nonpolar mobile phase, is designated as such because it was the first phase discovered. Normal phase (NP) is typically used for nonpolar to semipolar analytes. It is not generally chosen for polar analytes due to their high affinity for the surface, resulting in unacceptably long retention times. In general, in HPLC the samples

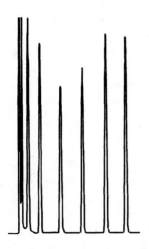

Figure 9.8. Typical gradient elution separation.

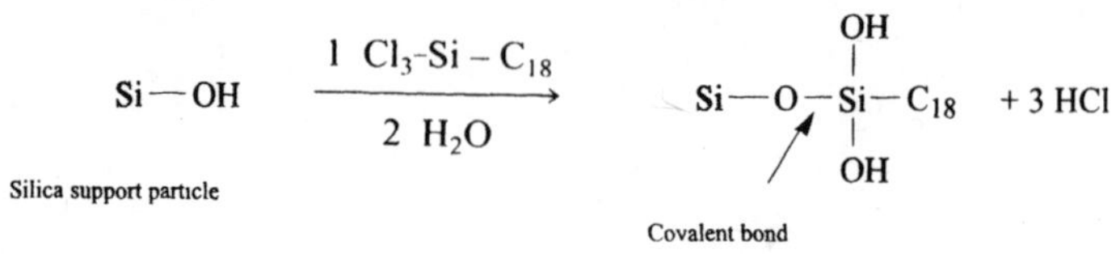

$$Si-OH \xrightarrow[\text{2 } H_2O]{\text{1 } Cl_3\text{-}Si-C_{18}} Si-O-\overset{\displaystyle OH}{\underset{\displaystyle OH}{\overset{|}{\underset{|}{Si}}}}-C_{18} \quad +3\ HCl$$

Silica support particle

Covalent bond

• **Covalent attachment of the stationary phase yields a thermally and hydrolytically stable *bonded phase*.**

Figure 9.9. Reactions for producing bonded C18 Phases.

have polarities more like the MP than the SP. This is just the opposite of the situation in GC.

There are two types of stationary phases used in the normal-phase mode: either raw silica gel [liquid–solid chromatography (LSC)] or a modified silica surface [bonded-phase chromatography (BPC)]. LSC is also referred to as adsorption chromatography (see Chapter 7). The mechanism of separation is the reversible adsorption of the analytes on the polar, weakly acidic surface of silica gel. The SP is high-surface-area silica, and the MP is a nonpolar solvent with a polar modifier added to control the separation. LSC is especially useful for the separation of positional isomers and "group-type" analyses (i.e., the separation of saturated, unsaturated, and aromatic hydrocarbons).

BPC is a newer approach to normal phase HPLC. In this approach, silica is used as a support material while polar functionalities are covalently bonded to the surface. Typical BPC normal-phase columns include cyanopropyl, aminopropyl, and diol phases. One might question why it is useful going through all of this chemistry to modify a polar surface just to end up with another polar surface. The reason is that this chemical modification yields a column that is durable and more stable than the original silica. Raw silica suffers from the surface activity being modified by adsorbed water. Even the humidity changes in the laboratory can lead to retention time shifts on raw silica. Therefore, it is usually recommended that raw silica or LSC methods be transferred onto a BPC packing such as cyanopropyl.

9.2.2. Reversed Phase

Reversed phase (RP) is designated as such because the polarities of the mobile and stationary phases are reversed, as compared to normal phase. Therefore, reversed-phase HPLC is defined as a nonpolar SP (i.e., C18) and a polar MP (i.e., methanol or acetonitrile and water). RP can be applied to the determination of polar, semipolar, and even nonpolar analytes. RP is the most popular mode of HPLC. The reasons for its popularity include its versatility and ability to handle polar analytes. The importance of polar compounds to the pharmaceutical industry was previously noted, but the fact that RP can handle such a wide variety of compounds makes it an especially useful technique.

Most RP packings are bonded (BPC) as just described. They are made by chemically bonding a hydrophobic molecule onto the silica surface. Figure 9.9 illustrates the bonding process of a trichlorooctadecylsilane onto the silica surface. The one chlorine atom acts as a good leaving group. It joins the H from the Si–OH (silanol) surface to leave as HCl. This results in a thermally and hydrolytically stable Si–O–Si bond, thus modifying the polar silica surface into a nonpolar C18 surface.

During the first level reaction, a certain percentage of silanols are left unreacted due to steric hindrance of the large C18 chains. These residual silanols are especially undesirable when acidic or basic compounds are the analytes. The silanols are often referred to as "hot spots" or "active sites" due to their polar, weakly acidic nature. Therefore, acids and bases may adsorb due to hydrogen bonding and acid–base interactions, resulting in tailing peaks. Bases are especially vulnerable to this interaction, which is why column manufacturers offer special RP columns referred to as "base-deactivated." See Chapter 10 for more discussion about deactivated columns.

The process of deactivation is complex, variable, and usually proprietary, but one of the most widely used approaches known as "end-capping" is worthy of noting. This technique utilizes the same chemistry as described above to react a C18 with the silica surface. But instead of using the trichlorooctadecylsilane, a trimethylchlorosilane (a silicon atom with three methyl groups and one chlorine) is used. In this case, employing a much smaller molecule minimizes the steric hindrance. This allows the trimethyl group to find its way to the surface and "cap-off" the remaining unreacted silanols. The resulting columns are called "end-capped" columns and are sometimes more desirable for analyses involving acids or bases as analytes.

The MP in RPLC is usually buffered to control the elution behavior of the solutes, many of which have acid/base functional groups. Normally the pH of the buffer is chosen relative to the solute pK values. Because stability is desired, the pH is ± 2 units from the pK so that the solute's ionization is fixed and not likely to change. If a pH is chosen which results in the formation of ions, the ionic solutes would be expected to elute very early, which can be a problem in the separation. In this case the techniques of ion pair chromatography can be useful (see the discussion later in this chapter). For the separation of ions, ion-exchange chromatography is preferred.

9.2.3. Ion-Exchange Chromatography

Ion-exchange chromatography (IEC) is an HPLC separation technique that takes advantage of the charge on analytes. Typically, one thinks of ion exchange as the mode of choice when dealing with inorganic anions (i.e., fluoride, chloride, nitrate, sulfate, and phosphate) or inorganic cations (i.e., lithium, sodium, ammonium, potassium, magnesium, and calcium). But ion exchange is also useful for organic compounds that can be ionized. Essentially, any organic acid or base can be ionized into its corresponding anion or cation by adjusting the pH of the mobile phase. Organic acids can be ionized into anions simply by raising the pH, thus removing a hydrogen ion and leaving a net negative charge on the molecule:

$$RCOOH \Leftrightarrow RCOO^- + H^+ \tag{9.1}$$

Likewise, organic bases can be ionized into cations by lowering the pH in order to protonate the molecule, leaving a net positive charge:

$$RNH_2 + H^+ \Leftrightarrow RNH_3^+ \tag{9.2}$$

Once the organic molecule has been ionized, it behaves just like an inorganic anion (for acids) or inorganic cation (for bases).

Now that our analytes have been converted into charged species, a stationary phase of the opposite charge is introduced. A positively charged stationary phase is chosen for anions, and a negatively charge stationary phase is chosen for cations. The stationary phases for IEC (called resins) are typically based on a copolymer of polystyrene-divinylbenzene. This polymer support material is then chemically modified to impart a positive or negative charge. The separation mechanism for ion exchange is the reversible adsorption of the charged analytes on a stationary phase of opposite charge. The ions are then selectivity removed from the stationary phase by a mobile phase that contains ions of the same charge as the analytes. This "competition" for sites is the basis of the separation in ion exchange. Therefore, the technique takes advantage of the coulombic interaction (attraction of opposite charges) between the analytes and the stationary phase [2].

9.2.4. Ion Chromatography (IC)

IC is an ion-exchange technique invented by Dow Chemical Company scientists Small, Stevens, and Bauman in 1975 [3]. Technically, IC is an ion-exchange separation followed by a chemical suppression of the mobile phase in order to allow detection with a conductivity detector. The use of a conductivity detector in ion exchange is desirable due to its universal response for ions. If an analyte is ionic enough to be separated by this approach, it will be detected. Conductivity detectors also are fairly simple in design, durable, and very sensitive. The only problem preventing their widespread use is the fact that the MP is also ionic. The ionic character of the MP is necessary for the separation because it competes for the same charged sites as the analytes, but the ionic character is undesirable when the MP reaches the conductivity detector. An ionic MP would produce an unacceptably high background in the detector, leading to high background noise and low sensitivity.

Mobile Phase Reaction			
$Na^+ OH^- + Resin^- H^+$	$\rightarrow$	$HOH + Resin^- Na^+$	$\rightarrow$ Detector
High Conductivity NaOH Enters	$\rightarrow$	Near Zero Conductivity Water Elutes	
Analyte Reaction			
$Na^+ Cl^- + Resin^- H^+$	$\rightarrow$	$HCl + Resin^- Na^+$	$\rightarrow$ Detector
Lower Conductivity NaCl Enters		Higher Conductivity HCl Elutes	

Figure 9.10. Suppression reactions used in ion chromatography (IC).

The elegant solution developed by those scientists was to simply follow the separator column with a "suppressor" column of opposite charge to the separator column. For example, in anion chromatography the separator is positively charged. The anion suppressor column is simply a high-capacity cation exchange column (negatively charged). The suppressor column removes the counterions, thus removing the conductivity. In the anion exchange example shown below, the MP is NaOH. The OH^- is the ion responsible for the separation, and the Na^+ is the counterion. If the MP were to enter the conductivity detector, the NaOH would produce an unacceptably high background. If, however, the MP (and column effluent) is first passed through a negatively charged cation exchange column in the hydrogen form (suppressor), the Na^+ would be exchanged for H^+, and the OH^- would join the H^+ to form HOH or pure water. Some commonly used suppression reactions shown in Figure 9.10 ensure that the ionic character of the MP is present for the separation but is removed prior to the detector.

The added feature of this procedure is what happens to the analytes. In this case the anionic analytes undergo the same exchange in the suppressor. The analyte, therefore, will travel through the detector as its H^+ salt instead of as its Na^+ salt. The resulting signal is approximately 350% greater due to the higher mobility of the H^+ ion as compared to the Na^+ ion. The ultimate result of this technology is high efficiency (>5000 theoretical plates), excellent linearity (six orders of magnitude), and high-sensitivity detection (parts-per-billion by direct injection). Figure 9.11 illustrates an impressive separation of 34 anions in 15 minutes.

9.2.5. Ion Pair Chromatography (IPC)

While IC requires a specialized column and potentially a dedicated system, IPC may be performed with existing reversed-phase columns and equipment. The steps involved in IPC are straightforward. The first step is to adjust the pH to ionize the analytes. This generally is done by adjusting the pH at least 2 units

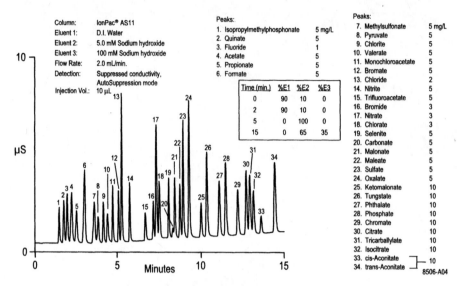

Figure 9.11. Ion-exchange separation of anions.

above the pK_a for acids, or 2 pH units below the pK_b for bases. The next step is to add an oppositely charged modifier to the aqueous portion of the MP. This modifier is referred to as the *ion pair reagent*. Typical ion pair reagents are (a) quartenary ammonium salts for negatively charged analytes and (b) alkyl sulfonic acids for positively charge analytes.

Once the ion pair reagent has been added to the MP, separation is carried out under typical reversed-phase conditions (C8 or C18 column, methanol/water or acetonitrile/water MP). The separation mechanism is a combination of analyte complexation and stationary-phase modification. The analyte complexation is due to the coulombic attraction of the oppositely charged analyte and ion pair reagent molecule. Once the analyte is introduced into the MP, the analyte complexes with the oppositely charged ion pair reagent, forming a neutral complex. This complex then travels through the column, as would any neutral compound.

The stationary-phase modification occurs due to the alkyl portion of the ion pair reagent partitioning into it. This leaves the charged head group exposed to the analytes, thus creating a pseudo ion exchange column. It is not clear which mechanism dominates, but the technique works.

9.2.6. Size Exclusion Chromatography (SEC)

SEC is a technique that separates analytes based solely upon their size in solution (hydrodynamic volume). It is common, but not entirely correct, to refer to SEC separations as "molecular-weight separations." The actual mechanism is

- Smaller molecules penetrate smallest pores, retained longest
- Larger molecules EXCLUDED
- Separation on basis of HYDRODYNAMIC VOLUME, not MW

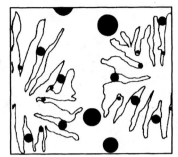

Figure 9.12. Schematic representation of the SEC mechanism.

quite straightforward. It is the exclusion of the *solvated* molecules from the pores of the stationary phase. Therefore the larger molecules elute early, and the smaller molecules elute late. The elution orders and even retention times are predictable due to the fact that, ideally, there is no sample–stationary-phase interaction (i.e., no adsorption, partitioning, ion exchange, hydrophobic interaction, etc.). Figure 9.12 illustrates the separation mechanism.

The large circles represent the largest molecules in the sample. These molecules are so large that they are unable to enter *any of the pores*. Therefore, they travel the shortest distance through the column and elute first. We call this the total exclusion limit of the column. Any molecule this size or larger will elute at the same time. The very small circles at the bottom of the pores represent the smallest molecules in the sample. These molecules are so small that they travel to the *bottom of each pore*. Therefore they travel the longest distance through the column and elute last. We call this the total permeation limit of the column. Any molecule this size or smaller will elute at the same time. The time between the total exclusion and total permeation limits of the column is referred to as the selective permeation area of the column.

There are two different types of SEC: gel permeation chromatography (GPC) and gel filtration chromatography (GFC), but their mechanisms are identical (see Table 9.5). The advantages and disadvantages of SEC are summarized in Table 9.6.

Table 9.5. Differences Between GPC and GFC

	GPC	GFC
Mobile phase	Organic solvents (THF, methylene chloride, etc.)	Aqueous buffers
Stationary phase	Rigid, cross-linked polystyrene-divinyl benzene	Soft, hydrophilic gel (Sephadex®)
Samples	Synthetic polymers (polystyrene, polyethylene, polypropylene, etc.)	Water-soluble biopolymers (proteins, peptides, oligonucleotides, etc.)

Table 9.6. Advantages and Disadvantages of SEC

Advantages	Disadvantages
Handles high-molecular-weight samples with short run times.	Limited peak capacity due to low resolution of SEC technique
Predictable separation times and elution order.	Cannot resolve similar sized compounds (~10% difference in molecular weight is required)
No on-column sample loss or reaction.	Different optimization strategy needed (pore size of column most important parameter).
Simple method development, no gradients.	Must be able to dissolve the sample.

9.3. PLANAR METHODS: TLC AND PC

9.3.1. Quick and Dirty Procedures

As we mentioned earlier, TLC and PC are generally thought of as old-fashioned versions of HPLC. Although PC is rarely used in the modern pharmaceutical laboratory, TLC is still a viable approach for the "quick and dirty" measurements, such as monitoring of organic reactions. TLC has the advantage of being much less expensive than HPLC and potentially faster for multiple samples.

9.3.1.1. Plates. TLC is similar to HPLC in that the separation depends upon the sample analytes distributing themselves between the MP and the stationary phase. In the case of TLC, the stationary-phase particles are coated as a thin layer onto a flat plate of glass, plastic, or aluminum. The most popular mode of TLC is normal phase, and the stationary phase is generally raw silica, although other materials are available. The MP is nonpolar (like hexane) with polar additives to adjust the elution strength.

9.3.1.2. Sample Application. Typical TLC applications may only require qualitative or semiqualitative information about the sample. Therefore, samples are not accurately quantitated, but simply "spotted" onto a plate with a small glass capillary or miscrosyringe. Sensitivity can be enhanced by spotting the sample several times onto the same spot. Semiquantitative analysis is further discussed in Section 13.3.7.1.

9.3.1.3. Visualization of Spots. There are several approaches for detecting the resultant "chromatogram" (spots). The simplest approach, which only works for colored compounds, is simply to use the naked eye to find the individual components. Colorless samples are often visualized with iodine vapors.

A second approach is to shine an ultraviolet (UV) lamp on the plate. If the components absorb UV light, they will appear as dark spots on the plate. If the compounds fluoresce, they will appear bright and colored. Alternatively, plates containing a fluorescent material can be used to visualize nonfluorescent solute spots.

Another approach is to spray the plate with sulfuric acid and bake the plate in an oven. This technique will "char" the organic compounds, resulting in black spots wherever an organic analyte elutes. There are yet other approaches involving, for example, coating the plate before or after the separation in order to derivatize the silica or the analytes.

9.3.2. Automation and Special Equipment

The TLC process can be made more reproducible and less labor-intensive by automating some of the steps. The sample application can be automated through the use of an autosampler that deposits a known amount of the sample onto a particular area or "lane" of the plate. The added advantage to this approach is the improved quantification.

One can also automate the detection portion of the procedure. Instead of simply looking at the resultant plate, a densitometer can be used to provide quantitative information about the position and intensity of the individual spots. A densitometer shines UV light onto the plate and then scans it, measuring the exact amount of light reflected by each small area of the plate. Therefore, if a sample component absorbs UV light, the reduction of the amount of light reflected will be proportional to the concentration of the analyte. The output of a densitometer resembles a chromatogram, with the x axis being distance from the origin (providing the qualitative information) and the y axis being intensity of light absorbed (providing the quantitative information). These data can be computerized and then handled the same as HPLC data.

9.3.3. High-Performance Thin-Layer Chromatography (HPTLC)

HPTLC is the ultimate form of TLC, combining the high-resolution advantages of HPLC and the multiple simultaneous sample capacity of TLC. The high resolution is accomplished by coating the HPTLC plates with HPLC packing material. These particles are much smaller with a tighter size distribution (5 ± 0.5 μm as compared to 40 ± 4 μm) as compared to traditional TLC. In order to take advantage of these better plates, the spotting process must be automated in order to provide narrower initial bands and reproducible sizes. The detection process is also automated with a high-resolution densitometer.

9.3.4. Advantages and Disadvantages of TLC

The advantages and disadvantages of TLC for the quick-and-dirty technique as usually practiced in the United States are listed in Table 9.7.

Table 9.7. Advantages and Disadvantages of TLC

Disadvantages	Advantages
Flow not constant and usually not controlled.	Inexpensive; no capital expenses.
Efficiency limited to several hundred plates.	All analytes are on the plate and detectable.
Temperature and solvent gradients exist.	Multiple samples run simultaneously.
Less accurate quantitation.	Shape can vary; two-dimensional possible.
Slower analysis (if only one sample).	Faster analysis (for multiple samples run simultaneously).
SP exposed to the elements and may require conditioning (drying).	Easy to learn and use.

9.4. USP

In the chromatography ⟨621⟩ section of the *United States Pharmacopeia* (USP), there are separate sections devoted to paper chromatography (PC), TLC, continuous development TLC, column chromatography, and HPLC. In general they discuss some definitions, apparatus, and procedures.

There is also a table listing 34 packings for LC. They are given in Appendix IV along with the GC lists. Some commercial equivalents are included in this Appendix, but it is impossible to include all possible manufacturers, so a few of the major ones have been used as examples. No endorsement of these companies over the competition is intended and, while we apologize that many are omitted, we do feel it is valuable to provide some information for the novice.

9.5. INSTRUMENTATION FOR HPLC

9.5.1. Pumps

In order for the theory of HPLC to become reality, stringent requirements are put on the individual components. The pump is given the difficult task of providing the force necessary to push the MP through the tightly packed column. The pressure requirement may be as high as 6000 psi due to the small-diameter packing material and length of the column. In addition to providing the force, the pump must also maintain an accurate and constant flow of MP in order to provide reproducible retention times. Finally the pump must deliver accurate MP compositions to establish the correct eluent strength for the HPLC separations. The pump accomplishes these tasks through a variety of designs.

Most HPLC pumps are based upon the reciprocating piston design. Although single-headed reciprocating piston pumps are still available, the majority of pumps employ one of two versions of a dual-headed reciprocating piston

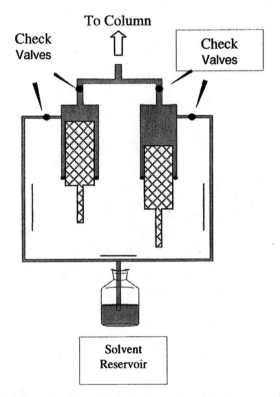

Figure 9.13. Schematic of dual-headed reciprocating piston pump, parallel design.

design (Figure 9.13). The first style of dual-headed design places the two pump heads in parallel and operate 180 degrees out of phase, allowing one pump head to deliver high-pressure MP while the other pump head refills with MP.

Another popular pump design is the series dual reciprocating piston pump (Figure 9.14). In this design the pump heads are assembled in series, rather than parallel. The first piston is typically twice the volume of the second piston, allowing the first piston to deliver high pressure flow onto the column while simultaneously refilling the second piston. When the first piston is empty, the second piston assumes the function of delivering MP while the first piston refills. Once again this design results in a pump delivering nearly pulse free flow.

Small pulsations are difficult to eliminate due to the differences in pressure between the flowing stream (high pressure) and the contents of the pump head at the beginning of its outward stroke. Even though liquids are only slightly compressible, this pressure difference is enough to cause a slight change in pressure each time the pump changes from piston #1 to piston #2. These pulsations are further minimized through the use of mechanical or electronic pulse dampers. They are noticeable only under the highest sensitivity settings of the HPLC system.

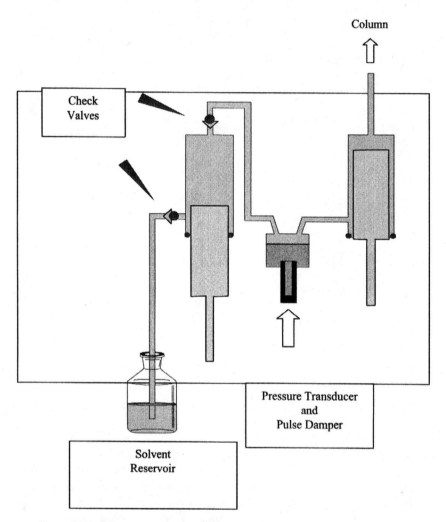

Figure 9.14. Schematic of dual-headed reciprocating piston pump, series design.

9.5.2. Sample Introduction Devices

HPLC presents more of a challenge than GC when it comes to introducing the sample. In GC, a small piece of rubber (the septum) is all that is required to isolate the sampling device (syringe) and the flowing stream. In HPLC the flowing stream is typically under several thousand pounds of pressure, and the flow path must be mechanically isolated from the sample. To accomplish this, a multiport switching valve is often used.

Whether the HPLC system has a manual or automated sampling system, the same approach is followed. The valve has two positions: load and inject. In the load position, the flow is diverted directly from the pump to the column, thus

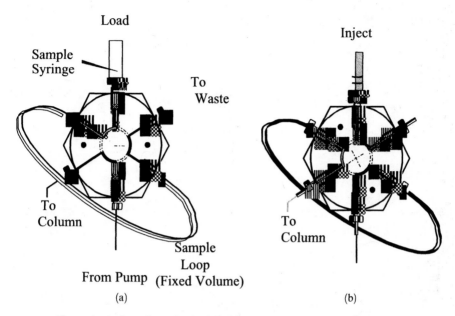

Figure 9.15. Sampling valve for HPLC. (*a*) Load position, (*b*) inject position.

leaving the sample loop at ambient pressure, as shown in Figure 9.15a. The sample loop is then loaded with sample via a manual syringe (manual injector) or through the use of an automated loading device (autosampler). The valve is then rotated into the inject position, redirecting the flow from the pump through the sample loop, and onto the column (Figure 9.15b).

In the case of manual injections, a fully loaded loop injection results in the highest reproducibility. This type of injection is accomplished by loading the sample loop with at least four times the volume of the loop. For example, a 20-μL loop should be filled with at least 80 μL. All of the excess volume goes to flush out the current contents of the loop and ensure a reproducible injection volume.

9.5.3. Tubing and Connectors

In order to minimize band broadening and maintain the efficiency of the column, the total volume of extra column tubing and connections must be minimized. The important section or "critical path" of the HPLC system includes everything from the injection loop to the detector cell. As the peaks travel along this path, they undergo longitudinal diffusion, leading to peak broadening and the loss of efficiency. If the peaks encounter any voids or unswept volumes, the peaks will tend to tail. In general, the length and volume of tubing and the number of fittings should be minimized in this critical path, and the total recommended length of tubing should not exceed that listed in Table 9.8.

Table 9.8. Maximum Allowable Length of Tubing[a]

Column Characteristics				Maximum Length of Tubing		
L (mm)	ID (mm)	d_p (μm)	N	0.007 in.	0.010 in.	0.020 in.
33	4.6	3	4,400	22	9	—
50	4.6	3	6,677	33	14	—
100	4.6	3	13,333	67	27	—
150	4.6	5	12,000	167	68	—
250	4.6	10	10,000	556	228	14
250	4.6	5	20,000	278	114	—

[a] In order to keep the extracolumn band broadening below 5%, the length of tubing used from the injector through to the detector must be less than the lengths listed in the table.

Source: J. W. Dolan and L. R. Snyder, *Troubleshooting LC Systems*, Humana Press, Clifton, NJ, 1989, p. 207.

9.5.4. Detectors

Detectors in general are discussed in Chapter 13, which should be consulted for background information and classification. The common HPLC detectors are listed in Table 9.9. They are not unique to HPLC, but are based on existing well-known principles of spectroscopy (see Chapter 6) and electrochemistry (see Chapter 5). To adapt them for use in HPLC, they need to be modified to accept small samples (small cell volumes required) and exhibit fast response times (small time constants).

Unfortunately, none of these detectors is both universal and sensitive as was noted earlier. The one that comes closest to meeting these requirements is the UV absorbance detector, and it is the most popular one. The mass spectrometer (MS) has ideal properties, but at present it is too expensive and complex for routine use.

Table 9.9. HPLC Detectors

Spectroscopic
Ultraviolet [visible (UV–VIS)] absorption
    Fixed-wavelength UV detector
    Variable-wavelength UV detector (VWD)
    Photodiode array detector (PDA), also called diode array detector (DAD)
Fluorescence
Mass spectrometer (MS)

Electrical
Electrochemical detector (ECD)
Conductivity detector

Other
Retractive index (RI)

9.5.4.1. Ultraviolet (UV) Absorbance Detectors. UV detectors are the most popular and are commercially available for HPLC use in a variety of configurations. Each of these will be discussed briefly.

Fixed-Wavelength UV Detector. The simplest version of the UV detector is the fixed-wavelength detector. This detector employs a UV light source—typically a low-pressure mercury vapor lamp. The mercury lamp provides several distinct lines of UV radiation, with the 253.7-nm wavelength (commonly referred to as 254 nm) being the most intense. The source light is then passed through a filter to remove the extraneous wavelengths.

Variable-Wavelength UV Detector. Because all compounds do not absorb UV light at 254 nm, variable-wavelength detectors are used to allow the option of choosing the wavelength. This is accomplished by adding a monochrometer to the detector design. The monochrometer starts with a continuum UV source, such as a deuterium lamp, producing a broad band of radiation from 190 to >800 nm. The light is then bounced off of a grating, which separates the light into a spectrum of its various wavelengths. The grating is placed on a moveable platform, allowing the user to chose any single wavelength from the spectrum.

Photodiode Array (PDA) Detector. The PDA detector takes the UV detector one step further than the variable-wavelength detector (VWD) by allowing the user access to all of the wavelengths simultaneously. This is accomplished by starting with a continuum source as in the VWD and passing the entire spectrum of light through the detector cell. The light is then bounced off of a grating as in the VWD. In this case the grating does not move and the single detector is replaced with an array of a multitude of individual detectors (photodiodes). These detectors are arranged on a single chip referred to as a *photodiode array*. A schematic is shown in Figure 9.16.

The advantage of this detector is that the user is given access to the entire UV spectrum all of the time. A variety of tasks can be performed such as peak purity (comparing UV spectra at various points along the peak), compound confirmation (adding spectral information to the retention time), and reprocessing any single wavelength as a chromatogram.

9.5.4.2. Fluorescence Detector. The fluorescence detector is one of the most selective and sensitive detectors in all of chromatography, allowing sub-ppb analyses. However, only a small percentage of molecules have the ability to fluoresce so its applications are limited.

9.5.4.3. Refractive Index (RI) Detector. RI detectors are the most universal and least sensitive of all the readily available HPLC detectors. They measure the minor refractive index changes that occur when the analyte concentration changes in the HPLC effluent. These detectors are universal because the RI of a solution will change if the temperature, density, or concentration changes.

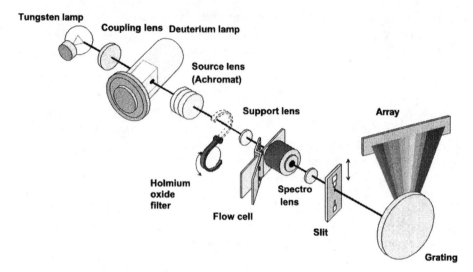

Figure 9.16. Schematic of a photodiode array UV detector.

Therefore, if any analyte passes through the detector cell, the change in RI will be recorded as a chromatographic peak. RI detectors are highly susceptible to slight variations in flow rate, detector cell pressure, and temperature. It is not unusual to wait several hours to allow an RI detector to reach temperature equilibration before beginning an analysis.

9.5.4.4. Electrochemical Detector (ECD). Along with the fluorescence detector, the ECD is one of the most selective and sensitive HPLC detectors. An ECD is chosen when the analytes can be oxidized or reduced at a reasonable voltage. To accomplish this oxidation or reduction (REDOX), the ECD employs three electrodes: the reference, counter, and working electrode. A potential difference is set between the working and reference electrodes. When an electroactive compound passes over the working electrode, oxidation or reduction takes place, resulting in the transfer of electrons between the working electrode surface and the analyte. This flow of electrons is measured as a current, and thus it forms the basis for the chromatographic peak. ECDs are commonly used in the determination of catacholamines at the low parts-per-billion (ppb) range.

9.5.4.5. Conductivity Detector. Conductivity is an extremely sensitive and universal detection mode for ions. This detector takes advantage of the ions' ability to carry a charge in solution. The conductivity is measured by placing two closely spaced electrodes in the flow path of the column effluent and measuring the resistivity of the solution. Some detectors also include a temperature-measuring device (thermister) and the associated circuitry needed to correct for the resistance variation due to slight temperature changes. Conductivity detectors combined with chemical suppression for use in IC have demonstrated sub-ppb detection limits and six orders of magnitude of linearity for common ions [4].

9.5.4.6. Mass Spectrometer (MS). LC/MS is regarded by some to be the most powerful analytical technique in the world. This fast-growing approach combines the versatility of the HPLC separation with the structural identifying power of mass spectrometry. The difficulty in mating these two techniques stems from the inherent incompatibility of a large liquid flowing stream (HPLC effluent) and the high vacuum requirements for the MS. This is a quickly evolving technique that is sure to have a major impact on the pharmaceutical laboratory, as well as on the rest of the field of analytical chemistry.

9.5.5. Troubleshooting

It is impossible to create an exhaustive list of all symptoms and problems that can occur in HPLC. There are, however, a number of common causes that account for a majority of the HPLC symptoms. Here is a list of those symptoms and the associated causes.

PROBLEMS

1. Periodic Pressure Fluctuation

Possible Cause: Leaky Check Valve. In order for the check valve to work correctly, the internal parts (ball and seat) must be able to form a tight seal. If foreign objects (air or dirt) are introduced into the check valve, it will leak. The leaks do not produce any visible signs such as drops of liquid, only regular or periodic pressure perturbations. After making sure that there is no source of air or dirt entering the pump, open the purge valve and purge the pump at a high flow rate. It may help to gently tap the check valves with a small wrench or screw driver. This will help to dislodge any trapped air bubbles. If this does not fix the problem, remove the check valve and either clean it through sonication in dilute nitric acid or replace it with a new check valve.

Another Possible Cause: Leaky Pump Seals. If the pump seals are leaking badly, liquid will be visible coming from the back of the pump heads. The seals have a limited lifetime and need to be replaced as a part of the routine preventative maintenance of the HPLC system. Although seals may last over 12 months, it is a good idea to replace them at regular intervals (every 6 months). Follow the manufacturer's direction for replacing the seals, along with any associated back-up seals, o-rings, or wear retainers.

2. High Pressure

There is a clog somewhere in the system. Disassemble the HPLC components starting with the last component—usually the detector. Keep removing components until the high-pressure problem goes away, then focus the troubleshooting efforts on that piece of equipment.

Possible Cause: Clog in One of the Following Filters:

1. Any filter in the pump (some pumps have a filter built into the purge valve)
2. High-pressure in-line filter (comes after the injector)
3. Inlet frit of the guard column
4. Inlet frit of the analytical column

Clean the filter by sonication for 10 min in dilute nitric acid (6 M) or replace the filter.

Another Possible Cause: Clogged Tubing Due to MP Buffer Precipitation. Check the entire injector and the tubing leading from the injector to the column. If you have been using buffer and have recently introduced a high concentration of organic solvent, the buffer may have precipitated in the lines. Wash the system with pure water. Start at a very low flow rate (0.1 mL/min) until the pressure decreases.

A Third Cause: The Sample or Buffer Has Precipitated in the Lines. Wash the system with warm solvent. Then make several large volume injections of that same solvent. This should dissolve any precipitate. It is "best practice" to dissolve the sample in the MP.

3. Low Pressure

Possible Cause: Leak. Concentrate on the high-pressure part of the HPLC. This includes everything from the pump heads to the top of the column. Find the leak (drips or puddle) and tighten or replace the fitting(s).

Another Possible Cause: No Solvent. This one happens to the best of us: You ran out of solvent! Refill the solvent container then purge the lines and pump with the fresh MP.

4. Poor Peak Area or Height Reproducibility

Possible Cause: Problem with the Autosampler. Concentrate you efforts on the injection system. There may be a clog in the syringe needle or a leaky line between the syringe (or metering device) and the sample vial. Replace the needle or tighten the leaky tubing.

5. Noisy Baseline. Evaluate further by shutting off the pump in order to rule it out as a cause. If the noise remains after the pump has been shut off, it is probably a problem with the detector.

Possible Detector Cause: Contaminated Detector Cell. Clean the cell with dilute (6 M) nitric acid. The is most easily accomplished with a 10-mL syringe and an adapter that allows the syringe to connect directly to the cell inlet. Inject 5 mL into the cell and allow it to stand for 10 min. Follow that with an additional 5 mL and rinse the cell with water. *Do not pump this Solution through your column.*

   If the baseline remains noisy, determine if the detector lamp needs replacement.

9.6. CAPILLARY ELECTROPHORESIS (CE)

All current forms of column chromatography (HPLC, GC, SFC, IC, GPC, etc.) are pressure-driven. This means that the techniques rely on a pressure differential across the column to drive the MP. This pressure-driven flow causes the undesirable effect of a parabolic flow profile. The parabolic flow profile is due to the resistance to flow along the walls of the tubing and causes peak spreading or band broadening. Even though we introduce the sample into the system as a narrow, straight band, the radial center of the band travels faster than the radial edges due to this phenomenon. Until recently, this undesirable effect of the parabolic flow profile has been is considered to be unavoidable. Capillary electrophoresis (CE) avoids this problem by driving the MP with a voltage differential instead of a pressure differential. This voltage differential results in electroosmotic flow (EOF). The first step in creating this type of flow is to start with a fused silica column. The surface of fused silica is similar to the silica support material used for HPLC columns, which means that the surface is covered with Si–OH or silanol groups. A basic buffer is passed through the column to remove the hydrogen ions, resulting in a net negative charge on the surface. The column is then filled with an aqueous buffer. The positive buffer ions align themselves with the negatively charged surface. When a high-voltage potential difference is applied across the column, the positive buffer ions move toward the negative electrode, dragging with them the aqueous buffer (MP). This establishes the EOF of MP through the column, resulting in a flat flow profile. Therefore, there is minimal band broadening due to the flow through the column. This ultimately results in a technique with the solvating advantages of HPLC and the capillary efficiency of GC. This technique in its simplest form is called capillary zone electrophoresis (CZE) and has already demonstrated efficiencies in excess of 1 million theoretical plates. This is compared to 20,000 for HPLC and 400,000 plates for capillary GC.

9.6.1. CE Systems

The CE instrument is similar to an HPLC. We start with a "pump." In the case of CE, the pump is actually a high-voltage power supply capable of delivering 30,000 V. The positive electrode is placed at the beginning of the system in a buffer reservoir. A piece of fused silica tubing (<75 μm ID) is placed in this reservoir. This end of the capillary also serves as the injection device when placed into a sample vial. The other end of the capillary is placed in a buffer container along with the negative electrode. When the voltage is applied, the resulting EOF drives the MP through the column.

There are several different modes of capillary electrophoresis that can be applied to a variety of separation problems. A few of the most popular techniques and their applications follow. Remember that all CE methods are not chromatrographic, because most CE methods do not have a stationary phase (SP).

9.6.1.1. Capillary Zone Electrophoresis (CZE). The EOF is adjusted so all analytes move toward the negative electrode. Even though negatively charged species will have the tendency to move toward the positive electrode, the EOF is adjusted to be greater than the negative flow resulting in a net flow toward the negative electrode. CZE, also called free zone electrophoresis, is the simplest form of CE and separates analytes based on their charge and size in solution. The higher the positive charge, the faster the analytes move toward the negative electrode. Also, the larger the molecule, the slower it travels due to the drag-induced resistance to flow. Therefore, positively charge analytes elute first (further separated by size), the negatively charge elute last (again further separated by size), but the neutral molecules all elute as a single peak. CZE is an electrophoretic method, not a chromatographic one.

9.6.1.2. Micellular Electrokinetic Chromatography (MEKC). Obviously a technique that cannot separate neutral molecules would have only limited application as an analytical technique. Neutral molecules can be added to the list of analytes by introducing a micelle to the buffer system. Micelles are formed when surfactants are added above their critical micelle concentration (CMC). The surfactants reach a high enough concentration that they start to interact with each other. The long alkyl chains group together, forming a nonpolar interior of the micelle. The polar head-groups form the hydrophilic exterior. As the analytes travel through the column, they will partition in and out of the micelle interiors, thus forming the basis of neutral separation. The micelles can be thought of as a "pseudo"-reversed-phase SP. The major difference is that the "stationary" phase is not stationary in the sense that a packed column contains an SP. However, it is classed as a chromatographic method, and the ultimate result is a separation of positives and negative analytes as described above for CZE, with the addition of neutral separation through micellular interaction.

9.6.1.3. Capillary Electrochromatography (CEC). If one uses a packed column in the CE mode, then the technique can be classed as a true chromatographic method, most commonly referred to as capillary electrochromatography (CEC).

CE is an evolving technique that has a great deal of potential (no pun intended). CE will not replace HPLC, nor will HPLC eliminate the need for CE. As CE continues to probe the advantages of EOF and HPLC continues to refine column technology and hardware, the two techniques will continue to complement each other.

Of the many references on CE, the recent one edited by Khaledi [5] contains separate chapters on CEC (Chapter 8) and the analysis of pharmaceuticals (Chapter 25). An older reference on the latter topic is worthy of note also [6].

REFERENCES

1. G. Lunn, *HPLC Methods for Pharmaceutical Analysis*, four volumes, Wiley, New York, 1999.
2. H. M. McNair and L. N. Polite, Recent Advances in Ion Chromatography, *Am. Lab.* **1988**, *20*(10), 116–121.
3. H. Small, T. S. Stevens, and W. S. Bauman, *Anal. Chem.* **1975**, *47*, 1801.
4. L. N. Polite, H. M. McNair, and R. D. Rocklin, Linearity in Ion Chromatography, *J. Liq. Chromatogr.* **1987**, *10*(5), 829–838.
5. M. G. Khaledi, *High Performance Capillary Electrophoresis*, Wiley, New York, 1998.
6. S. F. Y. Li, C. L. Ng. and C. P. Ong, in *Advances in Chromatography*, Vol. 35, P. R. Brown and E. Grushka (editors), Marcel Dekker, New York, 1995, Chapter 5.

General References

HPLC

B. A. Bidlingmeyer, *Practical HPLC Methodology and Applications*, Wiley, New York, 1993.

R. L. Cunico, K. M. Gooding, and T. Wehr, *Basic HPLC and CE of Biomolecules*, Bay Bioanalytical Laboratory, CA, 1998.

T. Hanai, *HPLC: A Practical Guide*, Springer-Verlag, 1999.

M. McMaster and C. McMaster, *HPLC: A Practical User's Guide*, John Wiley, New York, 1998.

V. R. Meyer, *Pitfalls and Errors of HPLC in Pictures*, John Wiley, New York, 1998.

V. R. Meyer, *Practical HPLC*, 3rd edition, John Wiley, New York, 1999.

L. R. Snyder, J. J. Kirkland, and J. L. Glajch, *Practical HPLC Method Development*, 2nd edition, Wiley, New York, 1997.

TLC

B. Fried and J. Sherma, *Thin-Layer Chromatography*, 4th Edition, Marcel Dekker, New York, 1999.

J. G. Kirchner, *Thin-Layer Chromatography*, in *Techniques of Chemistry*, 2nd edition, Vol. XIV, A. Weissberger (editor), Wiley, New York, 1978.

Capillary Electrophoresis

D. R. Baker, *Capillary Electrophoresis*, John Wiley, New York, 1995.

10

HPLC COLUMN PARAMETERS

Richard Hartwick

High-performance liquid chromatography (HPLC) was introduced in Chapter 9, including the classification of methods, the instrumentation, and the types of columns and packing. The basic theory of chromatography and many important equations were presented in Chapter 7. This chapter builds on these two chapters and provides more detailed information about many important aspects of HPLC. Because the column is at the heart of HPLC and because the selection of a column is a critical step, the material in this chapter is built around column considerations. The process of discussing columns will provide many opportunities to elaborate basic HPLC concepts of interest to pharmaceutical scientists.

Further justification for this approach is the fact that virtually every method of analysis in the pharmaceutical industry contains the sentence "Use column XXX, or *equivalent*." The answer to that directive is neither simple nor obvious, as will be shown in this chapter. Indeed, to make an informed decision as to what constitutes an equivalent column, one must utilize an appreciable depth of knowledge of the entire separation process. The purpose of this chapter, then, is to use the concept of column equivalency as a convenient framework within which to review those aspects of the HPLC separation process that are critical to successful daily bench work.

Ideally, the selection of an equivalent column should always take place during the method development cycle, while the chemist has available a variety of

Analytical Chemistry in a GMP Environment. Edited by J. M. Miller and J. B. Crowther
ISBN 0-471-31431-5 © 2000 John Wiley & Sons, Inc.

degraded samples representative of the worst-case separation problems. It is foolhardy to develop a new method around a *single* column type from a *single* vendor, without having an alternate, equivalent column identified. The history of HPLC is replete with examples of methods failing catastrophically when a vendor changed its column manufacturing process, without notifying the customer. Often, vendors themselves may not have sufficient control over their own processes to know when changes have occurred in their materials; and they will insist to the customer that the material is identical, when the separation clearly demonstrates it is not. In these situations, it is critical that a truly equivalent column had been previously identified and tested during method development. The alternative is chaos, as the laboratory scrambles to find a column that works, while the projects wait to be completed.

In practice, it is not uncommon to work with older methods for which the specified column is no longer available or for which the specified column is simply not working as it did originally (for a variety of reasons). This situation is another example of the need to identify columns that will give separations equivalent to those being achieved in a given method.

10.1. COLUMN EQUIVALENCY

The term "equivalent column" is in itself highly ambiguous. Any experienced chromatographer realizes that two nearly identical separations might at times be achieved on two columns of different particle sizes and lengths, for example. On the other hand, two columns that may have identical physical characteristics can produce wildly different separations. On an *operational level*, an equivalent column could be defined as one that produces results negligibly different from its antecedent column within the range of running conditions and samples for which it is intended. The keyword here is negligibly different. No two columns—whether of the same brand, type, or lot number—will be identical. Lot-to-lot variations of a given column can be severe, as all working chromatographers have discovered at some point in their careers. Thus, some type of reasonable acceptance criteria need to be set to decide whether a proposed column is equivalent to another.

On a *physical level* the selection of equivalent columns would appear straightforward. Simply choose another column of the same stationary phase type, particle size, and column length and diameter. In a superficial regulatory definition, columns with the same physical characteristics would appear to satisfy the regulatory demands of equivalency.

One faces the immediate choice then, of using columns that, operationally, produce the desired separation, or of choosing two columns that, technically, meet regulatory definitions, but which produce obviously different separations. Which choice should the chromatographer make?

The answer to this rhetorical question is obvious. The chromatographer must take both definitions into account, because the ultimate goal of this exercise is to find a second column that at once faithfully reproduces the desired separation, while not being radically different in its obvious physical parameters. No one would feel comfortable substituting a 5-cm, 3-μm C18 column for a 30-cm 10-μm C18 column used in an existing method and then calling these two columns "equivalent," even if the separations they produced happened to look very similar. The regulatory guidelines are virtually silent on the issue of what testing must be performed to prove legally that two columns are truly equivalent. When selecting an equivalent column, the burden of proof is on the analyst. Exactly what parameters one should investigate when trying to prove column equivalency will be discussed in more detail at the end of this chapter, but equivalent selectivity of the worst-case stressed samples is a good starting point.

Theoretically, if one had complete knowledge of (a) the most intimate physical and chemical aspects of a particular column and (b) their contributions to the separation, then choosing an alternate column that produced identical results to another would be a straightforward task. The fact that this is still not possible today underscores the lack of a true understanding of retention mechanisms in HPLC. Columns have improved enormously over the past 25 years, and the columns today must be regarded as generally very good. The so-called base-deactivated columns now offered by most vendors are really just regular HPLC columns, but with much cleaner, more uniform, and better-characterized silica gels and bonded phases as compared to older columns. However, until the chromatographic sciences can explain in detail the complex interactions occurring on the liquid chromatographic surface, the ultimate test for equivalency must remain at the bench, using our existing method and the worst-case samples we can find. Nevertheless, this does not mean that one cannot operate from a depth of understanding of the principles of chromatography in selecting either an equivalent column or a suitable column during the method development process.

To simplify this discussion, examples will be taken from the column type most frequently used in pharmaceutical laboratories—bonded reversed-phase on silica gel. These columns have a nonpolar moiety bonded to the solid support, usually silica. The most common one is a C18 hydrocarbon chain also known as ODS which stands for the octadecylsilyl group. It is important to distinguish between the term *bonded phase*, which describes how the stationary phase was prepared, and the term *reversed phase* (RP), which denotes that the stationary phase is nonpolar (and need not be bonded). Refer to Chapter 9 for more details.

10.2. REVIEW OF CHROMATOGRAPHIC PARAMETERS

Perhaps a good starting point in deciding upon equivalency is to ask the question, "What information is contained in a chromatogram? View the fictionalized chromatogram in Figure 10.1. Its characteristic features are listed in Table

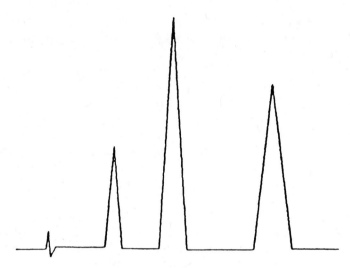

Figure 10.1. Typical fictionalized HPLC chromatogram.

10.1. Each of these characteristics is essentially completely independent from the others. That is to say, each of the physical and chemical processes leading up to each of these four characteristic features can be independently studied, understood, and optimized. The chromatogram, then, represents merely a composite representation of each of these processes, and in that sense it is not "real." The "real" events in a chromatographic separation are occurring on the column: the chromatogram is merely a time-based representation of those true events occurring within the column itself. This may sound pedantic; however, it is an essential starting point in one's dissection of a chromatogram, in order to better understand it.

Table 10.1. Characteristic Chromatographic Features

Feature	Expressed by the Parameter	Controlling Process
1. Retention or retardation	1a. Retention time, t_R 1b. Retention factor, k 1c. Retardation factor, R 1d. Separation factor, α	1. Thermodynamic
2. Peak width	2a. Plate number, N (efficiency) 2b. Plate height, H	2. Kinetic
3. Shape	3a. Symmetry or asymmetry 3b. Tailing factor	3. Differing rates
4. Response	4a. Peak height 4b. Peak area	4. Detector characteristics

Each of the parameters in Table 10.1 has been defined and discussed in Chapter 7. Now, we need to elaborate on them and apply them to the HPLC situation as we seek to define and characterize the equivalent column. We begin by reviewing the underlying physical–chemical processes behind each of the features. They are listed in the last column in Table 10.1; discussions of the first three can be found in Chapter 7, and a discussion the fourth can be found in Chapter 13.

While the first two features, retention and peak width, need no further elaboration, it is critical to remember that the two process of thermodynamics and kinetics are essentially independent from one another, where activation energies are not involved as limiting factors. In other words, one can manipulate and optimize the *retention* of a separation independently from the *efficiency* of a separation. This is accomplished every day at the bench by choosing a stationary phase type to produce a desired retention factor and then separately deciding upon the efficiency required to produce the necessary theoretical plates, by choosing particle size and column length.

The peak shape is a more complex problem, but is always related to differing rates of interactions for molecules within a column. A fairly broad definition of rates is used here, which can include both unequal flow properties, such as column channeling, and different rates of chemical interactions, as for example the effect of tailing of amine type compounds with acidic silanol sites on the silica gel. Whatever the cause though, *differing rates* of interactions will usually produce tailing or fronting.

Finally, the peak heights or areas—or, more generally, the responses of the molecules—are used for quantitative purposes, usually the overall goal of the separation. In chromatography (like all other analytical methods, with the exceptions of gravimetry and coulometry, which are absolute), it is not possible to quantitate without the use of reference standards. In HPLC, where optical absorbance detectors are widely used, the response of a given molecule will be a function of both its absorptivity and its concentration at the detector cell. Comparison against a known standard will yield the concentration of the unknown. The issue of peak response is really separate from the present discussion, and the principles of ultraviolet–visible (UV-VIS) absorbance spectrophotometer have been presented elsewhere in this book. Therefore, only the first three chromatographic features will be discussed further in this chapter.

10.3. PARAMETERS NECESSARY FOR EQUIVALENT COLUMNS

Given the information that a chromatogram contains, it is possible to define the basic chemical and physical characteristics that must at least be considered when investigating column equivalency. These are as follows:

1. Same selectivity (determined by chemistry of the system: stationary/ mobile phase *type*, silica gel *quality*)

2. Same retentiveness (determined within a phase type by bonded density and phase ratio)
3. Same efficiency (determined by particle size, column length, quality of packing)
4. Same peak shape (no additional tailing introduced)
5. When tested on the "worst case" samples, do they work?

The first four features defining equivalent columns form a necessary, but not sufficient, set of criteria. The bottom line for will always remain the operational definition of being able to do the job required of them (feature 5). However, it is difficult to argue that two columns are truly equivalent if the first four basic characteristics are not matched. Furthermore, in order for an analyst to work with the tool of HPLC in a deep and meaningful way, a full understanding of the theory involved in items 1–4 is essential. The remainder of the chapter will be devoted to understanding in detail each of these basic column characteristics, in preparation for selecting equivalent columns, which are then subjected to the final operational test: "Do they work?" The final section of the chapter will discuss what testing is required to prove the equivalency, or lack thereof.

10.3.1. Retentiveness and Selectivity

What gives rise to the general retentiveness of HPLC columns? Why should one C18 column produce retention times twice as long as another, ostensibly identical C18 column? As previously shown (Table 10.1), the controlling process is thermodynamic, and the parameters commonly used are t_R or V_R, k, and α.

Recall that the retention time is related to retention factors is shown in equation 10.1:

$$k = \frac{t_R - t_M}{t_M} = \frac{t'_R}{t_M} \tag{10.1}$$

Because the elution time for a nonretained peak is approximately (but certainly not exactly) constant during changes in mobile-phase composition, changes in retention time, t_R, are proportional to changes in k. Thus, either system of measurement is suitable. The retention factor has the advantage that it is independent of column length. However, it has the disadvantage that an accurate dead volume marker must be used. Routine pharmaceutical work is usually made more complicated by requiring that a certain dead volume marker be used for each method. In fact, the estimation of column dead volumes is a major issue in itself [1]. One can use markers ranging from those that are totally included within the stagnant intraparticle volume in the particles, as well as the interparticle volume between the particles (commonly called V_M), up to those totally excluded from the particle pores (often called V_0). Because on modern HPLC packing materials the ratio of V_M/V_0 is approximately 2, this means that capacity factors can vary by 100% or more, for a column and solute with a given retention time, depending upon the particular dead volume marker chosen. Obviously, if one is having problems trying to meet system suitability requirements for a method reporting retention factors, the nature of the dead volume marker should have been specified during method development and should be reported as part of the method. In too many cases, such methods are based upon a dead volume estimated by some type of disturbance peak(s) eluting somewhere near the supposed dead volume. These disturbance peaks may or may not be visible on other detectors. Given these problems, the use of retention times, rather than retention factors, is often a better way to specify system suitability requirements. In any event, the following discussion, which uses retention factor changes, obviously applies equally to retention times.

The starting point for a discussion of retentiveness can be the equilibrium distribution coefficient, or distribution constant, K_c (see Chapter 7):

$$K_c = \frac{C_s}{C_M} \tag{10.2}$$

where C_s is the equilibrium concentration of solute in the stationary phase and C_M is the equilibrium concentration of solute in the mobile phase. In chromatography, by convention, the stationary phase is always placed in the numerator, the mobile phase in the denominator. The distribution constant can be thought of as expressing the *chemistry* of the solute with the two-phase system. It relates the relative free energies of interactions (or more accurately, the chemical potentials) of a given solute in the stationary and mobile phases:

$$\Delta\mu_i^0 = -\mathscr{R}T \ln K_c \tag{10.3}$$

where $\Delta\mu_i^0$ is the difference of the standard-state chemical potentials of the solute in the two phase regions. It is similar to an electrical potential, because it expresses the driving force that forces the transfer of matter from one phase to another.

The distribution coefficient for a given solute solute is thus determined by the specific chemical interactions occurring between the solute molecule and each of the two phase regions, stationary and mobile. Anything that changes the energetics of these interactions will change K and, of course, the selectivity for two or more solutes. The energetic differences of, say, a dye molecule in an aqueous versus an organic solvent arise from a complex combination of hydrogen bonding, van der Waals, steric and many other factors. Even though it may be difficult or impossible to calculate or predict these forces beforehand, if the energies of interaction for a given molecule are different in each phase region, then the concentrations of solute in each region will adjust, until the chemical potentials become equal in the two regions.

It is interesting to note the logrithmic dependence of K (and thus the retention times) for linear changes in the $\Delta\mu_i^0$ term. While it is an *extrathermo-*

dynamic assumption, it is nevertheless often empirically observed that systematic changes to a given parameter in a chromatographic system will often result in linear changes to the free energies of that system. Thus, for example, when one linearly increases the percentage of organic modifier in a reversed phase system, the changes in retention are, to a first approximation, logrithmic, in accordance with the above equation for K. This is the basis for many of the computer programs that attempt to optimize separations by systematic changes to the mobile-phase composition while plotting the log of the retention times or retention factors of the mixture components.

On the other hand, if the type of organic modifier is changed—as, for example, changing from acetonitrile/water to methanol/water in a reversed phase separation while keeping all else constant—the *selectivity* between two solutes is likely to change significantly. This is because the $\Delta\mu_i^0$ terms for each solute are likely to change differently, although of course they may not, depending on the particles.

Because distribution constants are not generally known, they are not sufficient to describe retention. In chromatography, the retention is more often expressed by the total moles of solute in the stationary phase (SP) over the total moles of solute in the mobile phase (MP), at equilibrium:

$$k = \frac{\text{moles of solute in SP}}{\text{moles of solute in MP}} = \frac{K_c}{\beta} = \frac{K_c V_S}{V_M} \qquad (10.4)$$

where β is the ratio of the volumes of the two phases, V_M/V_S.

In modern HPLC, which commonly utilizes monolayer bonded phases, it is difficult to define the volume of stationary phase V_S. However, the retention factor in HPLC will be proportional to the specific surface area of the stationary phase, assuming equal bonded phase density. Thus, for HPLC the retention factor can be approximately expressed as

$$k = \frac{A_S}{V_M} \times K_c \qquad (10.5)$$

where A_S is the specific surface area of the packing material of the column, in terms of square meters per gram. For comparisons between similar families of materials, such as porous silica gels, the total number of grams of material within a given column size is, within broad limits, fairly uniform. For example, a 150 × 4.6-mm-type HPLC column will typically require about 2 g of material to pack it. Also, to a first approximation, the included dead volume, V_M, of a given column type, such as porous silica, is relatively constant, with a typical porosity on the order of 0.7 for the inter- and intraparticle volume combined.

Obviously, the stationary phase volume is also affected by the density of ligand groups bonded to the surface, commonly expressed as μmol/m². Monomeric bonded phases in HPLC typically have a maximum bonded phase density approaching 4 μmol/m² [2] for the most commonly used bonded ligand types, with more typical phase coverages falling in the range of perhaps 2–3.5 μmol/m².

Consider for the moment that the typical silica gel will have about 8–9 μmol/m^2 of available silanol groups, meaning that for many bonded phases in HPLC a significant number of unbonded silanol groups are available for interaction with the solute molecules, either via long-range interactions or by direct acid–base interactions with the silanols themselves, in the case of poorly bonded stationary phases.

Given these approximations—which, while not perfect or completely general, are realistic and accurate enough for decision-making at the bench—several generalization are possible. For example, for a given bonded phase density and assuming the same chemistry of the underlying silica gel, doubling the surface area should approximately double the retention time for a given separation. Suppose one was working with a separation on an HPLC material with, say, 150 m^2/g, with a bonded phase density of 3.8 μmol/m^2. Using a column of the same physical dimensions, but packed with a material of 300 m^2/g, at a bonded density of about 3.8 μmol/m^2, should produce retention times approximately twice as long as the original column. This assumes the same flow and MP composition. One would expect this, because the column dead volume typically does not change dramatically with surface area. Therefore, if a chemist were performing a method development for a given separation on a given column and then pulled a new one from the stockroom with twice the surface area, he or she would expect this to happen. However, in consulting at various times with many industrial chemists, this author has often observed that a chromatographer may have no idea of why a new column is giving twice the retention when it appears to be the same C18 material, packed into the same-size tube.

One can imagine many scenarios when a rational selection of a column type is critical. Suppose, for example, that after considerable method development, very weak retention of a mixture is still observed, even when lowering the organic modifier concentration to nearly 0, in a reversed-phase separation. Clearly, one needs all the retention possible to pull the molecules out of the dead volume, without having to change stationary phase type, such as going to a cyano (CN) column. In such an instance, one would select a column with the highest surface area, and greatest bonded phase density possible. Typically, manufacturers will mostly modify the bonding chemistry to include a polar group to prevent phase collapse and call such densely bonded, high surface area material "Aqua-" columns. Conversely, if too much retention is observed, and mobile phase strengths are very high, one might use a different mode of columns such as normal phase rather than reversed phase. However it is usually easier to try to reduce the column "strength" by using a material of lower surface area, thus weakening the retention factors. Because the pore size and surface area of silica gel are linked, this usually means going to a larger pore size material—for example, 120 Å rather than the 80 Å, which might have been the original choice.

10.3.1.1. Effects of Bonded Phase Density. According to the retention factor equation, one would expect the density of bonded phase ligands to have two effects on the separation. First, the phase ratio (β) would be affected, because increasing the bonded phase coverage, up to a point, will increase the phase ratio, although the effect of specific surface area will usually be much greater.

Of more importance is the effect of the silica gel (or other support) on the *selectivity* of the separation. Remember that numbers of perhaps 3.5 to 4 μmol/ m^2 of bonded ligand density represent well-bonded stationary phases. Even in these instances, the role of the silica gel support is never absent. The silica gel is always an active participant in the separation process. However, as the stationary phase bonding density is decreased, more and more of the underlying silica area is available for direct interactions with the solutes. This may, or may not, affect the distribution constant, depending upon the exact nature of the chemical interactions occurring, which are solute-specific. Thus, the bonded phase density, and of course the quality of the silica gel itself, will likely change the *chemistry* of the system, affecting K for a given solute and affecting α for two solutes.

It is difficult to underestimate the importance of the silica gel chemistry upon the separation. This is often the source of the selectivity differences observed between (a) similar columns from two different vendors and (b) the lot-to-lot variations observed for columns from the same vendor. This fact also underscores the point that an HPLC separation is far more complex than the type of stationary phase selected. The modern HPLC column represents a kind of complex phase system, comprised of both the silica gel and its bonded phase. This is actually a good situation, because otherwise all C18 columns would be essentially similar and would give similar selectivities for all separations. The problem of course is in understanding and controlling the silica gel and the bonding process, to produce uniform columns of a given selectivity from a given vendor—that is, to produce columns equivalent to one another, lot-to-lot.

A word about so-called *base-deactivated* columns mentioned in Chapter 9 is in order here. In these types of columns, which are now sold by virtually all the column manufacturers, the undesirable trace metal and activated silanol groups have been removed, either by using highly purified silica gels or by proprietary bonding processes that help shield the silica from interacting with the solute, or a combination of both. The peak tailing commonly observed when chromatographing basic solutes typically arises from activated, acidic silanol groups that have a slower rate of interaction with basic solutes than with other molecules in a mixture. This dissimilarity of the rates of interactions gives rise to tailing. The classic means of addressing this problem has been the addition of various additives to the mobile phase, to overwhelm and compete for these defective sites, preventing the solute of interest from interacting. Typical examples of such additives are triethyl amine (TEA) or diethyl amine (DEA), which are usually added to the mobile phase at about 1 g/liter. If the chromatographer finds such additives in the mobile-phase recipe, there is a good chance it was

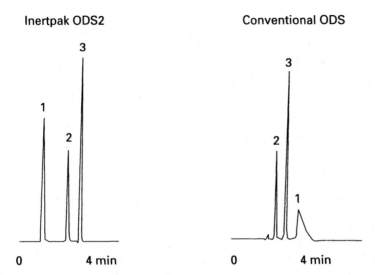

Figure 10.2. Comparison of test mixture separated on a base-deactivated C18 Column and on an older "Conventional" C18 Column of the Same Dimensions. Solutes: 1, pyridine; 2, phenol; 3, methylbenzoate. (Figure courtesy of Capital HPLC, Broxburn, Scotland.)

added during the method development stage to reduce a tailing or other peak shape problem in the method.

Contrary to popular belief, basic compounds will separate quite nicely, with symmetric peaks, even on bare silica in which such defects are absent. Acidic silanol and other defective sites can be created during the gel manufacturing or bonding process by (1) trace metals or other impurities in the silanes used to create the gels themselves, (2) improper heating of the gel before or after bonding, (3) fracturing of the material during packing, or a variety of other ways. Base deactivated materials are merely high-purity silica gel materials, in which precautions have been taken to prevent the introduction of defects. One could argue that the base-deactivated columns are merely the higher-quality columns that manufacturers should have been producing decades ago.

That such materials work is shown in Figure 10.2, where the highly basic compound pyridine, along with other neutral solutes, is separated on an older-style "conventional" C18 column and on a newer base-deactivated material. Not only does the pyridine lose its peak tailing, but it also switches elution order, eluting in the front of the chromatogram. This is because its retention on the conventional column was caused in part by acid–base interactions with the silica gel, as well as by the reversed-phase (hydrophobic) interactions with the stationary phase itself. Removing the acidic silanol sites from the gel permits the pyridine to separate the regular hydrophobic interactions only, reducing retention and improving peak shape.

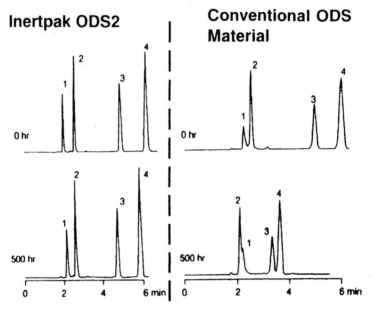

Figure 10.3. Separation of basic and neutral solutes on same columns as Figure 10.2 after 500 hours at pH 2. Note reversal of elution order on the conventional column. Solutes: 1, pyridine; 2, phenol; 3, methylbenzoate; 4, methylindole. (Figure courtesy of Capital HPLC, Broxburn, Scotland.)

Improving the quality of the silica gel and bonding process often improves column stability as well. The normal pH limits of mobile phase are 2–7. The lower pH limit arises from the hydrolysis of the silane ether holding the bonded phase to the silica. The upper limit is due to the disintegration and dissolution of the silica particle itself. The higher-purity silicas and improved bonding of the surfaces tends to improve stationary phase stability at both the upper and lower ranges, as shown in Figure 10.3. Here, several basic and neutral solutes are separated at pH 2 on both a conventional and a base-deactivated material. After 500 hr, the base-deactivated material maintains its selectivity. On the other hand, the lower-quality silica of the "conventional" column exhibits signs of stationary phase stripping. Notice how the imidizole peak begins to gain retention as the silica support becomes exposed, creating both additional retention and peak tailing as the column ages.

In summary, the retentiveness of a column arises from the specific surface area, in m^2/g, of the material used to pack the column, due to the phase ratio of the material. Retentiveness is also a function of the bonded phase density, in terms of $\mu mol/m^2$ of groups bonded to the surface. The upper limit, imposed by steric hinderence of the bonded groups, is on the order of 3.5–4 $\mu mol/m^2$ for the typical C8- and C18-type columns. Thus, bonded phase densities of 2 $\mu mol/m^2$ represent weakly bonded materials, with approximately only 50% of the maximum possible coverage of the gel surface by the bonded groups.

The selectivity of a column is affected by the bonded-phase ligand type, such as CN versus C18, by the density of the bonded phase (in terms of $\mu mol/m^2$), and by the chemical nature of the silica gel itself.

10.3.2. Peak Shape

Peak shape is also governed largely by the nature of the silica gel and the bonded phase, although peak tailing and fronting can have many other sources unrelated to the gel itself. For example, some methods exhibit peak shape problems by using exceedingly large injection volumes of solvents stronger than the mobile phase, or by using diluents of different pH values or simply too concentrated standards and samples. Obviously, the column chemistry cannot fix such problems. However, in those cases where peak shapes result from interactions of the solute with the stationary phase/support system, by definition the column becomes important. Selecting an equivalent column in such cases sometimes becomes a matter of trial and error, because the manufacturers themselves often cannot state exactly the properties of the gel that are causing the peak shape problems.

10.4. COLUMN EFFICIENCY

Of all the operating characteristics of an HPLC column, efficiency is probably the best understand. Since the first 1941 paper by Martin and Synge [3] describing the work for which they were later awarded the Nobel Prize, through the seminal work by van Deemter [4] and Giddings [5], considerable effort has been expended in understanding the basis for efficiency in chromatography. For a detailed discussion of the basic aspects of column efficiency, the reader is again referred to Chapter 7 of this book.

For HPLC, it is convenient to have a clear understanding of the relationship between column characteristics (such as particle size, column length and diameter, and flow rates) and the expected efficiencies of a separation. Fortunately, for the purposes for which we need this type of information, fairly broad but amazingly accurate estimations of theory can be applied which make the selection of column type straightforward.

In Chapter 7 the concept of plate number, N, was discussed in some detail as the common parameter to specify column efficiency. Related to it is the plate height, H. Their relationship is given in equation 10.6:

$$N = \frac{L}{H} \tag{10.6}$$

However, using the more general conceptual framework of Giddings, N is more accurately thought of a the total variance generated by a column operating on a peak that travels distance (or time, volume, etc.) X, thus,

$$N = \frac{X^2}{\sigma^2} = \frac{t_R^2}{\sigma^2} \tag{10.7}$$

Because X and sigma are both expressed in the same units, N is a dimensionless number. Imagine several 15-cm HPLC columns, all operated at a constant 1 mL/min with the same mobile phase and test mixture. The more efficient columns will produce narrower peaks for the same retention times, while the less efficient ones will have broader peaks. In terms of the above equation, X is the same (retention time), but the sigma-squared zone variance (σ^2) is less for the better columns, reflected in the higher N values. Thus, high N values are desirable, because the narrower peaks are both more detectable and allow room for the separation of more peaks within a given separation space (see resolution equation, Chapter 7).

When a chromatographer chooses a column to solve a given separation problem, he or she is choosing, in a sense, a prepackaged unit of a given efficiency. Separately, the chromatographer will choose a column based on the retention and selectivity that the column produces toward the separation mixture. However, as mentioned at the beginning of this chapter, the *retentiveness* and *selectivity* are thermodynamic terms, related to the free energies of separation of given solutes on a given phase system. The column *efficiency* is determined by the *rates* at which solutes can attain equilibrium within the column. These are two totally separate, independent processes, unrelated to one another. In practice, this means that one can select whatever column will produce the desired retention properties, such as CN, C18, base-deactivated C18, and so on. During the method development process, one can modify the efficiency of the separation by choosing the same packing material (but packed into longer and shorter tubes and, perhaps, on different particle sizes from the same manufacturer) without changing the general retentiveness (retention factors) or selectivity (alpha) of the separation. The column efficiency is simply another variable at the chromatographer's disposal to intelligently design the separation to meet the problem at hand.

If high efficiencies are good, then what is the penalty for producing large plate numbers? Time. Referring back to the relationship between retention factor and time, recall that

$$t_R = t_M(1 + k) \tag{10.8}$$

Using the same particle size, efficiently packed into the same-diameter column, at a constant flow rate, the above equation expresses that the retention time of a given compound will increase as t_M increases, if k is constant. Because the k value is determined by the chemistry of the phase system, to a first approximation, this will not change with any changes to the physical dimensions of the column or particle size. However, t_M will change with increasing column length; a column of the same internal diameter, twice as long as another, packed with the same material, will produce the same separations, but with twice the retention times for all the peaks. The longer column will also produce more theoretical plates—in fact, twice as many—if the particle size is kept the same and they are packed equally well. Therefore, the penalty for generating more efficiency, with a given particle size, is analysis time.

10.4.1. Resolution

Resolution (see Chapter 7) is a kind of bottom-line equation that expresses the degree of separation of two adjacent peaks.

A baseline resolution of 1.5 or greater (for equal-sized peaks) is usually desired to ensure a clean separation of the zones, and it is almost always a part of the system suitability requirements. Obviously, the resolution is a function of the selectivity of the column (t_{R_2} and t_{R_1}) and is also a function of the peak widths, w_2 and w_1, which result from the efficiency of the separation. Because the retention factors and thus selectivity are determined by the chemistry of the solute/phase system and not by column length or particle size, generating more plates by using longer columns packed with the same material increases resolution by decreasing the relative peak widths, w_2 and w_1, for given retention times. Of course, retention times are increasing with the longer column, because the column dead volume is also increasing, but the k values of the solutes remain the same.

What this means is that, ideally, one wants a column with *enough* efficiency to produce the desired resolution for a given separation. However, generating *too much* resolution is wasteful of time. If the selectivity of a separation is so great that one is essentially separating only a single peak, as is the case in many release assays, then the column can often be profitably shortened, and the analysis time reduced, without jeopardizing the robustness of the method. Conversely, for difficult separations, such as many stability-indicating assays, one generally wants a very selective separation and develops a method with the most efficient and selective column possible, regardless of the analysis times. The problem arises of how to rationally select the column efficiency, because one can choose among a wide variety of column lengths and particle sizes for any given packing material type by a given manufacturer. Probably the easiest way to understand this selection process is by using the concept of *reduced parameters*.

10.4.2. Reduced Plate Heights to Estimate Expected Column Efficiencies

J. C. Giddings, in his classic book, *Dynamics of Chromatography* [5], presented to concept of reduced plate heights and reduced velocities to express the intricacies of column efficiency in a broad manner. The reader is referred to that text and to Gidding's more recent text, *Unified Separation Science* [6], for a more detailed discussion of these concepts than can be presented here. Probably the clearest and easiest to understand presentation of both the concepts, and their application to solving realistic separation problems, can be found in Knox's classic textbook, *High Performance Liquid Chromatography* [7]. The purpose here is not to discuss the idea of reduced parameters in great depth but to present the ideas sufficiently to serve the purpose of using them to estimate anticipated column efficiencies to help select equivalent columns.

Definitions of the reduced parameters were given in Chapter 7. The reduced plate height, h, is given by

$$h = \frac{H}{d_p} \tag{10.9}$$

where d_p is the particle diameter and the reduced velocity, v, is given by

$$v = \frac{u d_p}{D_M} \tag{10.10}$$

where D_M is the diffusion coefficient for the solute in the mobile phase. In both equations the units cancel making the reduced parameters dimensionless. The reduced velocity expresses how many particle diameters the solute travels in the forward direction, via flow, versus its lateral movement via diffusion. The real value of using reduced parameters is that they can be used to normalize the van Deemter equation to a single universal curve. The resulting modified van Deemter equation is commonly referred to as the Knox equation and can be approximated as

$$h = \frac{B}{v} + A v^{1/3} + C v \tag{10.11}$$

where B, A, and C are the contributions to zone spreading due to longitudinal diffusion, eddy dispersion, and resistance to mass transfer, respectively. A typical Knox equation curve was shown in Figure 7.19.

The power of this equation arises from the fact that, within a given "type" of packing material, such as modern porous silica materials, generally similar A, B, and C coefficients can be expected for a well-made, well-packed material. For modern, silica-based HPLC packings, typical values will be $B \sim 2$, $A \sim 1$, or $C \sim 0.1$. To a first approximation, well-made and well-packed columns produced from C18 material, for example, of either 3-, 5-, or 10-µm particles, can generally be expected to exhibit similar performances, expressed in terms of A, B, and C. Furthermore, the Knox equation predicts that such similar materials will have a similar h value at about the same reduced velocity. For modern HPLC materials, the optimimum reduced velocity will typically be in the range of perhaps 3–5, resulting in minimum reduced plate height values of about 2–3. Again, these properties can be expected regardless of particle diameter.

It is important to note here that the effect of flow on efficiency losses is not nearly as severe in HPLC as it is in capillary GC. The portion of the curve to the right of the minimum is typically very shallow. The flow rate in terms of mL/min corresponding to the optimum is very low, typically about 0.05–0.3 mL/min for reversed-phase mobile phases, using 4.6-mm ID columns. However, there is not a severe penalty incurred by operating at the more normal 1–2 mL/min, because the curve is quite shallow. Thus, it is common practice to operate the HPLC column at these higher flow rates, because analysis times are directly reduced, while efficiencies may drop only 10–20% from the optimum.

If one substitutes equation 10.9 into the efficiency equation $N = L/H$, the result is

$$N = \frac{L}{hd_p} \quad \text{or} \quad h = \frac{L}{Nd_p} \tag{10.12}$$

Values of h for modern HPLC columns can be expected to fall somewhere in the range of perhaps 3–5 under most conditions. Values of 2–3 represent excellent, well-packed high-performance columns, while values approaching 10 or more indicate a fairly severe problem with the chromatography, either due to poor packing material or due to some unsuitability of the solute with the given phase system. If one measures N of a given column (which ideally should be approximately the same for all the peaks in the chromatogram) and calculates the h value using equation 10.12, one can readily estimate whether the column and HPLC system are working up to their full potential or if there is a problem somewhere.

The vast majority of separations in modern HPLC utilize more or less standard column lengths and particle diameters. It is interesting to calculate some of the typical expected efficiencies that these standard columns are capable of producing. Assume that a well-made reversed-phase column, 150×4.6 mm, operated at 1 mL/min, packed with 5-μm particles, is selected for a separation. How many plates can one reasonably expect this column to produce? Assuming a reasonable value of $h = 4$, one estimates the expected column efficiency as

$$N \approx \frac{15 \text{ cm}}{(4)(5 \times 10^{-4} \text{ cm})} \cong 7,500 \tag{10.13}$$

Is this a reasonable value? It would appear so, and the reader is invited to examine chromatograms from various methods using columns of this type to confirm this. Will every 15-cm, 5-μm column produce this value? Obviously not. Some columns of this size may be truly outstanding and produce N values of over 10,000, while others may have some type of peak shape problems and produce N values of 5000 or less. Nevertheless, this expected efficiency is sufficiently accurate to help plan reasonable method development strategies.

Table 10.2 presents typical expected column efficiencies for the various types of columns in common use, calculated using equation 10.12.

Table 10.2. Expected Efficiencies for Various Types of HPLC Columns[a]

Particle Diameter (μm)	Length (cm)	Approximate N
3	5	4,200
3	10	8,300
3	15	12,500
5	15	7,500
5	25	12,500
10	25	6,300

[a] Assumptions: $h = 4$.

It is interesting to group the data in Table 10.2 according to these classes of plate numbers: <5000 plates, 5000–10,000 plates, and >10,000 plates as shown in Table 10.3 on page 303. The under 5000 plate columns are best applied to very simple separations, with good resolution between peaks, where analysis time is highly important. Again, very simple release-type assays are good candidates for such columns. One must remember that the extracolumn dead volumes in many commerical instruments must be minimized in order to realize the efficiency of the short, highly efficient columns. Because connecting tubing must often be shortened and new low-volume flow cells installed, the popularity of this size column has been reduced in recent years.

The middle class of columns, those producing 5000–10,000 plates, are by far the most common. Within this class, there are three broad choices. The older-style columns, 25 cm (or 30 cm for the Waters μ Bondapaks), packed with 10-μm material, represents a column configuration dating back at least two decades. It produces low back pressures, and for a $k = 5$ peak, at 1 mL/min produces analysis times on the order of 15 min or so.

While there is often nothing wrong with these columns, perusal of the data shows that the same separation can be achieved using smaller particles packed into shorter columns, thus reducing the run times with no loss in efficiency. The more modern column style of 5-μm material packed into 15-cm columns will produce the same or better efficiencies as the 25-cm 10-μm columns, with analysis times reduced from 15 to 9 min, a reduction of roughly 40%. Similarly, 3-μm material in a 10-cm column will again produce roughly the same efficiencies, but with run times less than half that of the longer 25-cm columns. It is important to remember that these reduced run times come with no penalty to the separation quality and have many advantages. The resolution between two peaks will be generally similar using any of these three column types; however, the 5-μm, 15-cm, and 3-μm 10-cm columns will have significantly shorter run times, with sharper, more detectable peaks. The 15-cm, 5-μm type of columns represent the workhorses of the industry. They tend to produce enough plates for many routine separation problems, with moderate back pressures and analysis times and long lifetimes.

The final type of columns, those producing more than 10,000 plates, is limited to two column formats: the 15-cm 3-μm type or the 25-cm type packed with 5-μm material. Either column type will produce the most efficient separations routinely available using commercial instrumentation. The penalty one pays are in longer analysis times and higher column back pressures. When would one consider using such a column? These columns represent a good choice for difficult, stability-indicating assays. Conversely, if one were developing a routine quality control (QC) release assay, in the absence of specific complications, these would not be the columns of first choice. They would work, but the chromatographer would be developing a method that "overseparates." If the analyte of interest is very well resolved from any neighboring peaks, say with a resolution of 5 or more, there would be no need to utilize so many plates, and a shorter, less efficient, but faster column would be a far more rational choice.

Remember again that the efficiency, as determined by the particle size and column length, is *independent* of the retention and selectivity of the system. Thus, one may have completed a method development cycle, finding a particular mobile phase, stationary phase, and vendor whose columns are producing the desired selectivity, but the column initially tried was a 25-cm 5-μm type. If sufficient resolution is available, at that point the chromatographer would do well to review the data and perhaps try the same general material type, but with a 3-μm particle packed into a 10-cm column. Conversely, if a 15-cm, 5-μm CN column was found to give great selectivity, but a small degradation peak is very closely resolved from the active, some extra robustness could be designed into the separation by using a longer, 25-cm 5-μm material, or of course the 15-cm 3-μm format, to produce more than the minimum plates and resolution. Obviously, any such changes would be confirmed at the bench, prior to locking in the method development and before validation was started.

Finally, note that nothing in the above equations expresses column diameter. Column diameter, insofar as columns can be efficiently packed to a given h value regardless of diameter, is not a factor in determining column efficiency. Only column length and particle size determine efficiency. Larger- or smaller-diameter columns will require faster or slower volumetric flow rates, to maintain equal linear velocities (or reduced velocities) for different diameters, but adjusting the flow rates to the column diameter is otherwise not part of the efficiency equations. This is certainly true over the range of analytical column diameters from 1 to 5 mm. Larger preparative columns, often packed with 20-μm materials, are treated differently and are usually optimized in terms of mass of material purified, rather than simple plate heights.

Sometimes one needs to use columns of smaller or larger diameters, as for example when adapting a given separation for interfacing with mass spectrometry. Because the porosity of a given column is determined by the particle type packed into it, the linear velocities for different column diameters, packed with similar materials, are determined by the ratios of the cross-sectional areas. Because the area of a tube can be given as $\pi d^2/4$, the constants will all cancel, leaving

$$\text{Proportional flow rate} = \frac{d_1^{\ 2}}{d_2^{\ 2}} \qquad (10.14)$$

For example, suppose a given separation was currently performed on a 15-cm × 4.6-mm-ID column, operated at 1 mL/min, and that the separation must be adopted to a 2-mm column for interfacing with a mass spectrometer. The scaling factor becomes simply $2^2/(4.6)^2$, or 0.19. Thus, the flow rate for the smaller-ID column should be adjusted from 1 mL/min to about 0.2 mL/min. For the same-length column, packed with the same material, the elution times should remain essentially identical. Similarly, for older methods, specifying a 3.9-mm column that is to be adapted to a 4.6-mm-ID column, a 1-mL/min flow rate should be increased to 1.4 mL/min $[(4.6)^2/(3.9.)^2 \times 1]$ to maintain constant linear velocities in the column.

10.5. PUTTING IT ALL TOGETHER—SELECTING AN EQUIVALENT COLUMN

From this brief review, we are now in a better position to rationally select columns that are equivalent to one another. Examining the four major areas defined at the beginning of this chapter, it is clear that for two columns to be equivalent, they must be similar in terms of the following properties:

1. They first must be of the same particle size and physical dimensions, even though, as was shown above, similar efficiencies can be achieved using a variety of column formats.
2. The stationary phase type must be the same; that is, a C18 material with monomeric bonding must be matched to another C18 material, also with monomeric bonding.
3. The most difficult task in choosing column equivalency lies in matching selectivities. This is because the user, and sometimes the manufacturer as well, really has little control over the silica properties. It is hoped, however, that the manufacturer has some type of fairly rigorous quality testing that will maintain lot-to-lot uniformity of both the gel and the final bonded phase, over time. Every chromatographer who has worked for some years has at least several "war stories" about methods that have worked well for years, only find that the same-type column no longer will separate a given pair of solutes. Almost invariably, this type of problem can be traced to sometimes subtle changes in procedures at the gel manufacturing or SP synthesis.

The general retentivity of a phase is determined, we have seen, by both the specific surface area of a material and its bonded phase density, which also affects selectivity. This is not to be confused with % carbon loading. Percent

Table 10.3. Column Types as Defined by Efficiencies

Columns Producing N Values	Column Type	Approximate Analysis Time for $k = 5$ (4.6-mm-ID column)
<5,000	5-cm, 3-μm particles	3 min ($t_M \sim 0.5$ min)
5,000–10,000	10-cm, 3-μm particles	6 min ($t_M \sim 1$ min)
	15-cm, 5-μm particles	9 min ($t_M \sim 1.5$ min)
	25-cm, 10-μm particles	15 min ($t_M \sim 2.5$ min)
>10,000	15-cm, 3-μm particles	9 min ($t_M \sim 1.5$ min)
	25-cm, 5-μm particles	15 min ($t_M \sim 2.5$ min)

carbon values really give little information about the chemical nature of the bonded silica. Instead, it is the density of bonded ligands, in terms of μmol/m², that will say whether a given phase is very hydrophobic or if it is only lightly bonded, with extensive bare silica available for interactions.

In attempting to select an equivalent column, one must start by finding all of the above information. In the past, this was sometimes difficult, and manufacturers in general used to treat this type of information as proprietary. Fortunately, this attitude has changed significantly, and many column companies now routinely list many of these values for both their own and competitors' packing materials. Table 10.4 reports values taken from the current Phenomenex catalog.

10.5.1. Choosing Equivalent Columns: An Example

Suppose that a current separation utilizes a Shandon Hypersil ODS 5-μm material, packed into a 150- × 4.6-mm-ID column. Imagine that perhaps during method development, an alternative, equivalent column needs to be selected for testing, or that this particular column was discontinued by the manufacturer for some reason, and that an equivalent column must now be found for an older, existing method.

The first step in the process is to determine the properties of the column in use. Using Table 10.4, the following parameters are found for a Shandon ODS Hypersil, in addition to the data for several other columns which one might consider as candidates for equivalency.

The Hypersil ODS is a 120-Å pore-diameter material, with a specific surface area of about 170 m²/g. This can be considered a moderate surface area material. The bonded phase density is about 2.8 μmol/m². The phase is monomerically bonded and is end-capped. This material is available in 3-, 5-, or 10-μm particles, in a variety of column lengths. For our purposes, however, the column length and particle size have been fixed by the existing separation.

Of the various properties of the Hypersil ODS material, the ones of most interest are the specific surface area value of 170 m²/g, the bonding chemistry,

Table 10.4. Choosing Column Equivalency Candidates to a Hypersil ODS

Column:	Pore Size Å	Surface Area m²g⁻¹	Carbon Load %	Calc. Ligand Density μmol·m⁻²	Bonding/ End Capping?	Equivalent?
Hypersil ODS	120	170	10	2.84	Monomeric EC	—
LiChrosphere RP-18	100	350	21.4	3.9	Monomeric Not EC	No
Waters Resolve C18	90	175	10	2.76	Monomeric Not EC	Possible
Spherisorb ODS 2	80	220	12	2.72	Monomeric EC	Possible
Phenomenex PhenoSphere ODS (2)	80	220	12	2.50	Monomeric EC	Possible
Phenomenex Maxsil C18	65	500	12.5	1.25	Monomeric EC	Not in your dreams

Notes: All columns assumed to be in same physical format of column length, diameter and particle size. All data taken from Phenomenex 1999/00 Sales Catalog, Technical References section.

monomeric, and the bonded phase density of 2.8 μmol/m². The material is end-capped, meaning that a secondary bonding reaction was used with trimethylsilane (TMS) to attempt to "cap" any residual silica sites remaining after the primary bonding. End-capping is an older bonding technique that is not widely used today. The original idea was that the TMS groups would fit into places on the gel that the C18 groups, for example, could not reach. This approach has been proven flawed, because the two methyl side chains of the TMS have the same "molecular footprint" as the methyl groups on a monofunctional C18 or C8. In fact, if a bonded phase is well-bonded in the first place with the primary ligand, little or no extra TMS will be put onto the surface. Furthermore, the TMS groups that are bonded show much greater hydrolytic instability than the C18 groups, and they usually will hydrolyze back off the material fairly quickly [8]. Therefore, in this author's experience, the presence of end-capping is not considered to be an area of primary concern, and columns will be considered to be equivalent if they have the other necessary properties in common and if they produce the same separation, whether or not they are "end-capped."

Consider now each of the choices presented in Table 10.4. The Lichrosphere RP 18 is a spherical material, available in 5-μm particle size, and it has a long and venerable history as a good-quality material. The pore size is 100 Å. As a rule of thumb in all these comparisons, 1–2 significant figure type agreement is about the limit of accuracy for these values. Thus, 100 Å is generally considered to be essentially the same pore size as the 120 Å of the Hypersil. Of more importance is the surface area of 350 m^2/g. This is slightly more than twice that of the Hypersil ODS. The percent carbon is higher, but this really means very little. Instead, the bonded phase density of 3.9 μmol/m^2 is far more significant, indicating a tight, densely bonded surface, close to the maximum density allowed for a monomeric phase of about 4 μmol/m^2. The material is not end-capped, in part because it is so well-bonded that no TMS reagent could be put on the surface and is unnecessary anyway.

What is the signficance of these properties of the LiChrosorb material in determining column equivlanecy? Reviewing the basic equations governing retention, assuming all other properties were identical, the same-size column packed with the LiChrosphere RP18 would give twice the retention factors, and thus twice the retention times of the Hypersil ODS material, based on the phase ratio. However, the LiChrosorb's higher bonding density makes the material even more hydrophobic than the Hypersil ODS, resulting in probably even greater retentiveness for the LiChrosorb. All in all then, the LiChrosorb RP18 would not be considered equivalent to the Hypersil ODS. This is not to say of course that it is not a suitable packing material, or perhaps superior material, for other separations, but it is not equivalent to the Hypersil. In fact, given its generally high retentiveness, this would be an excellent candidate for a separation of very polar compounds with weak retention properties on a C18 phase.

Consider next the Waters Resolve C18, again assuming identical particle size and column lengths. The pore diameter of 90 Å is similar enough to the Hypersil ODS to be considered of the same general type. The surface area of 175 m^2/g is almost identical to the Hypersil ODS, as is the bonded phase density of 2.76 μmol/m^2 of monomerically bonded groups. The Resolve material is not end-capped, but again, in this author's experience, that should not be an overriding factor. According to the general principles of chromatography discussed throughout this chapter, one would consider the Resolve C18 material to be very similar to the Hypersil ODS.

The next candidate, the Spherisorb ODS 2, likewise has a similar pore diameter of 80 Å, a more or less similar surface area of 220 m^2/g, and a monomerically bonded stationary phase coverage of 2.72 μmol/m^2, with end-capping. Again, this would be considered another good candidate to consider for equivalency testing.

The Phenomenex PhenoSphere C18 material is almost identical to the Spherisorb material in its properties, with a pore diameter of 80 Å, a specific surface area of 220 m^2/g, and a monomeric bonded phase density of 2.5 μmol/m^2, with end-capping. For the same reasons, this could be considered as a candidate. In fact, if one were looking for a column equivalent to a Spherisorb ODS2, the PhenoSphere material would be an almost perfect match.

Finally, for contrast, consider the last candidate, the Phenomenex Maxsil C18. The pore diameter of 65 Å is becoming different enough from the wider pore Hypersil that it must be considered as a quite different material. This is reflected in the specific surface area value of 500 m^2/g—over four times higher than the Hypersil ODS. The bonded phase density, on the other hand, is an anemic 1.25 μmol/m^2, indicating a phase bonded only a little more than 25% of its available coverage of 4 μmol/m^2. It is end-capped, but this author would surmise that within a week on the bench under an aggressive mobile phase, most of the end-capping will have hydrolized off. Clearly, this material is about as unlike the Hypersil ODS as it is possible to be. Does that make the Maxsil C18 a "bad" material? Absolutely not. It would be a candidate for very hydrophobic molecules that do not have amines or other moieties capable of secondary interactions with the silica gel, although the high surface area and low bonding coverage indicate that some surface area is probably tied up in inaccessible micropores. Some people like vanilla, some chocolate, and then there are those days when we want tutti-frutti ice cream. For some applications, the Maxsil C18 would be the column of choice, but in no way could one defend that it was "equivalent" to the Hypersil ODS.

Thus, we now have three good candidates for equivalency testing, the Waters Resolve C18, the Spherisorb ODS 2, and the Phenosphere ODS-2. Perusal of the full table, along with careful review of the manufacturers' catalogs, will uncover others—this is not meant to be an exhaustive list. So, does that mean that one could substitute a Phenomenex Phenosphere ODS-2 for Hypersil ODS and begin running the method? Maybe, maybe not, but certainly not without further testing. It is quite possible that none of the above three replacement candidates will produce a separation identical to the one currently in use. It is also possible that one of the new columns will produce very similar separations, except for one tiny degradation peak, which inverts elution order with an excipient peak. On the other hand, one of these three candidates may very well produce the desired separation. Only testing will prove their suitability.

If there is no guarantee of success, why go through this excercise? Why not simply choose another C18 column and try it on the bench? The answer is obvious. The above selection process is a necessary, but not sufficient, test for column equivalency. Consider, however, that the alternative is an irrational, hit or miss approach, with no underlying understanding of the theory of operation of modern HPLC columns.

The need for a final testing to prove column equivalency brings us to the final segment, how to determine and prove equivalency? From the above discussions, it would seem self-evident that a realistic test mixture must be generated to evaluate equivalent columns efficiently. Normally, this might be either a stressed sample or some of the worst-case stability samples that one has available. It is not necessary that the equivalent column produce an identical separation in terms of absolute retention times, and so on. However, the elution order should certainly remain the same, and plenty of excess resolution between critical peaks must be evident to provide for adequate field robustness.

In our experience, small but significant quantitative errors can sometimes occur among different columns types, sometimes due to irreversible adsorption of solute onto the silica surface or due to other effects that cannot readily be isolated. As such, a study of the quantitation of samples on the new candidate column is highly recommended. A general scenario might be to perform an analysis on at least three lots of a drug product in triplicate, using both the existing method column and the candidate column. Some reasonable acceptance criteria must be set, which typically is agreement to $\pm 2\%$ for accuracy, between the two columns. In most real-world cases though, one would expect to find agreement of well under 1% in such an experiment. The analyst would also carefully compare all the usual chromatographic parameters, such as efficiency, peak tailing, resolution of critical pairs, and pressure drops in documenting equivalency.

Choosing an equivalent column is, in a sense, one of the hardest jobs a chromatographer has. To be done properly, it requires a great depth of both theoretical and practical knowledge, and it is limited by our general ignorance of the very subtle features of a column, which give rise to the selectivity of a given separation. Also, one should always expect to be surprised, and one should approach changing columns with some caution. Sometimes, months after a decision has been made to change a column specified in a method, some slight but significant change in quantitation is observed, sometimes in a stability program, which can be traced back to a column not being quite identical to another. Of course, all these arguments apply equally to lot-to-lot variations of columns within a single manufacturer, so one cannot escape this problem simply by refusing to change columns. Likewise, in the method development cycle, utilizing this type of knowledge and information will lead to the selection of columns appropriate to the separation. It goes without saying that only the most foolish of chromatographers would ever select a replacement equivalent column, or specify a column in a method, based on examination of only a single lot of chromatographic material sole-sourced from a single vendor. The cries of anguish arising from the laboratory a year or two later when the column changes will be loud and quite disturbing.

REFERENCES

1. J. H. Knox and R. Kaliszan, *J. Chromatogr.* **1985**, *349*, 211.
2. J. G. Dorsey and W. T. Cooper, *Anal Chem.* **1994**, *66*, 857A–867A.
3. A. J. P. Martin and R. L. M. Synge, *Biochem. J.* **1941**, *35*, 1358–1368.
4. J. J. van Deemter, F. J. Zuiderweg, and A. Klinkenberg, *Chem. Eng. Sci.* **1956**, *5*, 271.
5. J. C. Giddings, *Dynamics of Chromatography*, Part I, Marcel Dekker, New York, **1965**.
6. J. C. Giddings, *Unified Separation Science*, J. Wiley, New York, **1991**.
7. J. H. Knox, *High Performance Liquid Chromatography*, Edinburgh University Press, Edinburgh, **1980**.
8. T. R. Floyd, N. Sagliano, Jr., and R. A. Hartwick, *J. Chromatrogr.* **1988**, *452*, 43–50.

11

DISSOLUTION

Ross Kirchhoefer and Rudy Peeters

11.1. INTRODUCTION

This chapter will primarily focus on practical discussions of dissolution testing of solid dosage forms for oral administration. Dissolution tests are designed to monitor the rate at which the solid, semisolid, suspensions, etc. dosage forms release the active drug substance(s) into a uniform liquid medium at controlled temperature under standardized conditions at the liquid–solid interface. In practice, dissolution testing can be used for the following [1]:

1. *Determination of Bioequivalence:* "*... ensuring the bioequivalence of different batches of solid-dosage forms when a correlation between dissolution characteristics and bioavailability has been established.*" It is important to ensure continued product quality and bioequivalence upon changes in the formulation, manufacturing process, site of manufacture, and scale-up of the manufacturing process.
2. *A Release and Stability Test to Monitor the Manufacturing Process.* "*... quality control personnel as a convenient and reliable method of monitoring formulation and manufacturing processes.*"
3. *A Research Tool in Drug Development.* "*... establishing the intrinsic dissolution rate which is useful in screening new compounds being considered for new drug application.*"

Analytical Chemistry in a GMP Environment. Edited by J. M. Miller and J. B. Crowther
ISBN 0-471-31431-5 © 2000 John Wiley & Sons, Inc.

Earlier in Chapter 1 of this text, the drug development process was briefly reviewed and the importance of the bioavailability of the drug discussed. Over the years, scientists have attempted to directly correlate bioavailability *in vivo* with *in vitro* (IVIV) dissolution testing with limited success, particularly when product dissolution is the rate-limiting factor in absorption of the drug. For the most part, however, regulatory agencies rely on dissolution testing primarily to examine the quality of the lots of drug product; therefore much of the discussion of this chapter will focus on these quality control aspects of dissolution testing.

11.1.1. History

11.1.1.1. Background and FDA Involvement. To put the dissolution test in perspective, it may be useful to review the short history of the techniques, particularly from the FDA perspective because one author (R. K.) was a member of the FDA's St. Louis Laboratory Staff. In a *Pharmaceutical Technology* editorial [2], Gene Knapp states, "The agency (FDA) feels that *in vitro* dissolution testing can help pinpoint formulations that may present potential bioequivalence problems. The agency further believes that once a formulation has been shown to be bioavailable, dissolution testing is of great value in assuring lot-to-lot bioequivalence." William Hanson states in his useful and highly regarded *Handbook of Dissolution Testing* [1], "Dissolution testing, of course, is a regular quality control procedure in good manufacturing practice. Whether or not its numbers have been correlated with biological effectiveness, the standard dissolution test is a simple and inexpensive indicator of a product's physical consistency." The growth and value of the dissolution test was spurred by the realization of Wagner and Pernarowski [3] that the disintegration test, which predated dissolution testing, had little correlation with biological activity or bioequivalence while some dissolution *in vitro* results correlated with the *in vivo* results.

In the late 1960s and early 1970s, there were two general dissolution techniques in use: One was the rotating-basket procedure employing a three-necked round-bottomed flask, originally proposed by Pernarowski et al. [4] and also used by Levy et al. [5]. This technique was eventually modified and became the first official method in USP XVIII and NF XIII in 1970. The other was the paddle procedure originally developed by Levy, employing a beaker [6], and Poole [7], who also utilized a three-necked round-bottomed flask.

The FDA's St. Louis laboratory initially investigated two drugs, digoxin and prednisone, using the Pernarowski basket procedure and the Levy–Poole paddle procedure. Immediately it became clear that many factors influenced the dissolution test results. Some of these factors included variability in dissolution equipment (both home made or commercial), glassware, mechanism for stirring, analytical procedures for sampling, the dissolution medium, plus filtering and analyzing the test solutions.

11.1.2. Early Improvements in Dissolution Equipment

Detailed investigation into the physical considerations of the test indicated that the basket and paddle methods had some drawbacks. Concerns with the basket procedure in early equipment included the following:

- Physical contact with the drug formulation
- Lack of constant speed of rotation
- Wobble of the shaft and basket
- Variation in mesh size of the basket weave
- Construction materials of the basket
- Practical issues with fitting the basket into the three-necked round-bottomed flask

The paddle procedure suffered from either the difficulty of fitting a paddle into the center port of the three-necked round-bottomed organic reaction flask or being able to center the test formulation directly under the paddle when using a beaker. Both techniques suffered from the fact that only one tablet at a time could be investigated.

In either approach the major concerns were:

1. To find an appropriate vessel that could accommodate both the basket and the paddle
2. To be able to test more than one tablet at a time

Upon evaluation, the FDA's St. Louis Laboratory and Kimble Glass collaborated on making a round-bottomed kettle. This development was subsequently published by Kirchhoefer in 1976 [8]. The development of this vessel opened the door to using both the basket and the paddle easily in commercially produced equipment that could accommodate six vessels. Results could be obtained on six units at a time, and a better understanding of the influences affecting the test could be evaluated.

The FDA investigators published the results of these investigations in *Pharmaceutical Technology* [9]. Additional testing and experience led to the publication of an addendum in *Pharmaceutical Technology* [10]. Hanson [11] and Thakker et al. [12] also published information on factors affecting the test and test equipment.

11.2. DISSOLUTION BASICS

11.2.1. Disintegration Tests

The USP ⟨1088⟩ "*In Vitro* and *In Vivo* Evaluation of Dosage Forms," recognizes that unless an oral-dosage form disintegrated into small aggregates, it could not be absorbed into the body. Due to this concern, the disintegration test was made official in 1950 (USP XV). This technique continues to be a compendial procedure ⟨701⟩ in USP 24, where a complete description can be found. The disintegration test was mandatory for over 30 years, but is gradually being eliminated and replaced by the more meaningful dissolution test. As previously mentioned, the elimination of disintegration testing can be attributed to the realization that disintegration testing had little correlation with biological activity [3]. In fact, the solubility of a drug can be discribed in two discrete and general processes; disintegration and deaggregation into descrete particles. Figure 11.1 outlines the total process, indicating that at first approximation the dissolution process incorporates disintegration information.

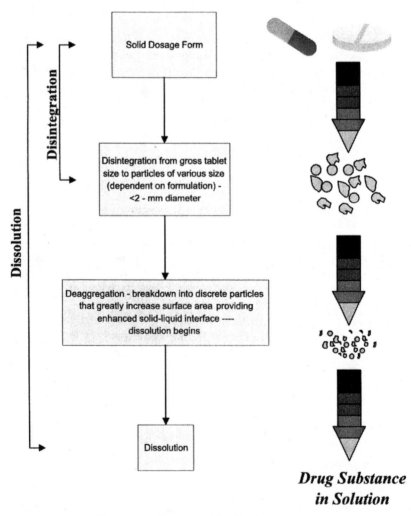

Figure 11.1. Overview of the dissolution process.

11.2.2. Elementary Theory

In the case of an orally administered solid dose, the steps of the adsorption of a drug product are a function of the following [13]:

1. Release of the active drug substance from the drug product
2. Dissolution or solubilization of the drug substance under physiological conditions
3. Permeability across the gastrointestinal tract

Dissolution testing is subsequently used to evaluate the first two steps of the three. The dissolution test, if appropriately performed, can experimentally determine the dissolution rate. The *dissolution rate* can be defined as follows [4]:

> ... the amount of active ingredient in a solid-dosage form dissolved in unit time under standardized conditions of liquid/solid interface, temperature, and the media composition.

Of the many mathematical approaches to describe dissolution (at sink conditions*), equation 11.1 is a reduced equation in its simplest form:

$$\text{Dissolution rate} = \frac{dW}{dt} = kS \tag{11.1}$$

where S is the surface area of the solid and k is a characteristic of the substance at a constant temperature in a defined solvent (includes shear rate).

This equation describes the dissolution rate as a function of the release of the active ingredient rather than the solubility in the dissolution medium. Sink conditions can be achieved by appropriate selection of dissolution apparatus and dissolution medium. Upon examination of equation 11.1, it is noted that as the surface area of the solid dosage increases (for tablets which disintegrate), the dissolution rate increases with time; conversely, homogeneous pure tablets (nondisintegrating) rate of dissolution will decrease with time as the surface area gradually diminishes. It should be noted that there are unique circumstances where dissolution rates decrease upon disintegration, but this is not typical [14].

11.2.3. Practical Aspects

The dissolution test can be useful for ensuring bioequivalency of different batches of solid dosage formulations where correlations have been established; this characteristic is fundamental to enable straightforward comparison of early

---

*Sink conditions are established when the volume is at least 5–10 times larger than the volume required to form a saturated solution.

Table 11.1. *Q* values Expected for Immediate Release Products of Various Classes

Classification	Suggested Dissolution Test Specifications
Case 1: High-solubility high-permeability drugs	$Q = 80\%$ in 60 min
Case 2: Low-solubility high-permeability drugs	Two points 1. 15 min 2. Later point (30–60 min) $Q = 80$
Case 3: High-solubility low-permeability drugs	$Q = 80\%$ in 60 min
Case 4: Low-solubility low-permeability drugs	Limited or no— IVIV correlation expected

pilot batches of drug product used in clinical trials to the scaled-up final production batches.

There are other purposes of the test such as testing new formulations, comparing older to new products, or evaluate/check dissolution release characteristics. The test can also be used as a diagnostic tool in tracking tablet/capsule production trends.

11.2.4. Dissolution Specifications

Dissolution specifications are established to ensure batch-to-batch consistency and to signal potential problems with *in vivo* bioavailability. For new chemical entities the specifications should be established based on data from acceptable clinical, pivotal bioavailability, and/or bioequivalence batches. In addition, the specifications should be based upon the drug development process. Three categories of dissolution test specifications have been described [15], and suggested specifications are given in Table 11.1:

1. Single-point specification as a routine quality control specification (for highly soluble and rapidly dissolving drug products)
2. Two-point specifications
   a. For characterizing the quality of the drug product
   b. As a quality control test for certain types of drug products (e.g., slow-dissolving or poorly water-soluble)
3. Dissolution profile
   a. For acceptable product sameness under SUPAC-related changes*
   b. To waive bioequivalence requirements for lower strengths of a dosage form
   c. To support waivers for other bioequivalence requirements

* SUPAC is discussed separately in this chapter.

Dissolution tests should be carried out under mild test conditions, basket method at 50/100 rpm or paddle method at 50/75 rpm; profiles may be generated by sampling at 10–15-min intervals. The dissolution testing is expected to be performed at physiological conditions when possible. Aqueous media are preferred over media containing organic solvents, such as alcohols. The pH range of these aqueous systems is typically between pH 1.2 and 6.8. The use of surfactants (e.g., sodium lauryl sulfate) is recommended for drugs insoluble in water. The volume of the dissolution medium is generally 500, 900, or 1000 ml; sink conditions are desirable but not mandatory.

Specifications can also be based upon drug solubility and permeability considerations; the Biopharmaceutics Classification System (BCS) is recommended in literature [15].

11.3. USP/NF PHARMACOPEIA GENERAL CHAPTER ⟨711⟩

The USP 24/NF 19 basket and paddle procedures have been in existence for many years and have survived as the primary techniques for testing solid dosage formulations. Differences with other pharmacopeias will be noted where applicable. The discussion will also include potential problems with the equipment and test procedure as well as current techniques used throughout the industry.

11.3.1. Apparatii

The USP 24/NF 19 [16] discusses the dissolution test for apparatus 1 (Figure 11.2) and apparatus 2 (Figure 11.3) in General Chapter ⟨711⟩. All the necessary information about the specifications for equipment, criteria, tolerances, sampling requirements, deaeration of the medium, and a procedure for stage testing are given; these are outlined in Table 11.2.

The USP/NF describes a sampling area from which the aliquot of sample is taken. The *European Pharmacopeia*, 3rd edition; the *Japanese Pharmacopeia*, 13th edition; and the *British Pharmacopeia*, 1998 edition, all use the same apparatus and tolerances.

11.3.2. Parameters Affecting the Dissolution Test

11.3.2.1. Physical–Chemical Properties of the Drug Substance. There are several physical–chemical properties of the drug substance that can be controlled to influence dissolution characteristics:

1. Solubility in the medium
   a. pH
   b. Ionic strength
   c. Additives (surfactants, enzymes,)
   d. Overall volume

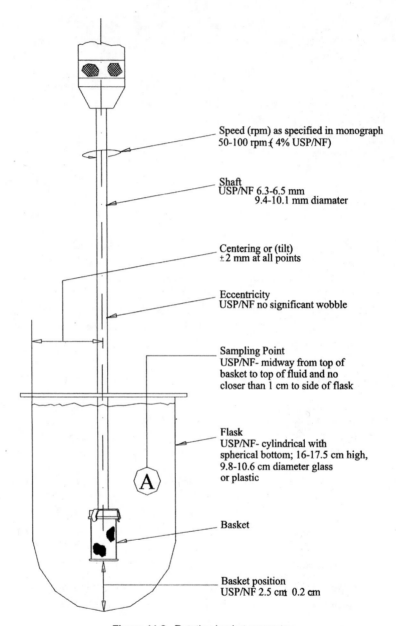

Figure 11.2. Rotating-basket apparatus.

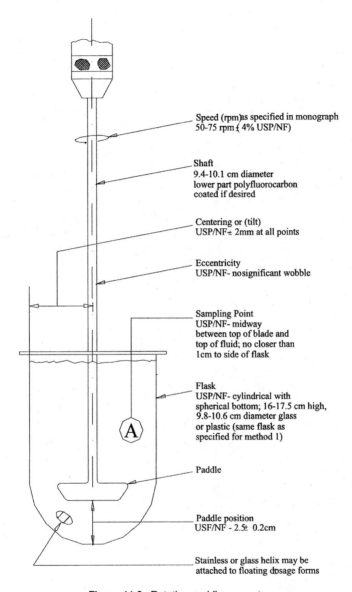

Speed (rpm)as specified in monograph
50-75 rpm { 4% USP/NF)

Shaft
9.4-10.1 cm diameter
lower part polyfluorocarbon
coated if desired

Centering or (tilt)
USP/NF± 2mm at all points

Eccentricity
USP/NF- no significant wobble

Sampling Point
USP/NF- midway
between top of blade and
top of fluid; no closer than
1cm to side of flask

Flask
USP/NF- cylindrical with
spherical bottom; 16-17.5 cm high,
9.8-10.6 cm diameter glass
or plastic (same flask as
specified for method 1)

Paddle

Paddle position
USF/NF - 2.5± 0.2cm

Stainless or glass helix may be
attached to floating dosage forms

Figure 11.3. Rotating-paddle apparatus.

2. Temperature (37°C as a standard)
3. Particle size/surface area
4. Crystal form

Other parameters and equations describe the dissolution properties of semi-solids (ointments and creams), suspensions, and transdermal formulations. The dissolution properties of the dosage form can be greatly influenced by the for-

Table 11.2. Nominal Settings for Dissolution Parameters

Parameter	Rotating Basket USP 24/NF 19	Rotating Paddle USP 24/NF 19
Water bath temperature (°C)	36.5–37.5	36.5–37.5
Dissolution medium	As specified in monograph, or 900 ml Dissolved gases must not interfere with the test	As specified in monograph, or 900 ml Dissolved gases must not interfere with the test
Required samples	6 + 6 + 12 sequenced until specification is met	6 + 6 + 12 sequenced until specification is met
Shaft speed (rpm)	As specified in monograph 50–100 4%	As specified in monograph 50–75 4% Lower speed preferred
Shaft diameter (mm)	6.3–6.5 9.4–10.1	9.4–10.1, lower part polyfluorocarbon coated if desired
Centering	2 mm at all points	2 mm at all points
Eccentricity	No significant wobble	No significant wobble
Sampling point	Halfway from top of basket to top of fluid; no closer than 1 cm to side of flask	Halfway from top of basket to top of fluid; no closer than 1 cm to side of flask
Flask	Cylindrical with spherical bottom 16–17.5 cm high, 9.8- to 10.6-cm-diameter plastic or glass	Cylindrical with spherical bottom 16–17.5 cm high, 9.8- to 10.6-cm-diameter plastic or glass
Basket position	2.5 ± 0.2 cm	2.5 ± 0.2 cm
Sinkers		Stainless steel or glass helix

318

mulation excipients, surface tension of the dosage form, granulation properties, compression force, and the storage/packaging conditions.

11.3.2.2. Variability of the Equipment.
The USP/NF allows certain variations, such as (a) gold plating of the basket to protect the stainless steel from corrosion in acid media and (b) coating of the paddle/shaft with Teflon for the same reason. The USP also allows the use of a wire helix to hold down floating dosage forms. This technique has not been standardized and still is a controversial topic with certain types of formulations. The wire helix may cause artificially high results by rubbing on the formulation or may cause artificially low results by not allowing the formulation to come apart properly. This is obviously an area where more study is required.

Geometry, Alignment, and Verticality. The axis of shaft is positioned so that it must coincide at all points with the vertical axis of the dissolution vessel to within 2 mm at any point. The shaft must also be at 90 degrees to the fluid in the vessel, and this straightness in verticality is measured with a level.

Stirring Rate. The stirring rate of the basket or paddle is usually between 50 and 100 revolutions per minute (rpm) with a tolerance of $\pm 4\%$. In general, the paddle is rotated at a slower speed than the basket. Paddle methods should normally be run at about 50 rpm and basket methods should be run at 100 rpm, but other speeds may be used. *In vivo* correlation experiments will ultimately determine which *in vitro* dissolution methods will be favored. Results obtained at 50 rpm with the paddle can usually be duplicated at 75 or 100 rpm with the basket. The paddle and basket, rotated at the same speeds, may give similar results if the drug is very soluble in the dissolution medium. Rotation speeds are measured by the apparatus internally and are displayed on a digital readout gauge. Rotation speeds can be quickly checked by attaching a piece of tape to the shaft, setting the speed of rotation, and allowing the tape to touch your finger by counting the revolutions using a calibrated stopwatch. Automated systems to check rotation speed are also currently available.

Temperature. The temperature of the medium in the dissolution vessel must be maintained at $37°C \pm 0.5°$. This temperature must be maintained during the entire test time. The temperature is measured with a thermometer that has been checked with an NIST calibrated thermometer. It is a good idea to check the temperature of the medium not only at the start of the test, but also at the end of the test time. Circulating water baths are used to maintain the temperature of the water slightly above 37°C, usually at 37.5°C. This is done because there is some heat loss through the glass or plastic vessels. Covers are used to prevent evaporation and maintain the temperature of the medium. The covers usually have a cutout for the shaft, sample, and thermometer ports.

Eccentricity. The USP/NF states "... no significant wobble ... or ... without significant wobble." The USP seems to constrain the outside wobble limit at 2 mm by defining the shaft axis as being no more than 2 mm from verticality with the vessel axis [1]. "No significant wobble" has been translated by engineers to mean that the stainless steel shafts can be manufactured and held in rigid chucks so that when rotated at the selected speed, 50 or 100 rpm, the measured wobble, near the bottom of the shaft, is 0.5 mm or less [1].

There is one more measurement required by the USP/NF, and that is the wobble measurement on the rim of the basket. The USP specification is not more than ±1.0 mm or a total runout of 2.0 mm. This tolerance is included to prevent use of baskets that are severely deformed from over use and handling.

The chuck is also part of the system, which holds the shafts. A poor chuck design can contribute to significant wobble. Wobble is measured with a runout gauge or wobble meter. There are several manufacturers of these gauges; these gauges are often supplied by the manufacturer of the test equipment.

Depth. The distance from the bottom of the basket or paddle to the bottom of the vessel is specified by the USP/NF at 25 ± 2 mm. This is measured with a depth gauge, and there are different manufacturers as well as home-made gauges in use. The gauge must slide under and be easily withdrawn from below the basket or paddle with very little up and down play between the bottom of the vessel and bottom of the paddle or basket.

When using apparatus 1, the basket must be attached to the shaft with the use of some force. This force is exerted upwards against the shaft. If the shaft is not held rigidly enough by the chucks, or if the chucks are worn, this upwards force may move the shaft slightly upwards. Eventually this movement will change the depth of basket bottom with respect to the bottom of the vessel.

Vessel. The dissolution vessel can be manufactured of glass, plastic or other transparent material and is available from the manufacturers of test equipment or glass catalogs. The tolerances and specifications for the vessel are given in the USP 24. If odd results are found with one particular vessel, check the flask for specifications. Newer vessels have tighter manufacturing tolerances and are more uniform in manufacture.

Care must be taken that the vessel material should not sorb, react, or interfere with the drug specimen being tested. This should be checked during method development before deciding what vessel is to be used for the test.

Centering. The shaft must be centered in the vessel. The USP/NF specifies that center of the shaft must be within 2 mm of the center or vertical axis of the vessel at all points. This is accomplished by use of a centering tool or an auto-centering template. Centering tools can be home made, or they can be purchased from equipment manufacturers. The use of a centering template, available with some manufacturer's equipment, removes the need for checking this variable.

Sampling Point. The sampling point has been defined by the USP/NF as midway between the side of the vessel and shaft and midway between the top of the medium and top of the basket/paddle, not less than 1 cm from the vessel wall. Sampling must be done as close to the same time as possible for all vessels. The USP allows a ±2% tolerance for sampling time when multiple sampling times are needed.

Sampling should be done so flow over the formulation will not be disturbed and to prevent influencing the rate of dissolution of the formulation. Sample probes are allowed but may not interfere with dissolution results. Cox et al. [9, 10] discuss this in detail.

Vibration. The test equipment, and more importantly the dosage formulation, must be kept in a vibration-free environment. Most modern test equipment meets this adequately, but there are no specifications given for vibration. Vibration is subjective and incumbent upon the user to maintain a vibration-free environment. If vibration is suspected, confirmatory testing must be done to isolate the vibration problem and demonstrate that test results as low as possible are obtained. There may be a vibration tolerance adopted by USP in the future.

Dissolution Medium. Properly designed and prepared dissolution medium is important for a successful dissolution test. Even from the beginning days of dissolution testing and research, dissolved gases in the medium were found to play a significant role in results obtained. The USP 24/NF 19 states that "... dissolved gases can cause bubbles to form which may change the results of the test. In such cases, dissolved gases should be removed prior to testing." This definition means that it is incumbent on the laboratory developing the dissolution test procedure to test the formulation in both deaerated and non-deaerated media. The *British Pharmacopeia*, *European Pharmacopeia*, and *Japanese Pharmacopeia* all say to "... use deaerated medium," thus taking this decision out of the analyst hands. There are several methods of deaeration, but there is no straightforward procedure for measuring the level of remaining dissolved gases.

Volume. The volume of the dissolution medium is usually 900 ml and measured by class A, 1000-ml graduate cylinders or by class A, 900-ml TD/TC volumetric flasks. The volumetric flasks are more expensive initially, but offer the advantage of accurate and repeatable measurements. Errors up to 2% may be introduced into the test procedure due to the inherent inaccuracy of the large-diameter graduated cylinder.

11.3.3. Test Equipment

There are many manufacturers of dissolution test equipment on the market. Some newer equipment uses a self-centering template to automatically center the round-bottomed vessel, making a centering tool and alignment unnecessary. Other equipment replaces conventional water baths with heating "mantels" with electronic temperature control and feedback. No matter what piece of equipment is in the laboratory, it should all have space for six to eight round-bottomed vessels and must meet the USP/NF requirements outlined in USP 24/ NF 19 General Chapter ⟨711⟩. The location of the test equipment must be vibration-free, and it may need a circulating water bath to hold the temperature of the test medium constant over the testing time and not contribute any vibration to the basket/paddle shafts or round-bottomed flasks. The speed of rotation must be held constant, and the shafts for the basket and paddle must meet the USP/NF specifications and should be equipped with a clutch to allow staggered starts for the individual flasks.

11.3.4. Stage Testing

The USP/NF uses a Q value as the specification for dissolution requirements. "Q value" is defined by the USP 24 as "the amount of dissolved active ingredient..., expressed as a percentage of the labeled content. The USP/NF gives a sequence of additional testing if any of the first six test units fail the USP stage 1 criteria. These criteria are given in General Chapter ⟨711⟩. The USP/NF testing sequence is 6 units + 6 units + 12 units until specification is met (Table 11.3). Stage testing may not be necessary if a value obtained in the first six units tested is outside the final stage 3 requirements. The lot would fail with the first six units tested. The *British Pharmacopeia*, 1998 edition, calls for testing six units with a release specification of not less than (NLT) 70%. If one unit fails, retest six units. The additional six units must all comply. The *Japanese Pharmacopeia*, 13th edition, calls for testing six units. If one or two units fail, retest six units; 10 of 12 must comply.

The individual drug monographs in the USP give the testing procedures for the drug formulation and the Q acceptance value.

Table 11.3.

Stage	Number of Vessels Tested	Acceptance Criteria
S_1	6	Each unit is not less than $Q + 5\%$
S_2	6	Average of 12 units ($S_1 + S_2$) is equal to or greater than Q, and no unit is less than $Q - 15\%$
S_3	12	Average of 24 units ($S_1 + S_2 + S_3$) is equal to or greater than Q, not more than 2 units are less than $Q - 15\%$, and no unit is less than $Q - 25\%$

11.3.5. Calibrators

To make the test sufficiently meaningful and give operators and managers some degree of confidence in the equipment and the test results obtained with the equipment, the USP introduced its calibrator tablets [16]. The disintegrating calibrator, prednisone tablets, and the nondisintegrating calibrator, salicylic acid tablets, were introduced with release specifications at 50 and 100 rpm using USP apparatus 1 and 2.

There has been much controversy about testing results obtained using the USP calibrators since their inception. One of the early major criticisms was that the prednisone calibrator was insensitive to dissolved air in the medium. Another early criticism, concerning the USP salicylic acid calibrator tablets, is that following USP paddle apparatus 2 technique instructions to not start the paddle rotation until after dropping the calibrator tablet, the tablets may stick on the sloping side near the bottom of the flask and not center themselves directly at the bottom of the round-bottomed flask. This normally causes higher results that will be outside the USP release specification.

On the other hand, even with the problems associated with the calibrators, they have brought a certain amount of uniformity to the dissolution test and elevated the test to a universally accepted technique. The USP has recently gone to a reduced testing posture with the latest lots of calibrators. In addition, the USP has expanded the release specification for the salicylic acid calibrator.

11.3.6. Sampling

11.3.6.1. Manual, Semiautomated, and Automated Procedures. Sampling may be accomplished by either manual, semiautomated, or fully automated procedures. The entire dissolution test can be automated, but lengthy validation and equipment costs must be closely considered before choosing this route. Most laboratories have settled on either a manual sampling procedure or a semiautomated procedure with sampling probes to remove aliquots for analysis. Both of these types of systems have been discussed in detail in references 9 and 10.

The USP 24/NF 19 specifies the area to take the sample specimen that was discussed earlier. It is defined as midway between the surface of the medium and the top of the blade (paddle) or basket and not less than 1 cm from the wall of the vessel.

Whether one samples manually or with a semiautomated arrangement will depend on the type of analysis, the number of drug formulations or similar samples to be tested, and the speed with which the information needs to be processed. A release test with a Q value and a single sampling time can be sampled at the end of the run by removing an aliquot of the test mix with a syringe–cannula arrangement. If this test must be done more than three times a day, a version of automated sampling may be useful. The USP/NF states that the test may be concluded in a shorter time than specified, if the requirement for minimum amount dissolved is met. If two or more sampling times are needed, the USP/NF allows sampling within $\pm 2\%$ of the stated times. The need to replace dissolution medium must be resolved during the method development of the test procedure.

Manual sampling is done with a syringe–cannula arrangement.

11.3.6.2. Staggered Start.
A finite time is needed to take aliquots and collect filtered portions of test units. If all six units are immersed at the same time, manual sampling of unit 1 will be at a different time than unit 6. Unit 6 is therefore in contact with the dissolution medium for a longer period of time. If rapid dissolution of the formulation is occurring at the sampling time, reproducible and variable results will be obtained. The USP/NF gives directions for starting the dissolution test. For apparatus 2, paddle, the USP/NF states, "The dosage unit is allowed to sink to the bottom of the vessel before rotation of the blade is started." For apparatus 1, basket, the USP/NF says to place the dosage unit in a dry basket at the beginning of each test. The USP means that the dosage units must enter the medium before rotation of paddle or basket starts.

A single operator is faced with a sampling dilemma without a staggered start option. The staggered start option is most easily accomplished when using apparatus 2, the paddle. Most, if not all, commercial equipment is equipped with a clutch arrangement that allows the operator to start and stop individual shafts. All the clutches on the paddle system can be disengaged. The preselected rpm is set, the first test unit is dropped into vessel #1, the run is started, and the clutch is engaged to start the paddle spinning. The remainder of the units is dropped at selected time intervals, usually about 1 min apart. When sampling time comes, the sample aliquot is withdrawn, filtered, and set aside. This is repeated with the rest of the test units as the sampling time arrives. A single operator has full control of the start and stop run time as well as adequate time to take the sample, filter, and prepare it for analysis.

The apparatus 1, basket, is not as convenient to stagger the start. In order to accomplish this, the shafts and baskets must all be removed from the medium in the vessel and held above the vessel by tightening the chuck. The dosage units are placed in the baskets. To start the run, all the clutches are disengaged. The first shaft chuck is loosened and the shaft and basket must be lowered to a stop mark previously made on the shaft to set the basket depth 25 mm from the bottom of the vessel. The chuck is tightened and then the clutch is engaged. The other units follow similarly at a preselected time interval usually longer than 1 min apart.

11.3.6.3. Positioning of the Dosage Unit.
The position the dosage unit assumes during the dissolution test can affect dissolution results no matter which apparatus is used. It is highly recommended that the operator *observe* what happens to the dosage formulation during the test. A minimum require-

ment would be to record your observation at the start, middle, and end of the test. If the basket is used, examine the residue left in the basket after the test.

When using USP/NF apparatus 1, basket, the dosage unit is enclosed in a confined space. When the basket is lowered into the dissolution medium, the lowering action forces the unit to tumble. Occasionally a bubble of air is trapped under the top of the basket and the dosage unit may become attached and trapped at the top of the basket. Test results for these units are usually lower than for the dosage units that remain at the bottom of the basket. Some dosage units swell in the basket and will not allow fluid to flow uniformly through the wire mesh. Some formulations break up rapidly and fall out of the basket to form a mound at the bottom of the vessel under the basket.

When using USP/NF apparatus 2, paddle, the dosage unit should come to rest on the bottom of the vessel directly under the paddle and centered with respect to the vertical axis of the vessel. Occasionally, the dosage unit sometimes gets stuck on the sides of the vessel before reaching the center bottom of the vessel. The velocity of the fluid stirred by the paddle is generally not sufficient to dislodge the dosage unit from its off-center position. However, the formulation is stuck in a higher fluid flow region, and results from these rogue positions are generally higher than dosage units, which are at the center of the vessel. The USP salicylic acid disintegrating calibrator tablet occasionally follows this rogue pattern, and higher results are obtained.

Some dosage units tend to float and may be carried along with the dissolution fluid flow. The use of sinkers is allowed by the USP/NF, but the sinker has not been standardized and caution must be exercised when using sinkers. They must not contribute to quicker release or retard the dissolution characteristics of the dosage unit.

11.3.6.4. *Filtering.*

Sample preparation prior to measurement of the sample often contains a sample filtration to remove suspended formulation components. This is one of the most problematic issues facing dissolution procedures and test methods. Using plastic syringes and filter discs has seen a dramatic increase and has become the norm for filtering dissolution aliquots. Filter discs constructed with nylon as the filter membrane are most commonly used for aqueous solutions; suitable filter materials and pore size are evaluated during dissolution method development.

Small ceramic porous discs, of various pore sizes, are also in use. The location of these ceramic discs can be critical. Some operators place the disc on the end of the sampling probe and insert the entire arrangement into the dissolution medium at sampling time. The sample aliquot is drawn through the disc, no filtrate is discarded and the sample is either measured directly or diluted and measured. This may be adequate for single time analysis but unsatisfactory for multiple sampling times. The position of the filter disc in the dissolution medium can disrupt the fluid flow and influence the dissolution results.

Plastic syringes of various sizes from 1 ml to 50 ml are common; these syringes usually have silicone lubricant in them along with a black rubber (plastic) gasket in the plunger portion of the syringe. The syringe is fitted with a stainless steel cannula, the dissolution aliquot is withdrawn, the cannula is removed, and a filter disc is attached. The first portion is filtered and discarded, and the remainder is filtered and used for further manipulation and/or measurement.

Regardless of the type of syringe or filter used, they need to be evaluated during method development/validation to make sure they are neither adding interference to the measured solution nor causing adsorption of active material from the solution to be measured.

The procedure of Cox et al. [9] can be used to validate the standard preparation, and the method used is essentially described as follows. Several standard solutions of different known strengths should be prepared in the water-miscible solvent. Each of these standard solutions should be diluted in the dissolution medium such that the final solution has the same drug concentration but different concentrations of the water-miscible solvent. The analysis of these solutions will demonstrate any problems associated with the water-miscible solvent. The lowest concentration of water-miscible solvent should be employed for the standard preparation. Finally, the standard solution used for calculating the sample dissolution results should be filtered identical to the sample filtration. The measured absorbance of the filtered standard should be identical to the unfiltered standard, thus showing no added interference or adsorption of the active drug component by the filter or syringe–cannula arrangement.

11.4. MEASUREMENT OF THE PHARMACEUTICAL ACTIVE

The detection of the active drug component in the dissolution medium can be accomplished by many and varied analytical techniques. Detailed descriptions of the more common techniques, HPLC and UV absorption, are reviewed in other chapters of this text.

After taking a sample aliquot of the dissolution medium, the sample aliquot is filtered if required and diluted to a concentration to allow proper measurement. In some instances, interfering substances in the drug formulation may be present which make the determination of the analyte difficult. Because the use of placebos is not acceptable to the FDA or USP/NF, the interfering substance can be made available as a standard and added to the analyte standard at the concentration in the measured solution when using UV assays. However, a specific and rapid HPLC assay method, which eliminates interferences from the formulation ingredients, is preferred. If a single wavelength is used for measurement, one should insure that the formulation excipients interfere minimally (typically less than 1.5%). Analytical wavelength must be chosen properly to maximize the sensitivity and minimize matrix and excipient interference.

There is a major problem when using HPLC for measurement of the pharmaceutical active since you are often required to inject aqueous, buffered aqueous, or rather strong acidic aqueous solutions onto sensitive HPLC re-

Table 11.4. Random Input Variables Checklist

Variable	Maximum Allowable	Excess Commonly Seen	Effect of Excess	Methods of Control
1. Eccentricity	±2 mm (compendium) ±3/4 (optimal)	2–5 mm	+4–8%	Straighten shafts, use wide shaft-guide points
2. Vibration	0.1 mils displacement at vessel	0.2–0.9 mils	+5–10%	Eliminate source
3. Alignment	1.5° to perpendicular	2°–7°	+2–25%	Adjust alignment in field
4. Centering	±2 mm (compendium)	±2–6 mm	±2–13%	Center individual flasks
5. Agitation rate	±4%	±10%	Linear	Use better, smoother control or use synchronous drive
6. Dissolved gas	Deaerated	Bubbles form	±50%	Dearate media by various methods
7. Media pH	0.00 accuracy	±0.05	±10%	Check buffers or deaerate, calibrate the pH meter
8. Media contamination	ppm	Ions, surfactants	Substantial	Carefully control media
9. Evaporation	None	2–5%	Linear	Use flask covers
10. Temperature (°C)	±0.05 (compendium) ±0.01 (optimal)	1°–2°	Linear	Monitor individual flasks, allow adequate equilibrium
11. Flow pattern	No interference	Turbulence from probes	Substantial	Remove probes
12. Sampling position	Compendium	±0.5 cm	Little	Use care
13. Filters	No sorbing	Considerable blockage	Significant	Use bidirectional filter flow, check sorbing
14. Detection	Use standard	Interference	Considerable	Use standard
15. Sorption	None	Considerable	Significant	Check materials

versed phase columns. Acidic solutions can degrade the column slightly or cause a large change in the mobile phase component thereby causing retention time shifts of the analyte peak. These retention time shifts may be changing slowly with each injection and may not be noticeable until after a series of injections has been made and the data examined. Dilution of the sample may be required to reduce the matrix effects.

11.5. ANALYST CHECKLIST

In conclusion, Hanson recommends an analyst checklist of parameters when using and validating the dissolution test procedures [1]. This checklist is of great value to operators who only occasionally run a test sample, and it is also a good reminder for operators who run tests frequently but may lose track of what needs to be checked. A similar checklist can be found in Table 11.4.

One final note: When using the basket procedure, a check of basket, rim, wobble, and depth should be made more frequently than at the recommended 6-month or yearly calibration schedule because the fragile baskets can distort with use.

REFERENCES

1. W. A. Hanson, *Handbook of Dissolution Testing*, Aster Publishing Corp., Eugene, OR, 1991, p. 2.
2. G. Knapp, Dissolution Testing: It's Increasing Role, *Pharm. Tech.* **1977**, *1*(4), 12–13.
3. J. G. Wagner, and M. Pernarowski, *Biopharmaceutics and Relevant Pharmakokinetics*, Drug Intelligence Publication, Hamilton, IL, 1971.
4. M. Pernarowski, W. Woo, and R. O. Searl, *J. Pharm. Sci.* **1968**, 1419–1428.
5. G. Levy, J. R. Leonards, and J. Procknal, *J. Pharm. Sci.* **1967**, 1356.
6. G. Levy, D. Pharm, and B. A. Hayes, *N. Engl. J. Med.* **1960**, 1053.
7. J. Poole, Some Experiences in the Evaluation of Formulation Variable in Drug Availability, *Drug Inform. Bull.* **1969**, *3*, 8–16.
8. R. D. Kirchhoefer, *J. Assoc. Off. Anal. Chem.* **1976**, *59*, 367.
9. D. C. Cox, C. C. Douglas, W. B. Furman, R. D. Kirchhoefer, J. W. Myrick, and C. E. Wells, *Pharmaceutical Technology, Guidelines for Dissolution Testing*, **April 1978**.
10. D. C. Cox, W. B. Furman, T. W. Moore, and C. E. Wells, Guidelines for Dissolution Testing: An Addendum, *Pharm. Tech.* **1984**, *8*(2), 42–46.
11. W. A. Hanson, Solving the Puzzle of Random Variables in Dissolution Testing, *Pharm. Tech.* **1977**, *1*, 5, 30–41.
12. K. Thakker, N. C. Naik, V. A. Gray, and S. Sun, Fine Tuning of Dissolution Apparatus, *Pharm. Tech.* **1997**, *1*, 5, 30–41.

13. Guidance for Industry, *Dissolution Testing for Immediate Release Solid Oral Dosage Forms*, United States Department of Health FDA (CDER), August 1997, http://www.fda.gov/cder/guidance.htm.

14. T. Macek, Formulations, in *Remington's Pharmaceutical Sciences*, 15th edition, A. Osol (editor), Mack Publishing Co., Easton, PA, 1975.

15. G. L. Amidon, H. Lennernas, V. P. Shah, and J. R. Crison, A Theoretical Basis for a Biopharmaceutical Drug Classification: The Correlation of *In-Vitro* Drug Product Dissolution and *In-Vivo* Bioavailability, *Pharm. Res.* **1995**, *12*, 413–420.

16. United States Pharmacopeial Convention, *United States Pharmacopeia*, USP 24/NF 19, 12601 Twinbrook Parkway, Rockville, MD, Rand McNally, Taunton, MA, 2000.

12

ANALYTICAL METHOD DEVELOPMENT FOR ASSAY AND IMPURITY DETERMINATION IN DRUG SUBSTANCES AND DRUG PRODUCTS

Jonathan B. Crowther, Paul Salomons, and Cindi Callaghan

12.1. BACKGROUND

Chapter 1 provided a thorough review of the requirements of analytical methods throughout the process of discovery and development (clinical, toxicology, and manufacturing/controls), through regulatory approval, and final manufacture and sale of the pharmaceutical product. The analytical methods that are ultimately filed in registration documents are used in the pharmaceutical quality control laboratory to do the following:

1. Identify the drug substance or product.
2. Quantitate the pharmaceutical active ingredient.
3. Determine level of purity.
4. Guarantee the overall quality of the product.

Analytical Chemistry in a GMP Environment. Edited by J. M. Miller and J. B. Crowther
ISBN 0-471-31431-5 © 2000 John Wiley & Sons, Inc.

These filed methods are typically run at initial release and during stability analysis; the results of the tests are compared against established specifications (regulatory and/or in-house). Additionally, in-process analysis may also be used to ensure and monitor quality during the manufacturing process.

Once the test methods are developed and validated, they may be transferred to a control lab, perhaps at a remote manufacturing site, or contract laboratory. Methods destined for a control laboratory [1] must be accurate, reliable, robust, and precise—preferably uncomplicated versions of research "screening-type" methods used in earlier stages of development.

In brief, the pharmaceutical method development scientist has the task to develop reliable yet simplified analytical procedures that are robust and able to stand the scrutiny of validation and method transfer. This challenge is confounded by industry requirements to continuously expand and improve the capabilities of pharmaceutical testing methods; yet current business trends often require accelerated product development timelines while resources committed to these efforts are often minimized to control costs.

This chapter will examine the blend of good organizational and technical skills required for successful and straightforward development of analytical methodology for the modern pharmaceutical-testing lab. The technical discussions are focused on the development of a stability-indicating high-performance liquid chromatography (HPLC) assay and purity release methods, because these represent some of the more difficult challenges common to most pharmaceutical laboratories.

12.2. INTRODUCTION

Chapter 1 of this text reviewed the variety of analytical methodology applied during the drug discovery and development (support to clinical trials, formulation development, pharmacokinetic and safety studies, support to regulatory filings) phases. The present chapter, however, will focus its discussion on methods required to produce, release, and support marketed products, specifically HPLC methods. Clearly, the objectives of each method development project will differ based upon end-user, method purpose, regulatory, and international requirements.

As previously stated, analytical methods that are destined to be used in a quality control environment require an additional degree of refinement [1] compared to research methods; methods for multilaboratory use are required to be robust. The general goal described in this chapter is to develop validatable, transferable, robust, reliable, accurate, and precise (V-TR^2AP) methodology.

From later stages in development to when the drug application is filed and approved, end-user laboratories typically test samples against the filed (or in-house) specification for the drug substance or drug product. Ordinarily, these methods (or able precursors to these methods) were used to determine and set the specifications presented to the regulatory authorities during development.

Tables 12.1 and 12.2 illustrate the variety of tests that may be required for release and monitoring drug substance and drug product. Indicated in each table are the typical validation parameters required for each test method type.

12.2.1. Specifications and Their Influence on Method Development

The International Conference on Harmonisation has released draft guidance on specifications, entitled "Test Procedures and Acceptance Criteria for New Drug Substances and New Drug Products: Chemical Products" [2]. A separate guideline addresses specifications, tests, and procedures for biotechnological/ biological products [3]. As defined in the document for chemical products,

A specification is defined as a list of tests, references to analytical procedures, and appropriate acceptance criteria that are numerical limits, ranges, or other criteria for the tests described. It establishes the set of criteria to which a drug substance or drug product should conform to be considered acceptable for its intended use.

Also noted in the same ICH guidance:

Specifications are chosen to confirm the quality of the drug substance and drug product rather than to establish full characterization, and should focus on those characteristics found to be useful in ensuring the safety and efficacy of the drug substance and drug product.

Specifications remain a binding standard of quality between the regulatory agency and the applicant. In general, a drug product or drug substance conforms to a specification when:

1. The article is tested according to the listed analytical procedure.
2. Values obtained are within the listed acceptance criteria.

Stated differently, the method and the specification limits are related; a change in one (i.e., method improvement or simplification) may justify change in another; this may require notification to regulatory agencies.

For excipient materials or active drug substances that are tested using compendial procedures, methods and specifications derived from pharmacopeias must be determined to be "suitable" for the given product or drug substance. Chapter 3 of this text reviews pharmacopeia requirements in detail.

12.2.2. International Guidelines and their Influence on Method Development

Clearly, all quantitative analytical methods used to support regulatory filings (setting of specifications, etc.), toxicology testing, release of clinical or marketed

Table 12.1. Typical Analytical Test Methods Required in Specifications—Drug Substance

Test (Specification)	Validation Requirements[a]	Requirement of the Method
Appearance—description		A qualitative statement about the state (e.g., solid, liquid) and color of the drug substance.
Identity (I.D.)	5	Able to differentiate the drug substance from compounds of closely related structure that are likely to be present.
Assay	1, 2, 3, 4, 5, 8, 9	Quantitate the amount of drug substance; highly desirable for the assay to be specific and stability indicating.
Impurity analysis Related substances	1, 2, 3, 4, 5, 6, 7, 8, 9	Levels determine through the development process coupled to toxicology information. Compare against ICH limits.
Residual solvents	1, 2, 3, 4, 5, 6, 7, 8, 9	Compare against known toxicity and ICH limits.
Inorganic—heavy metals	1, 2, 4	Detect and quantify heavy metals such as lead.
Physiochemical properties Melting point	1, 2, 4	Accurately determine a temperature at which a solid melts.
Refractive index	1, 2, 4	Determine the ratio of velocity of air in 589-nm light to velocity of a substance.
Viscosity	1, 2, 4	Ability to determine the resistance to flow or shear stress strain.
Optical rotation	1, 2, 4	Determine the rotation of polarized light.
Particle Size	1, 2, 4	Determine the size of small particles on the micron and submicron scale, and determine distribution of particles.
Solubility	2, 4	Determine the amount a substance dissolves in a specified solvent.

Table 12.1. *(continued)*

Test (Specification)	Validation Requirements[a]	Requirement of the Method
Bulk density/volume	2, 4	Determine the ratio of mass of a substance per a specific volume.
pH of aqueous solution	2, 4	Determine the pH of solution at a given temperature.
Chiral purity	1, 2, 3, 4, 5, 8, 9	Accurately detect and quantitate optical active isomers.
Residue on Ignition (ROI)	1, 2, 4	Determine the amount of organic material present.
Water content/loss on drying	1, 2, 4	Accurately determine the amount of water or moisture present.

[a] 1, Accuracy; 2, repeatability; 3, intermediate precision; 4, reproducibility; 5, specificity; 6, detection limit; 7, quantitation limit; 8, linearity; 9, range.

materials, and methods used in stability studies require some form of validation. Requirements for validation of various analytical methodologies are outlined in both *United States Pharmacopeia* (USP) [4] and International Conference on Harmonization (ICH) guidelines [5]. Drug registration agencies have accepted these guidelines; a review of the analytical method validation process and requirements are outlined in Chapter 15 of this text.

The harmonization of the validation procedures for pharmaceutical test methods directly influences the requirements for method development. In fact, ICH guidelines are reasonably specific. For instance, the guides suggest that impurity methods provide sufficient sensitivity to detect recurring impurities at 0.1% and above, (depending on dose) with reasonable precision and accuracy. In developing HPLC methodology, these validation requirements stipulate that stability-indicating impurity methods be designed and validated to do the following:

1. Simultaneously separate, identify, and quanititate degradation compounds/impurities from the "active" drug substance
2. Be free from interferences from the excipient materials.

On the other hand, content uniformity and dissolution methods may not be required to be "stability-indicating" because separation of the active compound and impurities may not be critical to these tests, particularly when impurities are determined through separate validated methods. These non-stability-indicating methods often offer advantages of simplicity and high sample throughput, along

Table 12.2. Typical Analytical Test Methods Required in Specifications—Drug Product

Classification	Test (Specification)	Validation Requirements[a]	Requirement of the Method
Solid dosage forms	Description		A qualitative description of the dosage form should be provided (e.g., size, shape, color).
	Identification	5	Able to differentiate active ingredient in a drug product from compounds of closely related structure that are likely to be present.
	Water content	1, 2, 4	Determine amount of moisture present is a given material or substance.
	Composite assay	1, 2, 3, 4, 5, 8, 9	A specific stability indicating assay.
	Degradation products—related substances	1, 2, 3, 4, 5, 6, 7, 8, 9	Able to quantitate impurities or degradants from an active.
	Solvents	1, 2, 3, 4, 5, 6, 7, 8, 9	Determine levels of solvents used during manufacturing and chemical process.
	Content uniformity	1, 2, 3, 4, 5, 8, 9	Uniformity of content and uniformity of mass.
	Dissolution/disintegration	1, 2, 3, 4, 8, 9	Refer to ICH Guidance Decision tree #7. Determine the rate at which an active ingredient is released (*in vitro*) into a liquid medium.
	Hardness and friability (tablets) friability (capsules)	2, 4	Typically an in-process control unless hardness and/or friability have an impact on product quality.
Solutions, injectables, lyophilized for reconstitution	Identification	5	Able to differentiate active ingredient in a drug product from compounds of closely related structure that are likely to be present.
	pH of solution	1, 2, 4	Determine the pH of solution at a given temperature.
	Particulate matter	1, 2, 4	Detect extraneous particles.
	Assay (potency)	1, 2, 3, 4, 5, 8, 9	Determine the amount of active present.

Degradation	1, 2, 3, 4, 5, 6, 7, 8, 9	Able to quantitate impurities or degradants from an active.
Uniformity of dosage unit	1, 2, 3, 5, 8, 9	Uniformity of content.
Dissolution/disintegration	1, 2, 3, 4, 8, 9	Refer to ICH Guidance Decision tree #7. Determine the rate at which an active ingredient is released (*in vitro*) into a liquid medium.
Osmolality	1, 2, 4	Able to detect flow of molecules from an area of greater concentration to an area of lesser concentration through a solution or membrane.
Reconstitution time	1, 2, 4	Ability to dissolve a powder in the diluent.
Particle size	1, 2, 4	Determination of the particle size distribution.
Redispersibility	2	Ability to resuspend a solution.
Specific gravity	1, 2, 4	Determination of mass versus water at 25°C.
Extractables	1, 2, 4	Ability to detect compounds leaching from a container or closure system.

[a] 1, Accuracy; 2, repeatability; 3, intermediate precision; 4, reproducibility; 5, specificity; 6, detection limit; 7, quantitation limit; 8, linearity; 9, range.

with less validation rigor. Similarly, methods used for drug substance/drug product identification are relatively straightforward to develop and validate.

In summary, many of the requirements of a method development project are being dictated through regulation, internationally accepted guidelines, and knowledge of current "good practice" accepted in the pharmaceutical industry.

12.3. THE METHOD DEVELOPMENT LIFE CYCLE—OVERVIEW

The process of method development "life cycle" parallels established approaches to analytical method validation [6] and computer validation [7]. Clearly, analytical methods have finite lifetimes and should be reviewed periodically, and then they should be revised or changed if required. Figure 12.1 provides an overview of the organizational components of the pharmaceutical method life cycle for filed methods. Successful implementation of the method development process requires careful planning and development of the requirements of the method, excellence in laboratory work, qualified instrumentation, and proper documentation from beginning to end.

Indeed the goals of any analytical development project should include the "scope," designing flexibility and robustness into the procedure, understanding the needs and environment of the lab where the method will ultimately transfer, and considering validation requirements. The method developers need to strive to minimize the effort required to use the method and simplify processing and interpretation of the data.

12.4. PLANNING

12.4.1. Review Company Policy on Method Development/Validation

Initially the method development scientist should review the in-house policy (or client's policy in a contract development laboratory) for method development, when available. The development policy sets standards that are required and sets goals to be met (with "reasonable" effort) for development of each type of analytical method and should consider current regulatory requirements. Because the ultimate objectives of the method development process are validation and transfer, these objectives should be incorporated into the policy document. A properly drafted policy document systemizes the approach to method development enabling different labs or contract facilities working on separate projects to develop analytical method of similar quality and capability. Thus the policy document requires a careful balance between practical considerations (resources and timelines) and target goals. The document could be quite specific, indicating the type of preferred analytical instrumentation and column vendor, or it could be very general. In either case, some flexibility should be in the policy. The group responsible for establishing the policy should consist of

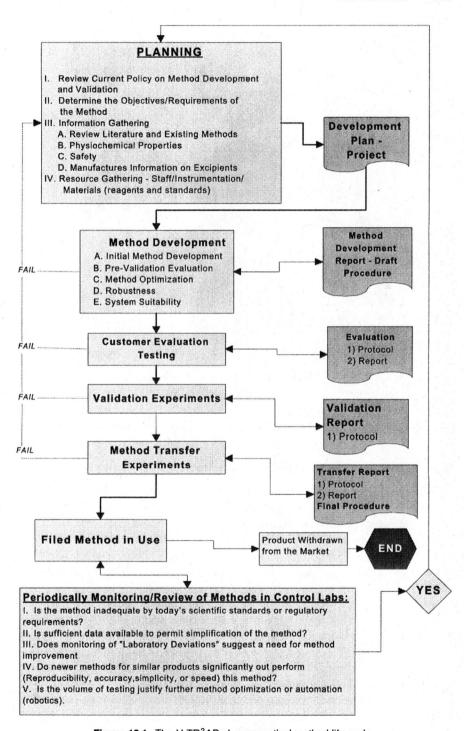

Figure 12.1. The V-TR^{2}AP pharmaceutical method life cycle.

Figure 12.2. Active participants in defining analytical method objectives.

developers as well as end-users whose decisions rely on the data generated by the analytical methods or may ultimately be responsible for the methodology.

12.4.2. Defining the Objectives/Requirements of the Method

Because many decisions will ultimately be made from the results generated from the methods, the scope and requirements of the method must be carefully defined up front. Indeed, these objectives ultimately define the extent of development plus the degree of validation, robustness, and optimization required.

Recent changes in business practices often require development scientists to be members of cross-functional project teams. They may work closely together with members of other departments (Figure 12.2) to ensure sufficient up-front planning, information gathering, and project organization; this approach attempts to ensure general agreement with regard to the requirements of the analytical method. Once the method goals are defined and documented, implementation of the laboratory portion of the method development process can begin. The new method, or "method improvement" if working on improvements of existing methodology, may require regulatory or other department support prior to beginning the project. Such support may be justified, especially if the updated methods require changes to the established specifications or influence product shelf life.

As stated above, the regulatory and compliance requirements must be considered early in the initial planning phase of a method development project. In addition, method validation parameters and possible method transfer requirements must also be considered when defining the requirements of the method.

12.4.3. Illustration of Method Requirements

Method requirements may be divided into several functional areas. These may include the following:

1. Regulatory requirements
2. Technical requirements
3. Practical requirements
4. Validation requirements
5. Considerations when transferring analytical methodology

12.4.3.1. Regulatory Requirements. The regulatory requirements of the method are generally straightforward, and they are often outlined into the "scope" or "purpose" section of the test method. In general, these requirements are directly related to the specification the method supports—for example, establishing assay, purity, or identification of the drug substance or product. Validation and method transfer requirements for these methods are reviewed below.

12.4.3.2. Technical Requirements. In general, the technical requirements are often suggested as target goals in a company policy or practice on method development. These technical requirements that may be considered in the development plan might include the following:

- Are the impurities and/or degradants to be separated known and available as standards? What is the justification for their choice? *References to summaries of stressed stability study reports are helpful* in addition to data available from ongoing stability studies.
- List the target method capabilities (i.e., to be able to separate all impurities, have a limit of quantitation (LOQ) of related substance < 0.05%).
- List the required validation parameters (USP and ICH) with target acceptance criteria (i.e., achieve linearity between 50% and 150%).
- Upon review of the formulation ingredients, what interferences might be a concern due to excipients?
- Will there be any anticipated sample preparation issues?
- For chiral separations, will capillary electrophoresis (CE) be preferred over HPLC using chiral stationary phases or additives?

· Can this HPLC method be designed to be compatible with liquid chromatography–mass spectroscopy (LC-MS) or Liquid Chromatography-Nuclear Magnetic Resonance (LC-NMR)?

These technical requirements must be ordered and ranked according to importance to enable streamlined and practical method development.

12.4.3.3. Practical Requirements of the Method. Once other requirements of the method are defined, practical issues should be incorporated into the development plan. In any method development process where the method is destined to be used in a control environment, the items in Table 12.3 need to be considered.

12.4.3.4. Validation Requirements of the Method. The validation requirements of the method are specifically outlined in regulatory guidelines. These requirements are coupled with proposed targeted acceptance criteria based on previous knowledge, company policy, or literature values for validation acceptance criteria [8, 9]. For general HPLC assay and purity methods the following validation parameters will typically be evaluated:

· Specificity
· Linearity
· Accuracy
· Range
· Precision (repeatability, intermediate, reproducibility)
· Limit of detection/quantitation
· Solution stability (recommended)
· Robustness (recommended)
· System suitability criteria

Details regarding definitions of validation terms, appropriate selection of parameters, and choice of acceptance criteria are reviewed in detail in Chapter 15. Method development activities should anticipate providing a method capable of meeting these validation criteria.

12.4.3.5. Considerations When Transferring Analytical Methodology. The following should be considered in the development plan if the method is expected to be used in a variety of laboratories with varying levels of expertise:

· Can we build simplicity into the method?
· Can we design the method to be robust?
· Can we review the listing of the proposed end-user (and end-user labs) that will use the method. What is the general level of instrument qualification and laboratory expertise?

Table 12.3. Examples of Practical Requirements to Be Considered During Method Development Planning

Considerations	Comment
Level of simplicity	What are the capabilities and qualifications of the target operating laboratory?
Degree of robustness	What type of laboratories will ultimately use the method?
Type of instrumentation	What instrumentation is available in the target lab? Testing such as particles size and viscosity is often instrument vendor specific; HPLC gradient "dwell" times vary among manufacturers and model. Would method development be simplified through standardization of laboratory instrumentation and equipment?
Cost considerations	Is the method cost effective (i.e., sample filtration versus the cost of a new column)? What are the number and frequency of samples to be tested? What are the costs associated with disposables required for sample preparation? Can the method design minimize the use of unique and possibly expensive materials or standards?
Sample preparation	What are the suspected difficulties in sample preparation? Is robotic automation required or anticipated, requiring specialized equipment or disposables? Are solvents impurities (residual peroxides), pH, temperature, or light known to cause stability concerns in sample solutions?
Safety/environmental	Are there avoidable safety hazards? Can concentrations/volumes be adjusted without affecting method accuracy or precision?
Run time, sample turn-around	Will the run time be critical? In-process analysis requires reasonable (sometimes "immediate") turnaround. HPLC assay/purity during release testing may permit 24-hr turnaround. Is the rapid method still a V-TR^2AP method?? What real benefit is realized in reducing the analysis time by 5 min if the method is less robust?
Testing volume	Will this method be used to an extent that significant effort is warranted to fully optimize the method?
Sample amount/ concentration	Bioanalytical analysis may permit only small sample volume. Potent drug products of low dose may require multiple samples to be tested or concentrations to be enriched. Will "recovery" be an issue?
Specialized reagents/ columns	Are the reagents and columns (HPLC/GC/CE) specified in the method widely and consistently available? Are the reagents of special purity? Is the water quality used in the HPLC method controlled?

Table 12.3. *(continued)*

Considerations	Comment
Resolution of impurities/degradants	Does the chromatographic method enable sufficient resolution of the impurities and degradants to permit straightforward integration that is repeatable from lab to lab?
Reference standards	How and in what quantities will reference standards be distributed and maintained?
Special considerations	Mass spectrometry compatibility for possible troubleshooting and for validation. HPLC-MS may also be useful during validation (specificity).

- Can we avoid unique instrumentation that may be unavailable/unpractical in the control lab?
- Can we minimize the method analysis time while maintaining robustness?
- Can we provide meaningful system suitability requirements?
- Can we incorporate automation (sample prep, etc.) into the design?
- Can we observe the end-user run the method. Can they perform the method as written?
- Can we avoid unnecessary safety risks or use of toxic solvents?
- Can we ask the receiving or "customer" laboratory to evaluate or "trial" the method before beginning validation studies?

12.4.4. Information Gathering

Often, there is a temptation to "jump in" to the experimental portion of the method development process. In general and prior to beginning laboratory method development activities, the method development scientist should consider checking resources for existing or similar methods, physical and structural information of the compound of interest, and information on the proposed formulation. Up-front knowledge is invaluable in streamlining the method development process. Table 12.4 generalizes information that may be useful in development activities.

Significant physiochemical information and stressed stability information should be available, because much of this can be found in the initial investigational new drug (IND) regulatory filing. Similarly, formulation development information should become available throughout the development process. Understanding the physical and chemical properties of the drug substance up front enables the developing scientist to successfully approximate initial method parameters. Characteristics such as structure, solubility, particle size, polarity, ultraviolet (UV) characteristics, and pK_a are invaluable to the developer. Unless some understanding of the properties of compound(s) exists, the experimental portion of the method development process becomes iterative, often resulting in

Table 12.4. Information Useful in Method Development

General Information[a]	HPLC Assay	HPLC Purity	Dissolution with HPLC	Residual Solvent	Identification
Chemical structure and molecular weight	√	√	√	√	√
UV spectra	√	√	√		√
Solvent solubility	√	√	√	√	
pH solubility	√	√	√		
pK_a	√	√	√		
Solvents used in manufacture		√		√	
Information in regulatory files (IND, NDA, IRF)	√	√	√	√	√
Specifications—set or proposed	√	√	√	√	√
Sensitivity to degradation	√	√	√	√	
Formulation and formulation components—use of placebo	√	√	√	√	
Number of impurities: Description of chemical synthesis of the drug substance	√	√	√		
Number of degradation products 1. What environmental conditions are these degs formed? 2. What mechanism is involved in their formation	√	√	√		
Safety Information— MSDS[b]	√	√	√	√	√

[a] Individual properties of the drug substance, degradants, and synthetic impurities may need to be considered.

[b] MSDS, material safety data sheets.

project delays. Chapter 10 of this text reviewed how adapting this information can lead to development of efficient and robust HPLC methods. Similar background information is desired when developing dissolution methodology (Chapter 11).

Information regarding formulation ingredients may be as useful as knowledge of the properties of the active drug substance. Members of a cross-functional team (i.e., formulation development) may be able to provide this information to the method development team. Otherwise, Material Safety Data Sheet (MSDS) sheets and manufacturer's literature are often the best sources for background information on excipient ingredients. An example of the importance of this information for method development is demonstrated by the fact that chromatographic peaks in impurity assay can often be attributable to colorings and flavorings used in the formulation but proper choice of wavelength or chromatographic condition may minimize the extent of the interference. Similarly, a prior knowledge of the level and type of antioxidants or preservatives in a formulation may enable simultaneous determination of their assay along with the assay of the active, there by eliminating the need for multiple methods. Knowledge of the results of stressed light stability studies may suggest that solutions of the active drug substance must be handled in amber glassware to protect them from light. Similarly, pH, heat, and moisture stressing (of both the drug substance and drug product) provide useful information to the development scientist.

Additional information that may be required if you are reworking or improving an existing method is listed below:

1. Input from laboratories that have used the method and have indicated their concerns
2. A review of the laboratory deviations with regard to use of the method

12.4.5. Resource Gathering: Resources/Instrumentation/Materials and Standards

In putting together a project plan for a method development and validation initiative, the necessary resources need to be assembled. Samples, reagents, and standards must be available along with laboratory instrumentation and staff; otherwise, the project may experience delay. Samples such as placebo formulations, standards of specified impurities, and certified reference standards of active material may require both commitment and resources in groups outside the development organization. In addition, the availability of the required instrumentation (which may require budgeting, purchase, and qualification) and trained staff may require significant up-front planning. Table 12.5 reviews some of the resource requirements that may be required.

12.4.6. Documentation: Development Plan

The requirements of the method are reviewed, prioritized, and documented in a method development plan. This document may serve as a reference for the method development report. To support the requirements, the justification to the choice of related compounds that must be separated by the HPLC method is documented. A project timeline may also be part of the development plan; this ensures that sufficient resources are allocated and will be available to enable the project to remain on schedule.

Table 12.5. Resource Planning—Method Development

Laboratory Staff

Method development scientist(s)
Validation scientist(s)
Transfer—scientist(s) at both sites; travel may be
  required
QA review—validation and Transfer protocols

Instrumentation and Equipment

Calibration/qualification instrumentation
Purchase additional instrumentation/equipment if
  required

Materials and Standards

Solvents used in methodology
Placebo materials
Individuals formulation components
Related substance standards
Reference standard of the active
Accelerated stress samples

12.5. METHOD DEVELOPMENT—GENERAL CONSIDERATIONS

12.5.1. Initial Method Development

Once the plan, the resources, and the objectives of the method are finalized, the laboratory portion of the method development process begins. In brief, at the initial stage of development of an HPLC assay/purity method, the mobile-phase conditions and HPLC column types are evaluated in order to provide resolution of the related substances from the active drug substance. In an example drug substance method development activity, ten related substances (0.25% level) are resolved from the active in the chromatogram shown in Figure 12.3. This group of eight related substances was chosen from reviewing stressed and real-time stability data (this may not be available for newer products in development) and knowledge of known impurities attributed to the synthetic process. For a drug product, placebo or individual ingredients would also need to be chromatographed to determine if they interfere.

The development plan further distinguishes that of the ten related substances, only three are "specified" impurities.* The method development plan further requires that the method accurately identifies and quantitates the speci-

---

* Specified impurity is defined as an impurity known and/or suspected to affect the quality and safety of the drug substance or drug product. Individual impurities are listed individually and limited to assure quality.

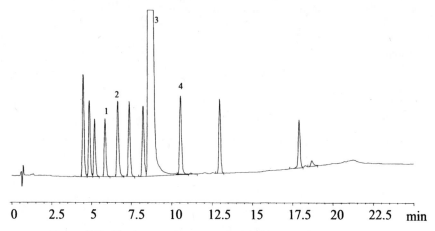

Figure 12.3. Chromatographic separation of all known related substances.

fied impurities, while the remaining seven related substances must be resolved in the chromatographic method. In further development and validation activities, less attention is given to the quantitation of the seven related substances because they are never observed in "typical" samples at levels above 0.1%. Using the same HPLC conditions as in Figure 12.4, the three are present at 0.25% relative to 100% of the active; furthermore, this sample will serve as a resolution check in system suitability evaluation.

Once basic resolution criteria are met (able to separate the ten related substances from the active), other laboratory method development activities are pursued including evaluation of sample preparation. An example later in this chapter further discusses the laboratory aspects of HPLC method development.

12.5.2. Method Optimization

If the method does not meet run time criteria or robustness requirements, it may require further optimization. Recently, many laboratories have found success using computerized HPLC method optimization software. Several commercial packages are available that combine classical chromatographic theory with statistical design to predict optimum separation conditions with a minimum number of experiments.

Methods development optimization is typically continued until the objectives outlined in the development plan are met. Method development experts often have the experience necessary to adequately optimize the methodology; others may benefit from the documented and organized development approach of available software packages.

12.5.3. Method Prevalidation Evaluation

Once the method is determined to be optimum, the method is evaluated to see if it will meet validation requirements. In this exercise, the method is challenged in some of the following areas:

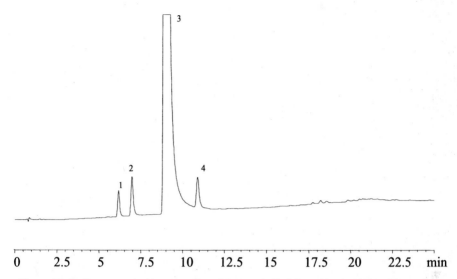

Figure 12.4. Chromatographic separation of three specified impurities in 100% of the active.

1. Can the target LOQ be met?
2. Is the method selective?
   - Placebo; stressed placebo
   - Degraded samples
   - Known impurities and degradants
   - Peak purity—diode array and/or LC-MS
   - "Collect" active—analyze on HPLC column with different selectivity
3. Linearity of the active (0–150%)
4. Linearity of the related compounds (0.05–2.0%)

Successful completion of prevalidation assessment suggests that the method is capable of entering the much more rigorous validation stage. The method should, however, be evaluated for robustness and assigned system suitability criteria before formal validation studies begin.

12.5.4. Robustness

The ability of a method to perform effectively in a typical lab environment and with acceptable variations is evaluated during robustness testing. Generally, if robustness is designed into the method development process, the methods should transfer more readily. In robustness testing evaluation, a variety of parameters are evaluated to determine the extent that they can be varied with the performance of the method unchanged. In a HPLC chromatographic experiment, the following representative robustness parameters (factors) may be evaluated:

- HPLC vendor
- Column supplier
- Flow rate
- Column temperature
- Mobile-phase pH
- Ionic strength
- Detector wavelength
- Gradient slope
- Injection size, sample concentration

Some but not all of these factors may need to be evaluated for robustness testing based upon general knowledge or experiences during method development. The range (levels) over which the various parameters are evaluated should be meaningful; that is, will the method perform successfully if the pH is adjusted ± 0.2 pH units from that specified in the method? Through HPLC experiments, typical "responses" (i.e., retention factor, resolution, peak tailing) are tracked while factors are adjusted. To optimize the evaluation of robustness, these factors can be evaluated simultaneously through an experimental design approach. Success has been achieved using a two-factor Plackett–Burman statistical approach [10]. In this approach, eight factors can be evaluated in 12 injections.

In most cases, a method will only be robust if the ruggedness of the method is considered as a requirement of the method development project. As an example in HPLC method development, at 5 mM and below, buffer ionic strength may be a critical parameter; raising the buffer concentration to 30 mM typically removes the sensitivity of the method to small variations in ionic strength. In another example, HPLC methods should be developed using column temperature control (typically 30°C to 40°C) to eliminate differences in chromatographic performance due to each laboratory's "ambient" environmental conditions.

Ruggedness evaluation should demonstrate that the method performs with variations that are typically classified as unintentional or random in nature. In contrast, a meaningful outcome of the *robustness* exercise is to quantitatively determine how a method performs during *deliberate* changes. On occasion when all practical approaches to improve robustness fail, a note should be included in the method description, such as "the pH must accurately be adjusted to pH 6.5 $\pm$ 0.2 for the method to meet system suitability." Obviously, with the goal to develop V-TR^{2}AP methodology, we need to ensure that robustness is *developed into* the method.

12.5.5. System Suitability

Generally speaking, it is expected that an analytical test method will perform in an acceptable manner each time it is used. System suitability testing ensures

Table 12.6. Using System Suitability Test to Monitor Validation Parameters

Validation Parameter	System Suitability Test
Linearity	Control sample or diluted standard (LOQ sample); response is expected to be within acceptable limits compared to 100% standard.
Accuracy	Control sample—evaluate standards as "samples"
Precision	RSD of five injections of a standard
Selectivity	Retention factor
	Number of theoretical plates
	Tailing factor
	Resolution (injection of a resolution mixture)
LOQ/LOD	Injection of a dilution of the standard to verify LOD/LOQ
Stability of solutions	RSD (area) of a series of injections or standards throughout a run

that the total system is functioning at the given time. System suitability testing, coupled with previous instrument qualification, periodic calibration, and method validation, provides assurance that the test method will provide accurate and precise data for its intended use (Chapter 15). Properly chosen system suitability criteria will fail just prior to the point where the system will begin to produce less acceptable data; however, criteria should not be so restrictive that *acceptable* data cannot routinely be used. It is the challenge of the method development scientist to develop realistic and meaningful system suitability criteria [11, 12]. Table 12.6 reviews some system suitability test and their associated validation parameter that is monitored.

As implied above, only meaningful system suitability criteria and those required by in-house or regulatory policy should be evaluated. During the method development process and robustness evaluation, marginal performance of the system can be observed. The experienced and attentive scientist will use these circumstances to suggest some of the system suitability criteria; minimum peak tailing and minimum resolution between "critical peak pairs" are typical examples. System suitability should be monitored over time to verify that the criteria remain realistic and achievable while continuing to provide assurance of the suitable performance of the method.

12.6. DOCUMENTATION

12.6.1. Method Development Report

Starting with the requirements outlined in the method development plan, the compiled method development report should summarize and provide justifications of how the method conditions, concentrations, reagents, instrumentation, wavelength, HPLC column types, and sample preparations were selected. An important component of this formal report is the data, or references to location of data, that provides justification of the selection of related compounds used in method development.

Often, examination of inadequate chromatographic data from early development can provide justification for method improvements (i.e., the pH was adjusted from pH 4.0 to pH 2.5 to improve peak shape; an example of both chromatograms should be recorded). In general, all useful data gathered during "information gathering" should be tabulated or referenced in the report. If the chromatographic method uses UV detection, the method development report should provide UV spectra of the active, related compounds, and any excipient materials in the formulation. Data incorporated in the method development report should support the final conditions and have thorough description of each experiment so future experimentation can be more readily performed. The report should summarize the results of method development, robustness, and prevalidation studies. The report, or excerpts, should be part of the documentation used in the method transfer.

12.6.1.1. Method Procedure.
The method procedure, description, or Standard Operating Procedure (SOP) should contain a complete rendering of the following information prior to validation:

- An introductory summary including the scope of the method.
- A list of reagents and their specification (HPLC grade, reagent grade, etc.). Any precautions should be included as well—for example, "test the bottle of THF to confirm the absence of peroxides, prior to use."
- A list of required standards.
- A listing of supplies: glassware, filters, and so on.
- A list of instrumentation and equipment.
- A description of solution preparation (mobile phases, standards, and samples).
- A list of method conditions (flow rate, wavelength, gradient profile, etc.).
- Procedure including sample preparation.
- System suitability criteria and how to calculate.
- If not company policy, a suggested sample sequence scheme including the order and number of injections of standards, blanks, system suitability, controls, and samples.
- Complete calculations, including calibration frequency and how calibration is performed.
- A table of analyte relative retention times and relative response factors, including excipient peaks and artifact peaks if present.
- A representative labeled chromatogram; a separate system suitability chromatogram may also be required.

The method is ready to proceed to validation. The method procedure may require updating following validation activities to include data [relative response factors (RRFs), LOQ, etc.] obtained during validation and any minor changes suggested by validation activities.

12.6.2. Completing Method Development

Prior to entering time-consuming validation activities, the method development progress should be reviewed by the team who initially drafted the method requirements. End-users of the method may desire to "evaluate" the method in their labs during "customer evaluation"; this should enable constructive dialog with the development scientist prior to time-consuming validation; this evaluation may also facilitate straightforward method transfer.

12.7. METHOD DEVELOPMENT—EXPERIMENTAL CONSIDERATIONS

12.7.1. Introduction

The purpose of this section is to review both practical and technical concerns that may arise during development of a V-TR^{2}AP HPLC method. To provide further understanding, practical "hints" that supplement the overall topic are presented in separate "comment" boxes. This section is not designed to be a thorough review of the HPLC method development process; interested readers should refer to specialized textbooks [13] or other chapters in this text.

Upon reaching the experimental section of the method development process, the following documents and resources should be available:

- Method development plan, which contains the method objectives and justification for the related substances to be tested, and project timeline
- Chemical, formulation, and safety information on drug substance and excipient materials
- Commitment of resources (human, instrumentation, samples/standards)

Once the objectives and resources are available, the method development experiments can begin.

12.7.2. General Components of HPLC Method Development

In addition to earlier discussions, the method development scientist should consider the following practical experimental components required in many pharmaceutical HPLC methods development activities:

1. Sample preparation
2. Reference standard materials, placebo formulations, stability lots, and so on

3. Chromatographic method development
4. Detection technique
5. Calculations/results—chromatography data system

These activities are typically not performed sequentially. As an example, the choice of detection technique can influence the choice of solvent used in sample preparation. The experienced HPLC method development scientist will recognize the interdependency of these parameters.

12.7.2.1. Sample Preparation. The separation of specified analytes in complex pharmaceutical matrices presents enticing challenges to chromatographic scientists. Often similar effort is required in evaluating suitable sample preparation schemes for these complex samples as in developing the chromatographic separation. When too little attention is paid to this important component of method development, the following can often be observed:

- Poor or irreproducible recovery
- Excessive column pressure
- Short column lifetimes
- Method transfer problems
- Safety issues
- Method robustness concerns

There are several excellent texts [13] that review the topic of sampling and sample preparation in detail. In the very basic sense, ensuring suitable "sample preparation" is essential to every HPLC development project. Proper sample preparation enables HPLC analyses that are free from unnecessary interferences and particulates, while requiring essentially 100% recovery of the active and related substances from the formulation matrix. The sample preparation step should be compatible with the proposed HPLC methodology.

Comment. It is not uncommon to use a different and/or "stronger" solvent to dissolve samples to ensure complete recovery of analytes. As an example, if a tablet is dissolved in 100% methanol and injected onto an isocratic HPLC system with a mobile phase of only 25% methanol, early peaks in the chromatogram may have distorted peak shapes or nonreproducible retention times; even worse, sample components may precipitate in the column. Try to dissolve the sample in the initial mobile phase, if possible.

Systems with lower injector "extra-column variance" are less forgiving when using a solvent that is chromatographically stronger than the mobile phase to dissolve the sample. This can result in split or poorly shaped peaks where none were observed in the less "efficient" system. This can be corrected by reducing the injection volume, lengthening the tubing between the column or injector, or using a weaker sample solvent when practical.

Figure 12.5 summarizes the optional steps of sample preparation for preparing pharmaceutical samples for analysis. Creams, ointments, suppositories, and extended release dosage forms represent dosage forms that present unique sample preparation challenges.

12.7.2.2. Reference Standard Materials, Placebo Formulations, Stability Lots, and so on. Reference standards, placebo formulations, and stressed stability lots should all be available prior to beginning method development experiments. These essential materials often take significant resources and time to obtain.

The development and validation of stability-indicating assay and purity methods require that studies be performed to verify the method's ability to resolve all possible degradants and synthetic impurities from active drug substance and accurately measure their concentrations in the presence of product excipients. Such studies are regarded as a minimum requirement for the development and validation of most any "stability-indicating" method [14]. Knowledge of degradants and synthetic impurities can be derived from the historical information that has accumulated for the drug substance/product. Ideally, drug substance and drug product degradants and impurities can be synthesized in sufficient quantity and purity to be characterized as reference standards. The availability of such materials streamlines chromatographic method development.

Stability-indicating methods should also be challenged with the analysis of real samples obtained from authentic stability studies. This challenge serves to "validate" the method's capacity to resolve unforeseen compounds that may result from the interaction of drug and excipients, particularly when stressed in ways that correlate well with realistic storage conditions in real packaging components.

Early in the drug development process, it is impractical that these standards are available. It is then highly desirable to perform studies to force degradation of the active compound in the presence and absence of placebo. One federal guideline [15] defines several conditions under which various drug product types should be stressed to elucidate potential problems that may occur during storage and handling. Depending upon matrix and packaging, these include extremes of pH, heat, high oxygen exposure, and light exposure. Excipient compatibility studies should also be undertaken for drug products in early development to determine any possible concerns regarding interaction of the drug substance with the inert ingredients. These stressed samples may also be useful in examining "specificity" during method validation.

12.7.2.3. Chromatographic Method Development. Factors critical to the development of an HPLC purity method include column selection, mobile-phase selection/development including buffer, and selection of a suitable detection scheme. These factors and their underlying theory were discussed in detail in Chapter 9.

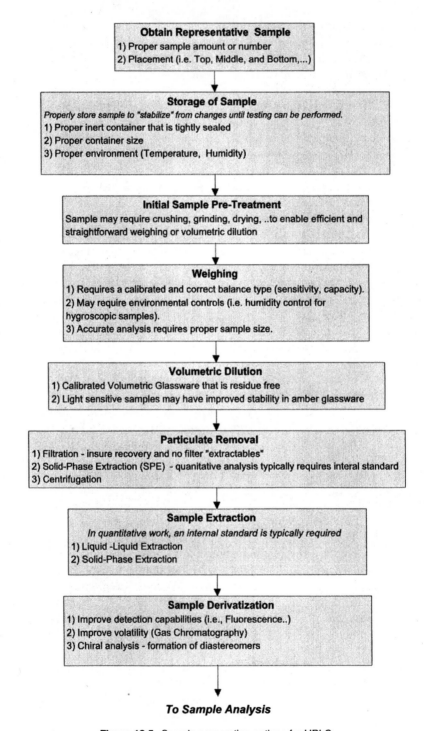

Figure 12.5. Sample preparation options for HPLC.

Table 12.7. General Guidelines for Selection of Chromatographic Mode

Mode	Analyte—General Characteristics
Normal-phase	Low to moderate molecular weight
	Low to moderate polarity
Reverse-phase	Low to moderate molecular weight
	Moderately polar
	Slightly nonionic
Reversed-phase/ion-pair	Ionic/nonionic mixtures of low- to moderate-molecular-weight moderately polar compounds
Ion-exchange	Low to moderate molecular weight
	Strongly polar
	Water-soluble
Size-exclusion	Large molecular weights ($>10,000$)

Choosing the Appropriate Separation Mechanism. Mechanisms and modes of HPLC separation have been examined in detail in Chapter 9. Briefly, the four basic modes of HPLC include normal phase, reversed-phase, ion-exchange, and size-exclusion (gel-permeation). Ion-pair chromatography is also used extensively in pharmaceutical testing and can be considered a mixture of reversed-phase and ion-exchange mechanisms. Table 12.7 is a general guideline to help select the modes that will most likely provide the desired separation of analytes.

Some separations can be performed in more than one mode. However, it is most desirable to perform the separation in the reversed-phase mode whenever possible because reversed-phase columns have been proven to be the most rugged and reproducible. Chapter 10 reviews columns commonly used in HPLC as well as some unique phases for specialized applications.

Developing the Mobile Phase for the Chosen Separation Mode. Mobile-phase selection is of course directly dependent upon the mode and column to be used. Solvent components commonly used in the preparation of various mobile phases are summarized in Table 12.8. Whenever possible, it is desirable to develop reversed-phase HPLC methods with volatile buffers that are LC-MS compatible. However, several factors should be considered in the selection of buffer type:

- Buffering range versus analyte pK_a
- Optical properties
- Volatility
- Reversed-phase, ion-pair, or ion-exchange mode

In reversed-phase mode it is desirable, although not essential, to choose a suitable buffer at the desired pH. In ion-exchange mode this *is* more critical.

Table 12.8. Common LC Solvents and Modifiers

Mode	Weak Solvent Component	Strong Solvent Component	Modifier/Comments
Reversed-phase	Water, buffer, dilute organic	Methanol, acetonitrile, tetrahydrofuran	Trifluoroacetic acid, acetic acid to suppress ionization of acidic compounds and residual silanols of the packing.
Reversed-phase	Water, salt water	Methanol, acetonitrile, tetrahydrofuran	Buffers of acetate, phosphate, citrate, trifluoroacetate—for selective pH control.
Reversed-phase/ion-pair	Water, salt water	Methanol, acetonitrile, tetrahydrofuran	Alkylsulfonic acid sodium salts for cationic analytes and tetraalkylammonium salts for anionic analytes (ion-pairing).
Reversed-phase/ligand exchange	Water, salt water	Methanol, acetonitrile, tetrahydrofuran	Ligands of copper, nickel, zinc (ligand exchange).
Normal-phase	Hexane, heptane	2-Propanol, ethyl acetate, methylene chloride	Acetic acid to suppress ionization of silanols on column.
Ion-exchange	Water	Buffer salts of potassium, ammonium, phosphate, lithium	Methanol, acetonitrile; often a constant amount of organic (<10% organic) is added to the mobile phase to minimize the "hydrophobic" mechanism of retention.

Ideally, the pH of the mobile-phase buffer should not be within ± 2 pH units of the pK_a values of the analytes. This ensures reproducible retention times and good peak shapes. The ionic strength of the buffer is not normally critical to achieving the desired separation in reversed phase, although separation robustness may be improved at higher ionic strengths. Typical concentrations include 20–50 mM.

Optical properties must also be considered for the separation of analytes that require low UV wavelengths for their detection. Phosphate, for example, can be used down to about 210 nm, whereas acetate has significant absorbance below 230 nm. Similarly, acetonitrile has superior low-wavelength transparency to alcohol or tetrahydrofuran.

One manufacturer of HPLC grade solvents has published an excellent guide to HPLC solvent selection, including UV spectra and tables of physical/chemical characteristics of each solvent [16].

Comment. When using gradient HPLC, it is highly desirable to maintain constant buffer ionic strength throughout the analysis. This requires that the buffers be premixed (and filtered) in each solvent bottle. Premixing of solvents (as opposed to 100% aqueous in solvent "A" and 100% Organic in solvent "B") is a "best practice" and often improves baseline stability and run reproducibility. With premixed solvents, helium sparging, and vacuum degassing should be minimized to limit evaporation of volatile solvents. An additional advantage of premixing solvents is that buffers will not precipitate in the inlet check valves of the HPLC pump as commonly observed when solvents are not premixed.

Comment. Short column life is a common complaint. Try to identify the cause—it could be due to any one or combination: pH is incompatible with column used, poor sample preparation exposes the column to excessive amounts of excipients that tend to precipitate upon injection, excessive pressure or temperature, poor solvent blending/buffer precipitation.

12.7.3. Obtaining Sufficient Resolution—Considering Method Requirements

Once the separation mode, column, and type of buffer have been chosen, the development chemist can apply various strategies to realize sufficient resolution of the required compounds in a reasonable timeframe. During evaluation of column and mobile-phase parameters, a practical time saver is the use of "diverse-concentration spiking" for tracking the elution of each of the analytes. Instead of injecting each one independently during given experimental conditions, binary or ternary analyte mixtures that contain the analytes at very dissimilar concentrations are injected and make analyte tracking both reliable and efficient.

Comment. Alternatively, the use of a mass-spectrometer can be a very efficient way to track analytes in complex mixtures provided of course the buffer system being investigated is volatile and mass-spec-compatible. Bench-top LC-MS systems are now becoming commonplace and affordable. For mass-spec incompatible mobile phases, a photo-diode-array UV detector can be useful to help track peaks through examination of their spectra compiled in a library.

12.7.3.1. Optimizing the Separation. The goal of separation science is to achieve the best possible separation that satisfies the method requirements, in the shortest possible time. The parameters α (selectivity), N (column efficiency),

and k (capacity factor or retention factor) that control resolution, R_s, are related by the master resolution equation as follows:

$$R_S = \frac{(\alpha - 1)}{4\alpha} \sqrt{N} \cdot \frac{k}{1 + k} \tag{12.1}$$

This classic resolution equation is discussed throughout the text. The α term relates to the actual chemistry of the conditions used in the separation. Often referred to as the selectivity factor, references to this as the "chemistry factor" can be found in the early literature on HPLC. Maximizing α will maximize the resolution between any two peaks. The success or failure of a separation is largely determined by selection of mobile phase components, including "modifiers" that offer the best compromise of chemistry-related forces (i.e., dipole moment, dielectric constant, etc.) to achieve a separation of many peaks.

The N term (Chapter 10) relates to the efficiency of the column and is often referred to as the plate count or number of theoretical plates. The value is determined by the physical configuration of the column. For example, size and pore depth of the packing particles largely predicts the value of N in units of plates/meter. Three-micron particles offer higher N values than do 5- and 10-μm particles. Finally, the form of the N term predicts that doubling the column efficiency will only increase resolution by a factor of 1.4.

The form of the k term imposes practical limitations on the retention times of analyte peaks because a plot of resolution versus k is an asymptotic function that levels out at about $k \cong 10$ or so. Thus, deliberate attempts to resolve peaks that are only partially resolved at k values between 7 and 9 through further increases in retention will be fruitless. In this case it makes more sense to improve selectivity and/or column efficiency as discussed above.

Method Optimization Versus Robustness. Analysis speed is important particularly in high-throughput applications. However, method robustness must not be sacrificed as the separation is optimized. Given the diversely equipped sites to which the method must ultimately transfer, along with the diverse levels of experience of the end-users that staff those sites, it is highly desirable to build as much robustness into the method as possible. As an example, in gradient chromatographic methods, the elution profile should begin with a short isocratic plateau. This provides a "window" to enable small adjustments to be made to compensate for differences between the "gradient delay" of instruments from different vendors or different models.

Comment. On some systems, early peaks will elute under isocratic conditions that provide adequate resolution from close neighboring peaks. When run on systems with lower delay volumes, they may elute under gradient conditions that can adversely affect the resolution and the transferability of the method. It can be very valuable to know how the delay volume of your systems compare to those in the operational laboratories.

Understandably, there needs to be a balance between method performance and the robustness of the method. It makes little sense to develop methods that are so finely tuned they cannot be transferred to the labs where they will be used.

12.7.3.2. Detection Techniques.
The variable-wavelength UV detector is the most common detector used in LC today. Other detectors such as refractive index and fluorescence are also in use. However, benchtop mass detectors such as single and triple quadrupole or ion-trap detectors have become more commonplace in recent years. In the future, CE-MS instruments may have a place in the analytical control testing laboratory.

As mentioned above, the choice of the detection method is interrelated with the chromatographic buffer and of course the nature of the analyte molecules. As stated earlier, a volatile buffer is prerequisite for straightforward mass spectrometry detection. For UV detection, the molecule must possess a good chromophore that can be measured against the background of the proposed mobile phase. In either case, the detection scheme must provide sufficient sensitivity to meet requirements for limit of quantitation (LOQ) and limit of detection (LOD). Typically, this applies to compounds related to active drug along with synthesis and/or degradation compounds.

12.7.3.3. Calculations and Results—Chromatography Data Systems.
Many methods may ultimately be used in a high-throughput environment by analysts that generally are not permitted privileges to manipulate data or customize reporting of the results. It is therefore essential to take complete advantage of available software so that data analysis and reporting can follow a template. To support that goal, the calculations specified in the method must be carefully defined by the development chemist and in such a way that they are compatible with the "logic" of the data system's software. The accuracy of this process must be carefully verified by comparing manually calculated results with those reported by the data system. Chromatography data systems will be discussed in Chapter 14.

12.8. VALIDATION ACTIVITIES

According to the ICH Guidelines [17], "The objective of validation of an analytical procedure is to demonstrate that it is suitable for its intended purpose." Validation requirements are depended upon the objectives of the method. Chapter 14 of this text discusses validation requirements, definitions, and approaches in detail.

Validation can be a time-consuming activity. Methods should not enter the validation phase unless they are fully developed. The following observation can be made about the relationship of validation and method development:

- When methods are properly developed, they readily validate.
- Validation is not a method development tool: Validation does not make a method good or efficient [18].
- A validated method doesn't necessarily imply a V-TR^{2}AP method.
- Validation acceptance criteria should be based upon method development experience.

The method development scientist should not begin the validation process unless she/he is confident of success. The validation process is "confirmation" that the method is suited for its intended purpose. Method validation is also a "holistic" process that requires suitable instrumentation and competence in laboratory techniques to ensure success.

12.8.1. Documentation—Protocol

In its simplest form, a validation protocol typically describes the requirements and the acceptance criteria that are used to evaluate a method. Validation protocols can be quite elaborate containing step-by-step procedural information, tables or test sheets to record and calculate data, with summary tables to compare results with acceptance criteria; others are more general templates. Regardless of the approach, in addition to the validation requirements and acceptance criteria, samples and lots that are tested should be carefully chosen and recorded in the protocol. It is also advisable to record in the protocol the required contents and form of the final validation report.

The protocol should be approved by lab management and quality assurance (QA). QA will ultimately approve the report and should be aware of the approach prior to initiation of experiments.

12.8.2. Method Validation—Experimental

The method should be fully developed with successful completion of robustness and prevalidation activities, before entering final validation. The details of validation experimentation are outlined in Chapter 15.

Validation is considered a current good manufacturing practice (CGMP) activity. Therefore, validation activities must be properly documented and performed on qualified and calibrated instrumentation and equipment.

12.8.3. Documentation—Report

The validation report can be a detailed report describing the results and validation activities. Alternatively, when using more elaborate protocols, the final report is simplified because the results are entered in protocol tables and a short executive summary is attached to the completed protocol. The final report may contain examples of the raw data; in addition, the location of the raw data and the laboratory documentation reference [notebook or Laboratory Information Management System (LIMS)] is also cited in the report.

Any deviations to the protocol are noted; changes to the acceptance criteria require justification and approval by laboratory management.

The final report and validation results should be reviewed by the development team; the validation report should also be reviewed by QA.

12.9. ANALYTICAL METHOD TRANSFER

The transfer of analytical methods may be part of the overall process of technology transfer of a product or process, or simply transfer of methodology from the research lab to a control or contract laboratory. The transfer of analytical methods is covered in detailed in Chapter 15. Method transfers are most successful when the receiving laboratories are somewhat involved in the development process (when practical) and are aware of the decisions and compromises made during development. On the other hand, analytical method transfer can be unsuccessful when:

- Methods are not sufficiently developed or documented.
- Methods are not sufficiently validated.
- Methods are not robust.
- Inappropriate transfer protocols/procedures are used.
- There is poor planning/scheduling.

Successful method transfers occur when an adequate transfer protocol is properly understood and executed. On occasion and when working with a new partner lab, on-site training of the methodology may be required. A possibly less expensive alternative would be to videotape critical components of the methodology (such as sample preparation) and forward to the receiving lab.

The sending or transfer coordinating lab is expected to test samples in a similar time period as the method-receiving laboratory.

12.9.1. Documentation—Protocol

Method transfers require a protocol that is similar to, but of lesser detail than, that of the method validation. The protocol should contain acceptance criteria, transfer requirements/procedures, and a listing of batches to be tested. The protocol should also include assignment of the responsibilities of the transfer and receiving laboratories. If the protocol contains fill-in tables and worksheets, the completed protocol can serve as the majority of the final documentation.

The transfer lab should include the following in the transfer documentation:

- A finalized method
- A method development report or summary
- Validation report or validation summary
- The transfer protocol—acceptance criteria

The transfer protocol must be approved by all laboratories involved in the transfer and by QA. For laboratories that are located nearby, an analyst from the receiving laboratory should consider visiting the transfer laboratory prior to the transfer to see the method performed in the originating laboratory.

12.9.2. Method Transfer—Experimental

The transfer experiments are initiated after acceptance of the protocol, review of the documentation, and specialized training completed, if required. Up-front planning is required to ensure that proper instrumentation, required supplies (volumetric ware) and standards, and sufficient quantities of the batches listed in the protocol are all available.

Because method transfer is considered a current good manufacturing practice (CGMP) activity, the laboratory analyst should review the documentation thoroughly to ensure that all concerns have been addressed prior to beginning the transfer experiments.

12.9.3. Documentation—Transfer Report

The transfer report should indicate whether the transfer was successful or not. The report typically consists of a completed protocol, a summary of the results, and references to both the location of raw data and items log into the laboratory notebook/LIMS. Any deviations to the protocol must be recorded.

12.10. PERIODIC REVIEW

The method should be reviewed regularly to consider the following issues:

- Is the method inadequate by today's scientific standards or regulatory requirements?
- Have there been recent improvements in technology that would enable improvements in the capabilities, speed, or simplicity of the method?
- Have there been any process or packaging changes that may affect stability or level of impurities?
- Does monitoring of "laboratory deviations" suggest a need for method improvement?

• Do newer methods for similar products significantly outperform (reproducibility, accuracy, simplicity, or speed) this method?

If method changes are warranted with a filed method, there may be required changes to the method requirements and subsequent consequences with regulatory agencies.

12.11. REFERENCE STANDARDS AND SAMPLES TO SUPPORT STABILITY INDICATING METHOD DEVELOPMENT

Any discussion of the method development process should consider obtaining a sufficient quantity of a qualified reference standard and appropriate samples to support the development and validation campaign. In most cases, the limit of the accuracy of the analytical test will be related to the correctness of the standard's assay. There is little consensus [19, 20] as to common requirements of reference standards as outlined by the regulatory authorities [21]; typical practice favors an approach of thorough analytical characterization of the standard supported by adequate documentation.

12.11.1. Types of Standards

Materials selected for use as a standard should be of high purity and stable. Practical and regulatory requirements for reference standards differ depending upon the objective and stage in the drug development cycle. For pragmatic reasons, reference standards used in early phases of development may be limited to a well-characterized portion of one of the early lots of material. In later stages, the primary reference standard may be a unique synthesis or a purified version of the material—sometimes in a different salt form to enhance stability or purity. The sections below will describe a process of qualification of a reference material.

Primary Reference Standard of the Active. The standard of an active is typically comprehensively characterized. Table 12.9 lists the details required.

Secondary Standard (Working Standard). A material of high purity that has been characterized against a primary or compendial standard. A secondary standard is often used to conserve the amount of primary standard.

Related Compound Standard. These standards are typically characterized, but not to the same extent as the primary reference standard of the active drug substance. Because quantities of these standards are typically limited, relative response factors for these standards are assigned to eliminate routine use in HPLC purity determination.

Compendial Standards. USP/NF/ASTM standards typically do not require further characterization. A certificate of analysis should be available.

Table 12.9. Reference Standard Characterization

Characterization Requirement	Measurement Technique
Physical Properties	
Proof of structure	NMR
	MS and MS/MS
	FTIR
	UV spectroscopy
	Functional group analysis
	Crystallinity
	Combustion analysis
Physical description	Appearance
Physical properties	Melting point
	pK_a
	Optical rotation
Purity	
Related compounds	HPLC purity
Chiral purity	HPLC, CZE chiral purity
Inorganic impurities	Residue on ignition, ICP
Solvents	GC
Water	Karl Fischer titration
Counterions	Ion chromatography
Verification—organic impurities	TLC analysis
Assay	
	Titration
	Assay by difference[a]
Verification	CH & N analysis
	Mass balance calculation
	Titration

[a] Assign assay by difference: % Assay = 100% − (% impurities/degradants including solvents) − (% water) − (% other, including counterions).

Internal Standard. Internal standards require a varying amount of characterization depending on the application. Often, a qualified commercially available standard is chosen.

Other Standards. Reagents and chemicals that are commercially available at high purity and have been characterized by the vendor can be suitable for standard materials. Prior to use, the vendor should provide sufficient documentation.

12.11.2. Handling of Standards

12.11.2.1. Storage. When possible, the standard material should be protected from light, heat, and moisture and housed in a controlled environment. Typical storage conditions for a stable solid standard would be a 2°C to 8°C stability room, protected from light, in a desiccator cabinet to protect from humidity. Reference stock solutions and working solutions should also be labeled with required storage conditions and expiration dates. The method procedure should specify the storage conditions and expiry dates should be supported by experimental data. There should be limited access to all standard materials. Although there is a risk of breakage, storage in tightly capped glass amber bottles is preferred. Containers should be sized to minimize headspace; small changes to assay have been observed in overly large containers as the solvents and water vapors in the solid standard materials equilibrate with the headspace.

12.11.2.2. Handling. Reference materials must be handled in such a way to preserve the integrity of the sample. As discussed above, the solid materials must be protected from heat, light, and high humidity. Standard materials that are stored at subambient temperatures must be equilibrated to ambient before opening to prevent water condensing on the materials, particularly in humid environments. If primary standard materials are used frequently, qualification of a secondary standard may be justified. Alternatively, primary reference standards can be subdivided into several smaller containers to limit contamination and changes in assay. The retest (recertification) date of the standard should be justified through stability monitoring in the actual storage container.

Certain standards require oven drying prior to use. These drying time/conditions should be adjusted to affect drying of the material but minimize any potential decomposition. Warm standards should be equilibrated to ambient conditions prior to weighing.

12.11.2.3. Documentation. A reference standard must have documentation to support its use as a standard, support its assigned assay, and defend the retest date (or expiry date for chemical standards).

Reference Standard Qualification Report. A detailed qualification report is drafted for a primary reference standard of the active, less comprehensive documentation is typical for other standards. Portions of this report may be submitted to the regulatory authorities. The report references the synthetic route for the standards and any purification schemes that were utilized in preparing the pure standard. Other components of this report may include:

- Description or reference to test methods used in characterization of the standard
- Report results of the characterization with representative data (UV, NMR spectra, HPLC chromatograms, etc.)
- Verification of data to establish purity and assay of the standard
- Suggested requirements for requalification interval and requalification tests

This report may be appended for additional requalification and stability data as it is obtained.

Certificate of Analysis. In general, all standards should have a certificate of analysis either generated from a qualified supplier or through analysis of in-house data. The certificate should contain the following information:

- Standard name
- Lot number and supplier
- Effective data and recertification date
- Purity
- Assay

Dispensing. A reference standard log is typically maintained to provide an inventory of the standard; it should indicate who, when, and how much standard was removed. When handling materials, care should be exercised to avoid contamination of the standard. Furthermore, observe safety cautions when handling potent drug substance materials.

12.12. SUMMARY

The objective of this chapter was to review the systematic but thorough process of developing analytical methods for stability and release testing of pharmaceutical drug substances and drug products. The components of the process are highly interrelated, and each must be considered to enable successful development → validation → transfer of the analytical methodology. The process enables consistent high-quality methods to be developed among several development laboratories. While the examples focused on HPLC methodology, a similar approach can be adapted to development of methods for other analytical techniques.

Thorough documentation is an essential regulatory component to the development process. Some additional development effort may be necessary to provide sufficient data to complete the final documentation. Similarly, extra effort will be required to provide justification for the selection of key parameters.

The benefits of the approach may enable the same methods to be used globally, in a variety of laboratory environments. The extra effort required in the development laboratory should provide a generous overall saving in costs, efficiency, and compliance by providing V-TR^{2}AP methodology to the operating and QC laboratory environments.

Analytical methods have a useful life cycle. Periodic evaluation of the capabilities of older methods is necessary to ensure that they remain suitable.

Acknowledgments

The authors thank Steven Scypinski, Ph.D., Ronald Erlich, Ph.D., and Prof. Jos Hoogmartens, Ph.D. for their input and review of the chapter.

REFERENCES

1. Bernard A. Olson and Paul K. S. Tsang, Considerations for the Development of Separation Methods for Pharmaceutical Quality Control, *Process Control and Quality* **1997**, *10*, 25–39.

2. International Conference on Harmonization of Technical Requirements for the Registration of Pharmaceuticals for Human Use (ICH), Q6A Specifications: Draft Guidance on Specification: Test Procedures and Acceptance Criteria for New Drug Substances and Drug Products: Chemical Substances. *Federal Register* **1997**, *62*(227).

3. International Conference on Harmonization, Guidance on Specifications: Test Procedures and Acceptance Criteria for Biotechnological/Biological Products, *Federal Register*, **1999**, *64*.

4. *U.S. Pharmacopeia* 24/*National Formulary* 19, United States Pharmacopeial Convention, Inc., Rockville, MD, Section ⟨1225⟩, 2000 pp. 2149–2152.

5. *International Conference on Harmonization of Technical Requirements for the Registration of Pharmaceuticals for Human Use (ICH) (Background and Status of Harmonization)*, Geneva, Switzerland, November 1996. ICH Secretariat, c/o IFPMA, 30 rue de St. Jean, P.O. Box 9 1211 Geneva 18, Switzerland or at www.pharmweb.net (check out FDA site).

6. Gerald C. Hokanson, A Life Cycle Approach to the Validation of Analytical Methods During Pharmaceutical Product Development, Part I: The Initial Method Validation Process, *Pharm. Technol.* **1994**, *18*(9), 118–130.

7. PMA's Computer Systems Validation Committee, Computer Systems Validation— Staying Current: Introduction, *Pharm. Technol.* **1980** *13*(5), 60–66.

8. J. Mark Green, A Practical Guide to Analytical Method Validation, *Anal. Chem. News Features*, **1996**, 305A–309A.

9. Dennis R. Jenke, Chromatographic Method Validation: A Review of Current Practices and Procedures. II. Guideline for Primary Validation Parameters, *J. Liq. Chrom. Rel. Technol.* **1996** *19*(5), 737–757.

10. M. Jimidar, N. Niemeijer, R. Peeters, and Jos Hoogmartens, Robustness Testing of a Liquid Chromatography Method for the Determination of Vorozole and Its Related Compounds in Oral Tablets, *J. Pharm. Biomed. Anal.* **1998**, *18*, 479–485.

11. D. E. Wiggins, System Suitability in an Optimized HPLC System, *J. Liq. Chromatogr.* **1991**, *14*, 1045–3060.

12. J. C. Wahlich and G. P. Carr, Chromatographic System Suitability Tests—What Should We Be Using? *J. Pharm. Biomed. Anal.* **1990**, *8*, 619–623.

13. L. R. Snyder, J. J. Kirkland, J. L. Glajch, *Practical HPLC Method Development*, 2nd edition, John Wiley and Sons, New York, 1997.

14. J. Mark Green, A Practical Guide to Analytical Method Validation, *Anal. Chem. News Features* **1996**, 305A–309A.

15. FDA Guidance Document, "Reviewer Guidance: Validation of Chromatographic Method," Center for Drug Evaluation and Research, Washington, DC, 1994.

16. Burdick and Jackson, *High Purity Solvent Guide*, Muskegon, MI, 1990.

17. International Conference on Harmonization, Guideline on Validation of Analytical Procedures: Definitions and Terminology; Availability, *Federal Register* **1995**, *60*.

18. James D. Johnson and Gale E. Van Buskirk, Analytical Method Validation, *J. Validation Technol.* **1998**, *2*(2), 88–105.

19. PMA Workshop on Reference Standards, sponsored by the PMA Analytical Steering Committee, Washington DC, 1992.

20. Paul K. Housepian, Bulk Drug Substances and Finished Products, in *Development and Validation of Analytical Methods*, Christopher M. Riley and Thomas W. Rpsanske (editors), Pergamon, Oxford, UK, 1996.

21. FDA Guidelines for Supporting Documentation in Drug Applications for the Manufacturing of Drug Substances, February 1987.

13

SOME PRINCIPLES OF QUANTITATIVE ANALYSIS

James M. Miller

The principles of quantitative analysis are basically the same for all the analytical methods used to test pharmaceutical samples, and they are dependent on the correct application of a variety of interdependent factors such as:

- Representative sampling
- Reduction of errors by careful sample handling
- Proper use of reference standards
- Correct application of the required test method, sufficiently validated
- Use of qualified (IQ, OQ, PQ) instrumentation, properly calibrated
- Instrument testing and validation
- System suitability testing
- Accurate and correct handling of data and calculations

Many of these topics are covered elsewhere in this book: sampling and quantitative laboratory manipulation in Chapter 5, statistics and error analysis in Chapter 4, data systems in Chapter 14, and validation in Chapters 15 and 2 (among others).

This chapter will focus special attention on the various methods for performing quantitative analyses and on the general characteristics of the detectors

Analytical Chemistry in a GMP Environment. Edited by J. M. Miller and J. B. Crowther
ISBN 0-471-31431-5   © 2000 John Wiley & Sons, Inc.

used for the analyses. Examples will be chosen from spectroscopy (see Chapter 6), but the focus will center on chromatography (see GC in Chapter 8 and HPLC in Chapter 9) because this technique is of greatest interest to pharmaceutical analysts. Chromatographic detectors have more variables to consider because they are being used to make measurements on flowing fluid streams, so they require special attention in this chapter. We will begin with a consideration of some classification systems that are used for them.

13.1. DETECTOR CLASSIFICATIONS (CHROMATOGRAPHIC)

13.1.1. Concentration Versus Mass Flow Rate

This classification system distinguishes between those detectors that measure the *concentration* of the analyte in the mobile phase and those that directly measure the absolute *amount* of analyte irrespective of the volume of mobile phase. One consequence of this difference is that peak areas and peak heights are affected differently by changes in mobile-phase flow rate.

To understand the reason for this difference in detector type, consider the effect on an ultraviolet (UV) signal if the flow is completely stopped. The detector cell remains filled with a given concentration of analyte, and its absorbance continues to be measured at a constant level. However, for a mass flow rate detector like the flame ionization detector (FID) in which the signal arises from a burning of the sample, a complete stop in the flow rate will cause the delivery of the analyte to the detector to stop and the signal will drop to zero.

This difference in performance has three consequences. First, the flow rate has different effects on these two detector types. For the concentration type, an increase in flow dilutes the sample and causes a decrease in peak area, but the peak height remains the same. For the mass flow rate type, an increase in flow causes an increase in peak height, but the peak areas remain unchanged. For most chromatographic analyses the flow rate is kept constant, but there are instances in programmed temperature gas chromatography (PTGC) where the flow decreases during programming.

Second, it is difficult to compare the sensitivities of these two types of detectors because their signals have different units; the better comparison is between minimum detectable quantities that can have the units of mass for both types. And finally, valid comparisons between detector types requires the specification of the flow rate, concentration, and analyte.

Classification of the most common chromatographic detectors is given in Table 13.1.

13.1.2. Bulk Property Versus Solute Property

Some detectors measure a property of the mobile phase (MP) and the same property of the analyte (dissolved in MP). An example is the refractive index

Table 13.1. Classification of Chromatographic Detectors

Detector	Concentration (C) or Mass Flow Rate (M)	Bulk Property (B) or Solute Property (S)	Selectivity: Universal (U) Somewhat (SS) Very (VS)	LOD (ng)
HPLC				
UV-VIS	C	S	SS	0.1
Fluorescence	C	S	VS	0.001
RI	C	B	U	1000
GC				
FID	M	S	SS	0.01
TCD	C	B	U	1

(RI) detector. The chromatographic baseline represents the refractive index of the MP, and the peaks represent the refractive index of the analytes in the mobile phase. Thus, the RI detector is a bulk property type of detector: It measures the bulk property (refractive index) of either the MP or the sample solution. Bulk property detectors cannot be used for gradient elution high-performance liquid chromatography (HPLC), and they tend to be less sensitive than the solute property type.

By comparison, the UV detector performs best when used with a MP that has no UV absorption; only the analytes (solutes) absorb UV. As such, it is classified as a solute property type. In the absence of a solute, the signal from this class of detector is very low (detecting only the background noise), and consequently solute property detectors are inherently more sensitive. Table 13.1 classifies some of the most popular GC and HPLC detectors according to this system.

13.1.3. Selective Versus Universal

This detector category refers to the number or percentage of analytes that can be detected by a given system. A universal detector such as the refractive index detector theoretically detects all solutes, while the selective type responds to particular types or classes of compounds. There are differing degrees of selectivity; the UV detector is not very selective and detects many organic compounds while the fluorescence detector is much more selective and detects only those species that fluorescence in solution. These general degrees of selectivity are given in Table 13.1: universal (U), somewhat selective (SS), or very selective (VS).

Both types of detectors have advantages. The universal detectors are used when one wants to be sure that all eluted solutes are detected. This is important for qualitative screening of new samples whose composition is not known, as well as when performing quantitative analysis by the area normalization method (to be discussed later in this chapter). On the other hand, a selective detector that has enhanced sensitivity for a small class of compounds can provide trace analysis for that class even in the presence of other compounds of higher concentration. It can simplify a complex chromatogram by detecting only a few of the compounds present and selectively "ignoring" the rest. A very common example is the analysis of pesticide residues in naturally derived pharmaceuticals whose matrix is very complex but not detected by the use of an electron capture (ECD) in a gas chromatographic (GC) analysis. Other examples include: the choice of wavelength of a UV detector to selectively detect aromatic compounds; the use of a MS detector in the selected ion monitoring mode whereby it will detect ions with only one given mass; or a fluorescence detector that is more selective than a UV detector. An example of the latter is shown in Figure 13.1, in which it can be seen that the fluorescence detector did not detect peaks 2, 3, 4, 6, 7, and 8 but was more sensitive (than the UV detector) for most other peaks, especially peak 10 [1].

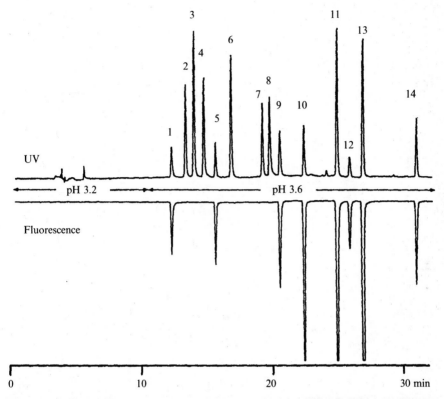

Figure 13.1. Dual detector chromatograms showing selectivity of the fluorescence detector. (Reprinted from *J. Chromatography* **1992**, *592*, G. A. Gross and A. Gruter, p. 271, with permission from Elsevier Science.)

13.2. DETECTOR CHARACTERISTICS

The most important characteristic of a detector is the signal it produces, of course, but three other important ones are noise, time constant, and cell volume. These will be discussed first to provide a background for the discussion about the signal.

13.2.1. Noise

Noise is made up of random signals produced by a detector in the absence of a sample. In chromatography, it appears on the baseline and is also superimposed on the analyte peaks. It is also called the background or dark current (in spectroscopy). Usually it is given in the same units as the normal detector signal. Ideally, there should not be any noise, but random fluctuations do arise from the electronic components from which the amplifiers are made, from stray

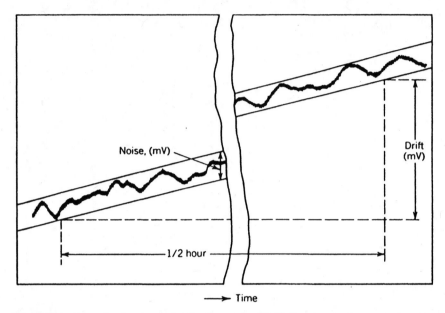

Figure 13.2. Example of noise and drift. (Copyright ASTM. Reprinted with permission. From H. M. McNair and J. M. Miller, *Basic Gas Chromatography*, Wiley, New York, 1998, p. 106. Reprinted courtesy of John Wiley & Sons, Inc.)

signals in the environment, from contamination, and from instrument malfunctions or problems [such as chromatographic leaks or entrapped air bubbles in liquid chromatography (LC) pumps]. Circuit design can eliminate some noise attributed to electronic components and connections; shielding and grounding can isolate the detector from the environment; and sample pretreatment and proper instrument operation and maintenance can eliminate some noise from contamination and malfunctions. Some additional suggestions for reducing noise can be found in reference 2.

The definition of noise used by ASTM is depicted in Figure 13.2. The two parallel lines drawn between the peak-to-peak maxima and minima enclose the noise, given in millivolts in this example. In addition, the figure shows a long-term noise or *drift* occurring over a period of 30 min. If at all possible, the sources of the noise and drift should be found and eliminated or minimized because they restrict the minimum signal that can be detected.

The signal-to-noise ratio is a convenient characteristic of detector performance. It conveys more information about the lower limit of detection than does the noise alone. Commonly, the smallest signal that can be attributed to an analyte is one whose signal-to-noise ratio (S/N) is 2 or more. An S/N ratio of 2 is shown in Figure 13.3 for a chromatographic run; it can be seen that this is certainly a minimum value for distinguishing a peak from the background noise.

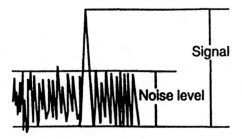

Figure 13.3. Illustration of signal-to-noise ratio (S/N) of 2. (Reprinted from D. W. Grant, *Capillary Gas Chromatography*, Wiley, New York, 1996. Copyright John Wiley & Sons, Inc. Reproduced with permission.)

13.2.2. Time Constant

The time constant, τ, is a measure of the speed of response of a detector. Specifically, it is the time (in seconds or milliseconds) a detector takes to respond to 63.2% of a sudden change in signal as shown in Figure 13.4. The full response (actually 98% of full response) takes four time constants and is referred to as

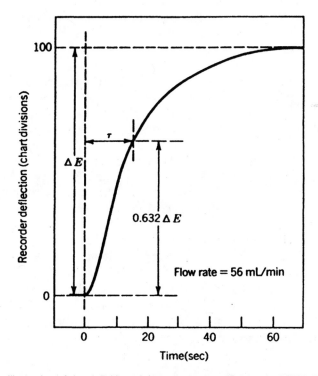

Figure 13.4. Illustration of the definition of time constant, τ. (Copyright ASTM. Reprinted with permission. From H. M. McNair and J. M. Miller, *Basic Gas Chromatography*, Wiley, New York, 1998, p. 108. Reprinted courtesy of John Wiley & Sons, Inc.)

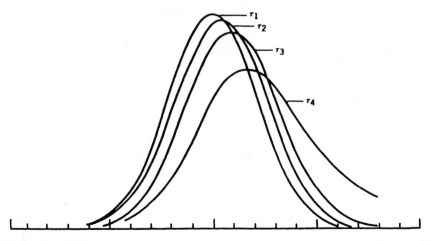

Figure 13.5. Effect of time constant on peak characteristics; $\tau_1 < \tau_2 < \tau_3 < \tau_4$. (From H. M. McNair and J. M. Miller, *Basic Gas Chromatography*, Wiley, New York, 1998, p. 108. Reprinted courtesy of John Wiley & Sons, Inc.)

the *response time*. One of these two parameters should be specified for a detector in every written procedure. However, because some instrument manufacturers have chosen to use other names such as *filter* or *rise time* without defining their relationship to time constant, some trial and error may be necessary to determine the optimum setting for a given instrument and to equate it to instruments from other vendors. In general, small time constants are most desirable, but they do result in increasing the noise level.

Figure 13.5 shows the effect of increasingly longer time constants on the shape of chromatographic peaks. The deleterious effects are the changes in retention time (peak position in the chromatogram) and on peak width, both of which get larger as the time constant increases. The area, however, is unaffected so quantitative measurements based on area will remain accurate while only those based on peak height will be in error. Consequently, peak height methods require proper optimization of the time constant to insure method robustness.

More dramatic is the effect of time constant on a chromatographic run. Figure 13.6 compares a five-component reversed-phase HPLC separation performed using time constants of 0.05 sec (Figure 13.6a) and 5 sec (Figure 13.6b).

The larger time constant produces wider peaks with decreased peak heights, decreased plate numbers, and decreased resolution—all undesirable.

Equation 13.1 [3] can be used to calculate an approximate maximum allowable time constant, based on a peak broadening of 1%:

$$\tau < 0.1 \left(\frac{t_R}{\sqrt{N}} \right) = 0.1 \left(\frac{w_b}{4} \right) = 0.1\sigma \tag{13.1}$$

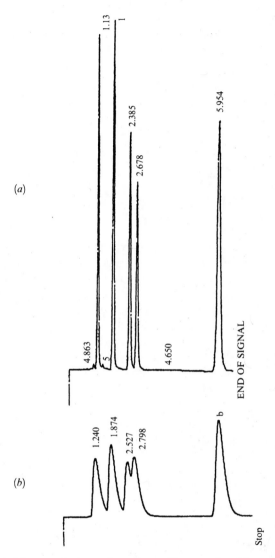

Figure 13.6. Effect of time constant on a chromatographic separation. (a) $\tau = 0.05$ sec, (b) $\tau = 5$ sec.

The parameters used for the calculation are measures of peak width or those related to it, namely the width at base (w_b), the quarter-peak width or standard deviation (σ), and the plate number (N). Note that column length is not important except to the extent it affects these other parameters; thus, plate height (H) is not relevant to this calculation. The peak used for this calculation should be the first one (or an early one) in a chromatogram because it will be the most narrow and the most critical one (in either isocratic HPLC or isothermal GC).

Either of the forms of Equation 13.1 can be used for the calculation. For example, a peak with a retention time of 10 min on a good HPLC column of conventional dimensions (4.6-mm ID and 25-cm length) and a plate number of 20,000 would require a time constant of 0.4 sec or less:

$$\tau = 0.1 \left(\frac{10 \times 60}{\sqrt{20,000}} \right) = 0.4 \, \text{sec} \qquad (13.2)$$

This same peak might have a peak width (at base) of 2.0 mm, measured on a chart at a speed of 1 cm/min. Its quarter-band width, σ, would be 3 sec, requiring a time constant of 0.3 sec or less. These calculations lead us to conclude that the detector time constant for use in HPLC should be on the order of 0.5 sec or less.

Another commonly used rule of thumb [4] is that the time constant should be less than 10% of the width of the peak at half-height, w_h. Thus, a peak with a width of 50 μL at a flow of 1 mL/min would need a time constant of 0.3 sec.

Many laboratories do not operate their HPLC detectors in this range. In fact, it is probably true that too many laboratories have no specification on the time constant in any of their methods! Problems often result during method transfer when time constants are not specified, and the receiving laboratory is using a detector whose default value for time constant is too large.

Peaks from capillary GC columns are likely to be even more narrow and they often require the smallest time constants, while GC peaks from packed columns are less demanding. Remember also that the narrowest peaks will be those with the smallest retention times and are the most critical. Finally, one needs to be aware that the overall time constant for the entire system is limited by the largest response value for any of the individual components: detector, amplifier, or data system.

In spectroscopy, the ideal magnitude of the time constant is dependent on the bandwidth and the scan speed:

$$\text{Response time} = \frac{(\text{bandwidth})}{(\text{maximum scan speed})} \qquad (13.3)$$

For example, if one has a sample that requires a spectrometer with a bandwidth of 8 cm^{-1} in order to achieve adequate resolution, and one wishes to scan at 20 cm^{-1}/sec, the response time required would be 8/20 = 0.4, or a time constant of about 0.1 sec.

Large time constants do have the advantage of decreasing the short-term noise from a detector. This effect is sometimes called *damping*. The temptation to decrease instrument noise and improve one's S/N by increasing the time constant must be avoided. Valuable information can be lost when the data system does not faithfully record all the available information, including noise.

13.2.3. Cell Volume

Cell volume is not usually an issue for normal spectroscopic measurements, although some dissolution methods may specify short path lengths (and smaller volumes) to avoid a dilution step, but for chromatography and for those samples of limited size, a small cell volume can be critical.

Operation of chromatographic detectors is optimized when their internal volumes are small, because band broadening is minimized. However, for concentration-type detectors, the magnitude of that volume has special importance. Suppose the cell volume of a concentration detector is so large that the entire sample could be contained in one cell volume. The shape of the resulting peak would be badly broadened and distorted.

Estimates can be made of ideal cell volume requirements, because the width of a peak can be expressed in volume units (the base width, 4σ, where the x-axis is in milliliter or microliter units). A narrow peak from a capillary GC column might have a width as small as 1 sec, representing a volume of 0.017 mL (17 µL) at a flow rate of 1 mL/min. If the detector volume were the same or larger, the entire peak could be contained in it at one time and the peak would be very broad. An ideal detector for this situation should have a significantly smaller volume, say 2 µL. When this is not possible, extra (make-up) MP can be added to the column effluent to sweep the sample through the detector more quickly. This remedy will be helpful for mass flow rate detectors but less so for concentration detectors. In the latter case, the make-up MP dilutes the sample, lowering the concentration as well as the resulting signal—not a satisfactory solution in many cases. Consequently, concentration-type detectors must have very small volumes if they are to be used successfully when the chromatographic peaks are very narrow.

Another consideration related to cell volume and peak width is the number of data points that are needed to provide a true representation of a chromatographic peak. Narrow peaks require fast data-sampling times as well as small cell volumes (and time constants). Like large cell volumes, too few data points will give peaks that are not symmetrical and Gaussian. It is generally recommended that at least 10 data points should be taken for each peak [5]. For more details, see Chapter 14.

13.2.4. Signal

The detector output or signal is of special interest when an analyte is being detected. The magnitude of this signal (peak height or peak area) is proportional to the amount of analyte and is the basis for quantitative analysis. For spectroscopy, the absorbance (A) is most often used for quantitative calculations; it is essentially the peak maximum or peak height. For chromatography, the peak area is most often used. An integrator (or data system) is needed to convert the analog signal into a digital peak area. Peak height can also be used; see the additional discussion in Chapter 14.

The signal specifications to be defined are sensitivity, minimum detectability, linear range, and dynamic range.

13.2.4.1. Sensitivity.

The detector sensitivity, S, is equal to the signal output per unit concentration or per unit mass of an analyte, depending on the detector classification. The units of sensitivity are based on area measurements of the peaks and differ for the two main detector classifications, namely, concentration and mass flow rate [6].

For a concentration-type detector, the sensitivity is calculated per unit *concentration* of the analyte in the mobile gas phase,

$$S = \frac{AF_c}{W} = \frac{E}{C} \tag{13.4}$$

where A is the integrated peak area (in units such as mV/min), E is the peak height (in mV), C is the concentration of the analyte (in mg/mL), W is the mass of the analyte present (in mg), and F_c is the MP flow rate in mL/min. The resulting dimensions for the sensitivity of a concentration detector are mV mL/mg.

For a mass flow rate type of detector, the sensitivity is calculated per unit *mass* of the analyte in the mobile phase,

$$S = \frac{A}{W} = \frac{E}{M} \tag{13.5}$$

where M is the mass flow rate of the analyte entering the detector (in mg/sec), W is the mass of the analyte (in mg), the peak area is in ampere-sec, and the peak height is in amperes. In this case, the dimensions for sensitivity are ampere-sec/mg or coulomb/mg. As noted earlier, the differences in the units of sensitivity between the two types of detector makes comparisons of the sensitivities difficult.

Figure 13.7 shows a plot of detector signal versus concentration for a thermal conductivity detector (TCD), a concentration-type GC detector. The slope of this line is the detector sensitivity according to equation 13.4. A more sensitive detector would have a greater slope. Because the range of sample concentrations often extends over several orders of magnitude, this plot is often made on a log–log basis to cover a wider range on a single graph. For example, a UV detector, having a large linear range, can enable 0.1% impurities to be quantitated simultaneously with the assay of the active ingredient (at approximately 100%).

As shown at the upper end of the graph, linearity is lost and eventually the signal fails to increase with increased concentration. These phenomena will be discussed later in the section on linearity.

13.2.4.2. Minimum Detectability or Limit of Detection (LOD).

The lowest point on Figure 13.7, representing the lower limit that can be detected, has been

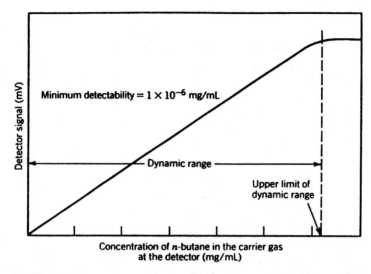

Figure 13.7. Plot of detector signal versus concentration. (Copyright ASTM. Reprinted with permission. From H. M. McNair and J. M. Miller, *Basic Gas Chromatography*, Wiley, New York, 1998, p. 110. Reprinted courtesy of John Wiley & Sons, Inc.)

called by a variety of names such as minimum detectable quantity (MDQ), limit of detection (LOD), and detectivity. The IUPAC report on chromatography [6] has defined the *minimum detectability*, *D*, as

$$D = \frac{2N}{S} \tag{13.6}$$

where N is the noise level and S is the sensitivity as just defined. Note that the numerator is multiplied by 2 in accordance with the definition discussed earlier that a detectable signal should be at least twice the noise level. The units of detectability are mg/mL for a concentration type detector and mg/sec for a mass flow rate type. The ICH definition of LOD is "the lowest amount of an analyte in a sample which can be detected but not necessarily quantified as an exact value" [7]. For many purposes the LOD is taken as 3× signal-to-noise ratio.

If the minimum detectability is multiplied by the peak width of the analyte peak being measured, and if the appropriate units are used, the value that results has the units of mg and represents the minimum mass that can be detected chromatographically, allowing for the dilution of the sample that results from the process. Some call this value the minimum detectable quantity (MDQ). As such, it is a convenient measure to compare detection limits between detectors of different types.

A related term is the limit of quantitation (LOQ), defined as "the lowest amount of an analyte in a sample which can be determined as an exact value"

[8] should be above the LOD and is widely taken as 6× the signal-to-noise ratio. Alternatively, the ACS guidelines on environmental analysis [9] specify that the LOD should be three times the S/N and that the LOQ should be 10 times the S/N. The definitions of the USP are similar and also state that the LOQ should be no less than two times the LOD [10]. Other agencies may have other guidelines, but all are concerned with the same need to specify detection and quantitation limits, and the relationship between them. Additional discussion about LOD and LOQ can be found in other chapters in this book; see especially the definitions in Chapter 15 and Appendix II.

13.2.4.3. *Linear Range and Dynamic Range.*
The straight line in Figure 13.7 curved off and became nonlinear at high concentrations. It becomes necessary to establish the upper limit of linearity in order to measure the linear range that is a fundamental parameter in the method validation process. Because Figure 13.7 is often plotted on a log–log scale, the deviations from linearity are minimized and the curve is not a good one to use to show deviations. A better plot is one of sensitivity versus concentration as shown in Figure 13.8. Here the analyte concentration can be on a log scale to get a large range while the y axis (sensitivity) can be linear. According to the ASTM specification, the upper limit of linearity is the analyte concentration corresponding to a sensitivity equal to 95% of the maximum measured sensitivity. The upper dashed line in the figure is drawn through the point representing the maximum sensitivity, and the lower dashed line is 0.95 of that value.

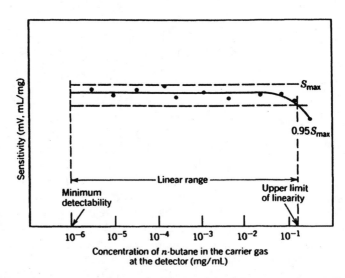

Figure 13.8. Example of a linearity plot. (Copyright ASTM. Reprinted with permission. From H. M. McNair and J. M. Miller, *Basic Gas Chromatography*, Wiley, New York, 1998, p. 111. Reprinted courtesy of John Wiley & Sons, Inc.)

Having established both ends of the linear range, the minimum detectivity and the upper limit, the linear range is defined as their quotient:

$$\text{Linear range} = \frac{\text{Upper limit}}{\text{Lower limit}} \tag{13.7}$$

Because both terms are measured in the same units, the linear range is dimensionless. Obviously, a large value is desired for this parameter.

Linear range should not be confused with *dynamic range*, which was indicated on Figure 13.7 as terminating at the point at which the curve levels off and shows no more increase in signal with increasing concentration. The upper limit of the dynamic range will be higher than the upper limit of the linear range, and it represents the upper concentration at which the detector can be used.

13.3. METHODS OF QUANTITATIVE ANALYSIS

Instruments can be divided into two classes: those that are very stable and not subject to variation from day-to-day, and those that are not. Most instruments for absorption spectroscopy (UV-VIS and IR) fall into the former category, and most chromatographs fall into the latter. Consequently, chromatographic measurements sometimes require the use of special methods for quantitative analysis, such as the Internal Standard Method or the Standard Addition Method. We will consider the more simple methods first, followed by these two special methods.

In discussing these methods, we will not distinguish between the use of peak areas and the use of peak heights (or absorbance in spectroscopy). The process of converting a peak to a peak area is covered in Chapter 14, which also discusses the advantages and disadvantages of each. We will simply refer to the data, be they heights or areas.

13.3.1. Standards and Calibration

All quantitative analyses and validations need to be based on standards. Usually, the standards are highly characterized chemicals, often with the same identity as the analytes to be run. Chemical standards must be pure and certified by an appropriate agency. The USP ⟨11⟩ lists about 1000 reference standards. They are dated with expiration dates, beyond which they must be recertified. The NIST is another official source of standards; although they have available over 1200 standard reference materials (SRMs), few are of use in the pharmaceutical industry. Some laboratory supply houses provide standards also. For example, Restek [11] offers a calibration mixture of organic volatile impurities for USP Method 467 as well as over 60 individual drugs. Additional information about standards and their handling was presented in Chapter 12.

The calibration procedure may vary somewhat depending on the instrument and the method, but basically the process is intended to establish a relationship between the calibration standard and the output signal from the instrument. A linear relationship is most desirable, and this discussion will be limited to that case. However, if the calibration is not linear, quantitative analysis is still possible, but additional precautions must be taken.

Two types of calibration are common, depending on the number of standards run. If one standard is used, prior investigation is necessary to establish that a linear relationship does in fact exist. In addition, it is necessary to show that a blank produces a zero signal, because a minimum of two points would be necessary to establish a linear straight line. Thus, this type of calibration is referred to as the two-point method, even though the zero point might not be confirmed for each analysis.

13.3.1.1. Two-Point Calibration.
The calculation for this method could be a simple ratio:

$$C_u = C_s \times \frac{S_u}{S_s} \qquad (13.8)$$

where C stands for concentration, S stands for signal, subscript u stands for unknown, and subscript s stands for standard. In absorbance spectroscopy, the ratio would be

$$C_u = C_s \times \frac{A_u}{A_s} \qquad (13.9)$$

where A is the absorbance.

Alternatively, a standard could be used to determine a proportionality constant, as for example the determination of the absorptivity constant, a, in Beer's law for absorption measurements:

$$a = \frac{A}{bC_s} \qquad (13.10)$$

where b is the cell length and C_s is the concentration of the standard. Then, for the analysis of an unknown, Beer's law is again used, but with the predetermined constant, a.:

$$C_u = \frac{A}{ab} \qquad (13.11)$$

The calculations are the same either way, and equations are used rather than graphs. To generalize, the equation is really that of a straight line:

$$S = mC \qquad (13.12)$$

Calibration Curve:

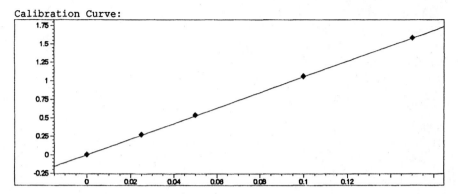

Figure 13.9. Typical HPLC calibration curve for dichromate. correlation coefficient = 0.99999.

where m is the proportionality constant (and slope) of the line. If the blank is not zero, then an additional term is needed to account for that fact:

$$S = mC + \text{blank} \qquad (13.13)$$

13.3.1.2. Multiple-Point Calibration. The multiple standards used in this method are different concentrations of the same standard. They might be dilutions from a single standard preparation, although an error in preparing the single standard will cause all of the dilution standards to be in error too. The signals produced by the standards are plotted versus their concentrations, and a straight line is fitted to the data by the linear regression process described in Chapter 4. Usually the data are entered into a spreadsheet program which then carries out the curve fitting. An example is shown in Figure 13.9. Visual inspection of the graph is usually desirable to confirm the fit of the data to a straight line even though the correlation coefficient, r, is most often used to express the quality of the fit. An r value of 1.0 represents a perfect fit, and typical calibration curves result in values of 0.99 or better. Clearly, the use of multiple standards should produce better accuracy than the two-point method because the errors associated with each standard will be averaged out by this process.

13.3.2. External Standard

External standard is the name sometimes used for the basic analysis procedure using a two-point calibration. Modern instruments usually include this method in the software of the data system. However, multiple-point calibration can also fall under this name.

For chromatographic analysis, a calibration curve is plotted for each peak (each analyte) to be analyzed. Because the standard solutions vary in concentration, a constant volume must be introduced to the column for all samples and standards. Manual injection is usually unsatisfactory and limits the value of this method for GC work. Better results are obtained from autosamplers that inject at least one microliter, and HPLC analyses are usually carried out with sample valves rather than syringes. Fixed-volume loops in HPLC provide very constant volumes, so the external standard is suitable for HPLC.

13.3.3. Area Normalization

Area normalization is a procedure that is often used in chromatography when pure standards are not available or when impurity peaks have not even been identified. As the name implies, it is really a calculation of area percent which is assumed to be equal to weight percent. If X is the unknown analyte, then

$$\text{Area}\%X = \frac{A_X}{\sum_i (A_i)} \times 100 \tag{13.14}$$

where A_x is the area of X and the denominator is the sum of all the areas.

For this method to be accurate, the following criteria must be met:

1. All analytes must be eluted.
2. All analytes must be detected.
3. All analytes must have the same sensitivity (response/mass).

These three conditions are rarely met, but this method is simple and is often useful if a semiquantitative analysis is sufficient or when (as mentioned above) some analytes have not been identified or are not available in pure form (for use as standards).

13.3.4. Area Normalization with Response Factors

If standards are available, the third limitation can be removed by running the standards to obtain relative response factors, f. One substance (it can be an analyte in the sample) is chosen as the standard, and its response factor f_s is given an arbitrary value like 1.00. Mixtures, by weight, are made of the standard and the other analytes, and they are chromatographed. The areas of the two peaks—A_s and A_u for the standard and the unknown, respectively—are measured, and the relative response factor of the unknown, f_u, is calculated:

$$f_u = f_s \times \left(\frac{A_s}{A_u}\right) \times \left(\frac{W_u}{W_s}\right) \tag{13.15}$$

where W_u/W_s is the weight ratio of the unknown to the standard.

Relative response factors of some common compounds have been published for the two most common GC detectors (FID and TCD) by Dietz [12]. These values are ±3%, and because they were obtained using packed columns they may contain some column bleed. For the highest accuracy, one should determine his/her own factors. For HPLC, compilations of response factors for the common UV detector have not been attempted, so one must determine one's own factors.

When the unknown sample is run, each area is measured and multiplied by its factor. Then, the percentage is calculated as before:

$$\text{Weight} \%X = \frac{A_x f_x}{\sum_i (A_i f_i)} \times 100 \qquad (13.16)$$

More information about the determination of relative response factors for HPLC can be found in Chapter 15.

13.3.5. Internal Standard Method

This method and the next are particularly useful for techniques that are not too reproducible, and for situations where one does not (or cannot) recalibrate often. The internal standard method does not require exact or consistent sample volumes or response factors because the latter are built into the method; hence, it is good for manual GC injections.

The standard chosen for this method can never be a component in a sample and it cannot overlap any sample peaks. A known amount of this standard is added to each sample—hence the name *internal standard* (IS). The IS must meet several criteria for chromatographic analysis:

1. It should elute near the peaks of interest, but it must be well-resolved from them.
2. It should be chemically similar to the analytes of interest and not react with any sample components.
3. Like any standard, it must be available in high purity.

The standard is added to the sample in about the same concentration as the analyte(s) of interest and prior to any chemical derivatization or other reactions. If many analytes are to be determined, several internal standards may be used to meet the preceding criteria. For more details on the calculations, see Magee and Herd [13].

Three or more calibration mixtures are made from pure samples of the analyte(s). A known amount of internal standard is added to each calibration mixture and to the unknown. Usually the same amount of standard (e.g., 1.00 mL) is added volumetrically. All areas are measured and referenced to the area of the internal standard, either by the data system or by hand.

If multiple standards are used, a calibration graph like that described in Section 13.3.2 is plotted where both axes are relative to the standard. If the same amount of internal standard is added to each calibration mixture and unknown, the abscissa can simply represent concentration, not relative concentration. The unknown is determined from the calibration curve or from the calibration data in the data station. In either case, any variations in conditions from one run to the next are canceled out by referencing all data to the internal standard. This method normally produces better accuracy, but it does require more steps and required additional effort to validate (see Chapter 15).

13.3.6. Standard Addition Method

In this method the standard is also added to the sample, but the chemical chosen as the standard is the *same* as the analyte of interest. It requires a highly reproducible sample volume, a limitation with manual syringe injection in GC.

The principle of this method is that the additional, incremental signal produced by adding the standard is proportional to the amount of standard added, and this proportionality can be used to determine the concentration of analyte in the original sample. The procedure can be performed with one standard (two-point) or multiple standards. In the former case, linearity is assumed as discussed earlier. Equations can be used to make the necessary calculations in the single standard method, but the principle is more easily seen graphically for multiple standards. Figure 13.10 shows such a typical standard addition cali-

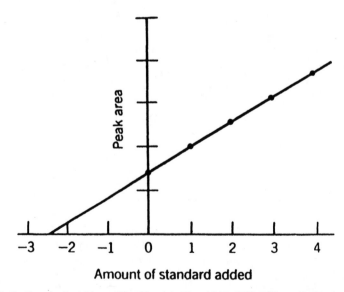

Figure 13.10. Standard addition calibration plot. (From J. M. Miller, *Chromatography: Concepts and Contrasts*, Wiley, New York, 1988, p. 107. Reprinted courtesy of John Wiley & Sons, Inc.)

bration plot. Note that a signal is present when no standard is added; it represents the original concentration, which is to be determined. As increasing amounts of standard are added to the sample, the signal increases, producing a straight-line calibration. To find the original "unknown" amount, the straight line is extrapolated until it crosses the abscissa; the absolute value on the abscissa is the original concentration. In actual practice, the preparation of samples and the calculation of results can be performed in several different ways [14].

Matisova et al. [15] have suggested that the need for a reproducible sample volume can be eliminated by combining the standard addition method with an *in situ* internal standard method. In the quantitative analysis of hydrocarbons in petroleum, they chose ethyl benzene as the standard for addition, but they used an unknown neighboring peak as an internal standard to which they referenced their data. This procedure eliminated the dependency on sample size and provided better quantitation than the area normalization method they were using.

13.3.7. Summary

Chromatographic results can be very precise, down to about 0.1% relative standard deviation (RSD) in the ideal case, and are typically less than 0.5%. However, for all quantative methods the precision and accuracy decrease as the concentration of the analyte decreases.

An early round-robin study of HPLC precision was conducted by ASTM in 1979 [16]. After removal of poor data (outliers) from four laboratories, the remaining 70 labs reported RSDs between 3.1% and 4.6%. These results were considered to be quite good for an interlaboratory study (good reproducibility) but not as good as they should be for a single lab (good repeatability). And, of course, a few of the labs studied showed that they had serious problems. This study also reported a highly satisfactory degree of overall accuracy.

13.3.7.1. Semiquantitative Analysis. The area normalization method described earlier is usually only semiquantitative at best because of the differences in response factors for most compounds in a given sample. One should clearly indicate this possibility when making analyses without standards, and one way to do this is to report data as area % rather than weight %. Or, area % data can be converted to weight % using relative response factors (RRFs) determined during the validation or method development process (Chapter 15).

Another common semiquantitative procedure is commonly used in thin-layer chromatography (TLC). As noted in the USP [17], TLC is most often used for qualitative anlaysis, but quantitative analysis can be performed by a visual comparison of the size of the spots, which is a fast semiquantitative method. More accurate analysis can be achieved by densitometry or video scanning, which results in the production of a chromatogram like those obtained by HPLC, and the data can be handled analogously.

13.4. ADDITIONAL TOPICS

13.4.1. Trace Analysis

Trace analysis, which is becoming increasingly popular, requires that all steps in the analysis be done with care. As an example of the guidelines that are common in trace analysis, the report of the American Chemical Society Subcommittee on Environmental Analytical Chemistry can be consulted [9]. It addresses the issues of data acquisition and data quality evaluation. For HPLC method development, Chapter 14 in Snyder et al. [18] contains valuable instructions for trace analysis.

13.4.2. The High–Low Method for HPLC

In the development of new methods of analysis to support investigations of a new drug substances, HPLC is commonly used to assay the main component and to detect minor impurities. Especially when combined with gradient elution operation, screening of drug mixtures is ideally accomplished by HPLC. However, depending on the dynamic range (linearity) of the HPLC detector, it is often difficult to get the required accuracy on the main drug substance and the desired detectivity on the minor impurities in the same run. For this reason, the method called the "high–low" method has been devised [19].

The basic idea of the high–low method is that two solutions are needed to achieve a satisfactory analysis—a low concentration solution to provide data for the main assay, and a higher concentration solution to provide adequate signals for detecting the trace components. Together they provide the data needed for determining the purity of bulk drug substances. For further details and a justification of this method, reference 19 can be consulted.

REFERENCES

1. G. A. Gross and A. Gruter, *J. Chromatogr.* **1992**, *592*, 271.

2. G. I. Ouchi, *LC-GC* **1996**, *4*, 472–476.

3. V. R. Meyer, *Practical High-Performance Liquid Chromatography*, 2nd edition, Wiley, West Sussex, England, 1994, pp. 309–310.

4. E. L. Johnson and R. Stevenson, *Basic Liquid Chromatography*, Varian Associates, Palo Alto, CA, 1978, p. 278.

5. E. Katz (editor), *Quantitative Analysis Using Chromatographic Techniques*, Wiley, New York, 1987, p. 43.

6. L. S. Ettre, *Pure Appl. Chem.* **1993**, *65*, 819–872.

7. Draft International Harmonisation of Pharmacopoeias, Text on Validation of Analytical Procedures, *Pharmaeuropa* **1993**, *5*(4), 341.

8. J. Fleming et al., *Accred. Qual. Assur.* **1997**, *2*, 51.

9. D. MacDougall et al., *Anal. Chem.* **1980**, *52*, 2242–2249.

10. *United States Pharmacopeia, USP 24*, United States Pharmacopeial Convention, Inc., Rockville, MD, 2000.

11. Analytical Refernce Materials, Catalog #59974, Restek Corp., Bellefonte, PA, pp. 85 and 90.

12. W. A. Dietz, *J. Chromatogr. Sci.* **1967**, *5*, 68.

13. J. A. Magee and A. C. Herd, *J. Chem. Educ.* **1999**, *76*, 252.

14. M. Bader, *J. Chem. Educ.* **1980**, *57*, 703.

15. E. Matisova, J. Krupcik, P. Cellar, and J. Garaj, *J. Chromatogr.* **1984**, *303*, 151.

16. ASTM Subcommitee E-19.08, *J. Chromatogr. Sci.* **1981**, *19*, 338–348.

17. *United States Pharmacopia* (USP 24/NF19), United States Pharmacopeial Convention, ⟨621⟩, Rockville, MD, 2000, pp. 1916–1917.

18. L. R. Snyder, J. J. Kirkland, and J. L. Galajch, *Practical HPLC Method Development*, 2nd edition, Wiley, New York, 1997, Chapter 14.

19. E. L. Inman and H. J. Tenbarge, *J. Chromatogr. Sci.* **1998**, *26*, 89–94.

General References

Quantitative Analysis

G. Guiochon and C. L. Guillemin, *Quantitative Gas Chromatography*, Vol. 42 in the Journal of Chromatography Library, Elsevier, New York, 1988.

E. Katz (editor), *Quantitative Analysis Using Chromatographic Techniques*, John Wiley, New York, 1987.

Detectors

H. H. Hill and D. G. McMinn (editors), *Detectors for Capillary Chromatography*, John Wiley, New York, 1992.

14

LABORATORY DATA SYSTEMS

R. D. McDowall

14.1. INTRODUCTION

Information systems and data systems are an important, but underrated, component of the analyst's tool kit, and they are usually taken for granted. Most analysts focus their attention on the analytical instrument attached to the data system, yet data systems are as important as the instruments they control and acquire data from.

In this chapter we will review the purpose of information systems such as Laboratory Information Management Systems (LIMS) and investigate chromatography data systems (CDS) in depth as a typical example of the data systems used within the laboratory. Other data systems such as near infrared (NIR), ultraviolet (UV) spectroscopy, and combined dissolution bath/UV spectroscopy have similar principles—for example, control of instrument operating conditions and parameters, data acquisition, analysis and reporting, and management of data—but they will be discussed only briefly.

14.1.1. Data and Information Management

Traditionally, laboratory data and information were kept in either laboratory notebooks or separate pieces of paper and then summarized in reports. However, with the requirement for rapid dissemination of information from the lab-

Analytical Chemistry in a GMP Environment. Edited by J. M. Miller and J. B. Crowther
ISBN 0-471-31431-5 © 2000 John Wiley & Sons, Inc.

oratory to the sample submitter, computerization is essential and has resulted in the need to manage data produced by analytical instrumentation (e.g., data file, instrument control parameters, etc.) and manage the information (calculated results) derived from the initial observations from a variety of instrumental and chemical analyses.

14.1.2. Purpose of Data Systems

Data systems are the interface between the analytical instrument and the user, and their purpose is one or more of the following:

- To collect data from instruments
- To control the instruments: either operating and/or data collection parameters
- To reduce data to information (process the initial data by calculation, analysis, integration, etc., into results)
- To organize and manage the data or information transferred or acquired
- To ensure the integrity and security of the data stored within it
- To report the analytical results and any associated quality information (e.g., system suitability test results)
- To interface with a LIMS for unidirectional or bidirectional communication

Data systems operating within a pharmaceutical laboratory must operate under regulatory requirements for good laboratory practice (GLP) or good manufacturing practice (GMP). In essence, this means that:

- System functions must be specified in a document known as a user requirements specification (URS).
- After installation, the system must be qualified to demonstrate its fitness for purpose; this will be based on the requirements in the URS (requirements traceability).
- Operation of the system must be consistent with GXP regulations: Train users with a controlled function to ensure data security and integrity, and any changes to the system must be controlled and the system requalified where necessary.

14.1.3. Types of Data System

14.1.3.1. Chromatographic Data Systems (CDS). A chromatography data system is probably the most common type of data system used within pharmaceutical laboratories. The purpose is to acquire analog data from chromatographic detectors, convert them to digital data, integrate them to peak areas, and calculate the concentration or amount of the analyte present in the sample and then report the results. Instrument control may also be a requirement for some laboratories; this can include control of flow rate, mobile-phase composition, column temperature, injection sequences and replicates, and detector wavelength. This information must be associated with a specific instrument and analytical run as it will form part of the documentation.

Owing to the emphasis on equipment qualification by the FDA, resulting from the Barr Decision [1, 2], it is important that chromatographers understand their data systems and be able to demonstrate their suitability for use. The principles for understanding and using a chromatography data acquisition system for a pharmaceutical laboratory are outlined in this chapter. The skills of the chromatographer are not redundant with a data system, but rather the automated system ensures an improvement in quality over manual methods.

14.1.3.2. Nonchromatographic Data Systems, Including LIMS. Other types of data system can be found connected to spectrometers such as UV, IR, NIR, and so on, and are being used for acquiring and manipulating data and reporting results in a manner similar to that of a CDS. However, many of these data systems are point solutions or islands of automation within the laboratory.

In contrast, a Laboratory Information Management System (LIMS) is a more encompassing information solution for the laboratory. It can interface with the customers of the laboratory and with other computer applications; it can also deal with the administration of samples, schedule workloads, manage results from many different types of instruments, and report results. Results can be reported on paper or transfered to spreadsheets or LIMS. An overview of LIMS will be discussed first before returning to an in-depth discussion of CDS.

14.2. LABORATORY INFORMATION MANAGEMENT SYSTEMS (LIMS)

The major function of most analytical laboratories is the creation and presentation of information quickly to facilitate decision making. A LIMS is one of the major laboratory automation tools at the disposal of analytical chemists to help achieve this aim. Although a LIMS does not undertake analysis, it can be pivotal in integrating both the laboratory operations and the laboratory itself within an efficient organization. A LIMS can provide a laboratory with the means to automate the processes of information creation and presentation, as well as being the platform for information dissemination to clients and senior management.

However, this is not always the case. A large number of systems fail to meet initial expectations, suggesting that LIMS are not fully understood [3]. A LIMS model was proposed [3, 4] that enabled the requirements of a system to be visualized conceptually [5]. This approach has been adopted, modified, and used as the basis of the LIMS concept model in the ASTM LIMS guide [6]. However, the LIMS model focuses primarily on identifying and visualizing the user functions within the laboratory environment and less on the strategic siting of a system.

To overcome this problem, a matrix for the development of a LIMS with a strategic focus [7, 8] has been proposed. It introduced the concept of three types of LIMS: operational, logistic, and strategic. The matrix is formed by plotting these three types of LIMS against the scope of laboratory and organizational tasks that can be undertaken by such a system. The purpose of this chapter is to discuss the impact that a logistic and strategic LIMS can make within an organization.

14.2.1. A LIMS Has Two Targets

A LIMS is one of the major tools of laboratory automation available to the analytical chemist [9]. However, to be effective and provide maximun benefit, it should be sited correctly. It is important to realize that a LIMS has two targets: The first is the laboratory that is responsible for generating information, and the second is the organization that uses that information to make decisions.

A LIMS should be sited to benefit both groups. However, many systems are conceived and implemented by laboratory managers who are only concerned with improving the efficiency of their laboratories, not benefiting the organization. When a LIMS is designed from the top down, the designer does not consider the laboratory and the laboratory does not benefit from the system. There is a balance to be struck between these two extremes.

Figure 14.1 shows an outline of the functions that a LIMS should undertake. The diagram shows a LIMS sited at the interface between a laboratory and an organization. Samples are generated in the organization and logged into the

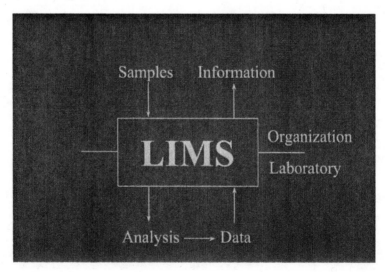

Figure 14.1. Functions of a LIMS showing the interface between laboratory and organization.

LIMS, the samples are analyzed within the laboratory, and data are produced and reduced within the LIMS environment to information that is transmitted back into the organization. Figure 14.1 represents the ideal siting of the LIMS: Both the organization and the laboratory benefit. The line dividing the organization and the laboratory show that the system is of equal benefit to both.

However, there are two other implementations that are possible with a LIMS, resulting in different positions of the interface between the laboratory and organization. These are discussed in more detail in reference 8, but the business benefits of any system implemented in these ways will be lost.

There is a balance to be found between the needs of the organization and the laboratory. The interface between the two must be carefully defined; however, the initial implementation should be toward the analytical laboratory, the information generator. Automating the information generator is the key to success for the whole LIMS, and this is best done by first understanding the process workflow, where necessary simplifying or eliminating processes and then automating the resultant flow with the information system.

14.2.2. Benefits of a LIMS

The benefits of a LIMS are dependent upon the laboratory in which it is installed and how it is connected with other computer applications within an organization. The majority of systems are implemented on a local level, and the real benefits from a system implemented according to business need are lost. They have little connection with the departments that submit samples, and therefore only meet one strategic requirement. The benefits are reduced administration, simplified work procedures, improved reporting, and increased laboratory productivity. The laboratory benefits, but the organization does not.

When sited properly, a LIMS will produce benefit to both. A system sited according to business requirements will enable information produced by the laboratory to be used to make decisions such as product release or rejection (using specifications stored in the LIMS) or calculation of formulation shelf lives.

Productivity increases achieved by an individual LIMS will be dependent on the aims and configuration of each system. No definitive study has been performed to assess the overall benefit of LIMS, but increases in productivity of 10–20% is the average estimate made by most laboratory managers who have installed such systems [10]. Golden also assessed the impact of LIMS in both research and development (R&D) and quality control (QC) environments. In the former, increases in productivity are the main benefit by removing the transcription and checking of data brought about by automatic data capture. This productivity is manifested by the speed at which a product can be perfected and brought to the marketplace. The rationale for a LIMS here is not only the improvement in productivity or in the speed at which specific analyzes are completed, but lies in the laboratory's ability to complete entire projects more quickly.

Specific benefits of a LIMS within a production environment are the quicker acceptance or rejection of raw materials and finished products, because data from individual analyses can be collated, checked, and reported more rapidly against specifications using a LIMS than with a manual process. This should enable lower stocks of both to be kept, thus saving and generating money respectively. In addition, information about product lines can be kept to comply with GMP regulations and generate product quality assurance data. Connection of the LIMS with the production department allows the production staff to interrogate the system. If the LIMS is connected to on-line process analyzers, the data generated by these instruments can be interpreted by the LIMS and fed into the manufacturing control systems. A LIMS sited like this will make the laboratory more effective within an organization.

Within a research and development environment, a LIMS is intended to provide information earlier, either to speed new compound to market or to terminate projects earlier and save money. This will help ensure the quality of the products emerging from a research and development pipeline. From a laboratory perspective, a LIMS should remove many of the administrative tasks associated with analysis. This will make the laboratory more efficient. Again, it is the connections that are made between a LIMS and existing computer applications that will help make the R&D laboratory effective within an organization.

14.2.2.1. Justification of a LIMS. The justification of a LIMS must be based upon business and organizational needs and not just laboratory benefits. The benefits of a LIMS, outlined above, can be divided into tangible, intangible, and unpredictable benefits [11]. Tangible benefits are those that can be assigned a monetary value such as cost savings, but intangible benefits (e.g., quality) cannot normally be assigned such a value. Intangible benefits, conversely, have no monetary value such as quality and image. If enough work is done, any intangible benefit can become a tangible one, but the extent of the work involved may not be cost effective. Unpredictable benefits are those that occur unexpectedly or are too infrequent to be assigned a value—for example, the prevention of mistakes such as the release of a batch of wrongly specified material or avoiding product recalls. For more detail on the justification of a LIMS, the paper by Stein [11] is the best on the subject.

14.2.3. Regulatory Issues

14.2.3.1. Quality of Operation. A LIMS is a tool for improving the quality of laboratory and organization operation in any area of the pharmaceutical industry: production, development, or research. To cope with changing environment in both development and especially research, a LIMS needs increased flexibility over a production system. In fact, a LIMS can be thought of as an essential compliance tool.

14.2.3.2. Regulatory Inspection. There are a number of implications of implementing a LIMS in a regulated environment, including system development documentation, system validation, log books, and change control. In many respects, regulatory authorities consider a computer system to be equivalent to an analytical system [12]. Therefore the system documentation and training records of the users should be available for inspection.

LIMS vendors can help users in validation by developing systems with full documentation and delivering them at the same time to the customer. Some vendors have chosen to apply for ISO 9001 accreditation while others have used independent companies to audit the whole system development. The responsibility for validation of the system still remains with the end-user and the laboratory management. Here vendors can help by providing test scripts for users to qualify their systems. For a more detailed discussion about development procedures and vendor audits, the reader is referred to the papers by Segalstad [13, 14] and McDowall [15, 16].

14.3. CHROMATOGRAPHY DATA SYSTEMS

The main discussion in this chapter concerns chromatography data systems (CDS). There are three main types of chromatographic data system available now:

- Integrator
- Single personal computer connected to a single or multiple chromatographs (single or multiple user)
- Multiuser client server networked systems

The advantages and disadvantages of each type are presented in Table 14.1, adapted from Darkin [17]. The principles outlined in this section are applicable to all three types of data system; however, the main discussion will concern multiuser client server chromatography data systems because these have the best approach for use in regulated pharmaceutical laboratories, such as version control of methods, sharing and control of data, overall data integrity, and data backup. Today it is not possible for a regulated laboratory to take any other course of action if security and integrity of data are to be maintained. Furthermore, chromatography data systems operating in pharmaceutical laboratories must be validated either retrospectively [18] or prospectively as outlined in Chapter 15 of this book.

The next sections will discuss CDS from a number of perspectives. First the principles of analog-to-digital (A/D) conversion and peak integration will be discussed as a prelude to a view of the workflow of a typical CDS. Please note that the terminology used here for some of the functions may not map exactly to all data systems; for example, a sequence file may be found both pre- and post-injection, so that for some systems, interpretation of what is written here may be needed for an individual example of a CDS. For readers wanting more information on the process, the book by Dyson [19] is highly recommended.

Table 14.1. Advantages and Disadvantages of Different Types of Chromatographic Data Systems

System Type	Advantage	Disadvantages
Integrator	Always available for use	Unintelligent
	Little downtime	Little scope for development
	Lowest training time and cost required	Little scope for reprocessing raw data
	Immediate local hardcopy	Limited scope for storage and archival of raw data
	Lowest cost per channel	Limited communication potential
		Single user
		Little or no security
Personal computer	Moderate cost per channel	Slower response for multiple users
	Moderate initial expenditure	Single keyboard: contention for access by multiple users
	Good graphics support	Little channel expansion
	Fast response for single user	Poor security as single PC
	Local storage of data	No immediate report unless requested
	Better user interface	Backup usually poor (user responsible)
	Provides hard disk storage for data and methods	Difficult to control methods between separate PCs
	Provides facilities for reprocessing and reporting of results	Multiple copies of software Minimum security
		Poor sharing of data
		Poor communication
		No network access to instruments
		LIMS integration requires interfacing to all PCs
Multiuser client-server	Highest computing power	Highest initial cost
	Multichannel	Significant liability if server fails: no further acquisition runs started
	Multiuser	
	Good networking facilities	
	Good graphics support	No immediate hardcopy unless requested
	Method standardization and control	Resilience in network required to ensure no or limited downtime of system
	Highest security and data integrity	
	Backup of data good	Backup of data may require full access to system limiting data acquisition
	Incremental growth of system possible	
	Single installed copy of the software	
	Central storage of data	

Table 14.1 (continued)

System Type	Advantage	Disadvantages
	System manager provides single point for administration and maintenance	
	Sharing of methods and data easily achieved through network	
	Data stored and organized either through directories or database	
	Single interface to LIMS required	

Source: Adapted from Darkin [17].

14.4. ANALOG-TO-DIGITAL (A/D) CONVERSION

14.4.1. Rationale for A/D Conversion

Because the heart of a chromatography data system is the A/D conversion unit, we will discuss this process starting with one of the early methods of recording analog chromatography data, a chart recorder. The chart recorder has two major advantages:

- A continuous real-time data plot
- Visual inspection to see any over-range peaks

However, the use of a chart recorder restricts quantification to manual (e.g., pencil and ruler) or semiautomated (e.g., planimeter, disk integrator) methods to quantify analytes. The introduction of automated A/D converters within CDS brought great relief from the tedium of physically measuring peak heights and widths followed by manually calculating the peak areas. It also eliminated the necessity to rerun samples whose peaks were offscale or too small in the original run.

14.4.2. Principles of A/D Conversion

Analog-to-digital conversion is a process by which a continuously variable signal (e.g., analog voltage) from a chromatography detector is converted to a binary number that accurately represents the original data. These data are then passed to a computer for processing and storage. It is necessary to convert the analog signal to a digitized form because computer systems can only handle

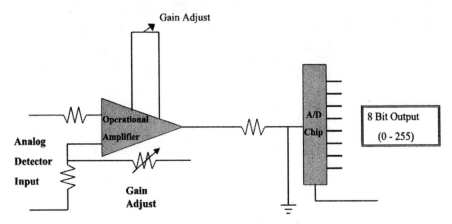

Figure 14.2. Diagram of an 8-bit analog-to-digital (A/D) converter.

numerical information in the form of a binary code comprising of a series of zeros and ones [20, 21].

A description of an 8-bit A/D converter device is shown in Figure 14.2, and the process of data conversion can be described in a number of simple steps. An 8-bit A/D unit can convert an input voltage into 2^8 bits or 0–255.

The first stage of the A/D process is the connection of the detector input to an adjustable gain amplifier. This amplifier takes the input voltage from the chromatographic detector and produces an output that is matched to the input of the A/D circuit. Many older high-performance liquid chromatography (HPLC) detectors have an output range of 0–10 mV intended for chart recorder input; however, the amplifier boosts the detector signal to the standard input range (typically 0–1 V) of an integrated circuit A/D converter. When the detector has an output of 0–1 V, connecting the detector lead to the 0–1 V input on the box eliminates the amplifier, and the detector signal is fed directly into the A/D unit.

The signal is sampled by the A/D unit when an appropriate control signal is applied to the A/D chip's "convert" pin, so that the analog voltage at its input (0–1 V) is converted into a digital value (in the range 0–255) and stored in an internal buffer. The digital value can then appear as an 8-bit parallel digital signal on the output lines. There are eight output lines, and each one carries one-eighth of the signal. By adjusting the amplifier's gain and offset voltage, an 8-bit A/D unit can be set to produce binary 0 (i.e., 00000000) output for 0-V input and binary 255 (i.e., 11111111) output for 10-mV input, binary 127 for 5-mV input, and so on.

The two main types of A/D unit used in CDS systems today are voltage-to-frequency (V/F) and successive approximation. Their advantages and disadvantages are compared in Table 14.2 [17].

14.4.2.1. A/D Converter Resolution. The two constraints as outlined above are the sampling rate (see the next section) and resolution of the A/D unit.

Table 14.2. Types of Analog to Digital Converters used in Chromatography Data Systems

	Voltage to Frequency (V/F)	Successive Approximation
Advantages	Single dynamic range	Lower cost
	Better signal-to-noise ratio	Higher data rates (up to 1000 Hz)
	Less sensitive to high-frequency noise	
Disadvantages	More expensive	Noisier than V/F
	Limited to data capture rates under 100 Hz	Can lose data during range switching
	Higher rates can create more noise	Can produce noise spikes that require software filtering

Source: Adapted from Darkin [17].

Resolution is the primary factor in determining the quality of the chromatography. It is defined as the number of bits that the A/D is capable dividing the signal into; the higher the bit length, the more sensitive the A/D. Table 14.3 gives some typical values to illustrate this relationship [21].

So why is resolution important? This depends on the type of analytical work that you do. If you are working within a narrow concentration range, a low resolution A/D may be suitable. There will be problems in discriminating between 97% and 100% of the nominal amount, but that's only minor. However, if you are determining peaks over a wide concentration range, say a main component and any related impurities, a low-resolution A/D will allow you to pass every batch of product, because you won't be able to see the impurities! For example, where the output signal accuracy is limited by the resolution of the converter is 8-bit, if the input is a maximum of 10 mV, then this can only be represented by 256 values—that is, in steps of approximately 0.039 mV. Hence it cannot discriminate between 0 and 0.01 mV for example, even assuming that the device is perfectly accurate and has no noise.

So how much resolution do we need? As much as we can get; however, there are cost penalties, such as longer readout times, software issues, and noise and drift considerations to take into account. Chromatography data systems typically employ a minimum of 16-bit A/D converters and some vendors use 21- or

Table 14.3. Relationship Between A/D Bit Length and Resolution

A/D Bit Length	Resolution Elements ($2^\wedge$ Bit Length)	Discrimination for 10-mV Input (μV)
8	256	39.0625
10	1024	9.765625
16	65536	0.152588
21	2097152	0.004768
24	16777216	0.000596

24-bit. The situation is complicated by the fact that some 16-bit devices can be mathematically enhanced by an oversampling technique.

14.4.2.2. Characteristics of V/F A/D Converters. Voltage-to-frequency A/D converters have the following characteristics:

- *Resolution:* As discussed above, the ability of the A/D unit to distinguish between peak size. Expressed in bits, the higher the number of bits, the more sentitive the A/D unit.
- *Noise:* The amount of short-term electrical noise signal that the A/D will add to the detector signal. Usually expressed in microvolts.
- *Linearity:* Linearity between the voltage output of the detector and the digital number that the A/D produced over the dynamic range.
- *Dynamic range:* Defined as the specified operating range for which the A/D is linear. There should be a message on the chromatogram that the input voltage was exceeded. However, the vendor should state what happens to a peak that is outside of the range of the A/D unit and how this is marked in the data file. Systems that have no over-range message should not be considered further—your analyses will be compromised.

14.4.2.3. Multiplexing. Some A/D chips are multiplexed, mainly in the less expensive single-user personal computer (PC) systems to reduce overall cost. Multiplexing involves one A/D sharing a number (2–4) of input lines from different detectors or chromatographs, and a multiplexer is used for multi-channel data acquisition from a single A/D converter. Its purpose is to sample each analog input line in a fixed sequence, allowing sufficient time on each channel for the A/D converter to acquire a portion of the analog signal, convert it to digital, and to pass the output to the computer. Because an A/D converter takes several milliseconds to acquire a reading, a single multiplexer can easily handle several analog inputs. However, the greater the number of channels connected, the smaller time the amount of available for data acquisition.

The greatest advantage that a multiplexer data collection system offers is one of economy, because several chromatographs can be connected to a system. However, the cost differential is eroded by the sophistication of the software required to acquire the data. The sampling rate and sequence of the multiplexer must be strictly controlled in order to maintain the integrity of the data from each individual channel. Whichever approach is used, the signal will have to travel a distance of several meters and background electronic noise that can degrade the quality of the signal will inevitably be generated.

In addition, there is the problem of ensuring that data being acquired on one channel does not interfere with the data being obtained on another. This problem is known as cross-talk.

14.4.2.4. Requirements for Effective Peak Integration. Definitions are important here, and the following terms are important for understanding the

process of peak integration: sampling rate or frequency, peak width, and threshold. Each will be defined below; for more detail please refer to the book by Dyson [19].

- *Sampling Rate or Frequency:* This is a key to understanding the operation of any data system and is the rate at which the analog signal is sampled by the A/D. The sampling rate is correctly expressed in hertz (Hz) because it is a frequency, but is usually more colloquially stated as points per second; however, regardless of the unit, the figures are the same. Therefore a 100-Hz sampling rate is 100 times per second. Normally an A/D samples at a constant rate and the data are bunched (averaged) using the peak width factor. There should be at least 15 points over the smallest peak of interest. Oversampling (faster sampling rate than needed) by the A/D unit can interfere with the peak integration and also increase the disk space of the stored data file. Undersampling will result in distorted peak shapes like those shown in Figure 13.5. Some data systems and integrators use a peak width setting instead of sampling rate, but the purpose is exactly the same.
- For capillary gas chromatography (GC), sampling rates of between 5 and 20 Hz (samples per second) are required. In contrast, for conventional liquid chromatography (LC) with a run time of about 10–20 min, a sampling rate of 1 Hz should suffice. For longer run times, a slower rate may be appropriate because later eluting peaks emerge from the column, and this may be found as the *time to double* used in some data systems. Capillary electrophoresis in various forms will require much faster sampling rates and may actually exceed the specification of many standard A/D units used for conventional chromatography.
- *Peak Width:* This is the value used to calculate the data bunching factor applied to the raw data during peak detection. Usually bunched data points are used to detect peaks, but all sampling points are used for integration. The product of the peak width is multiplied by the sampling rate divided by 15 to determine the number of points bunched together to produce a single point for integration. Bunched data are only used to detect peaks, in contrast to the fact that all raw data are used for integration. Data bunching has no effect on the acquisition of raw data. It is an internal calculation used to enhance the process of determining peak start and end. It is best to use the narrowest peak in the chromatogram to set the peak width initially, because optimum integration occurs when at least 15–20 data points per peak are available.
- *Threshold:* This defines the minimum rate of change required for peak detection at both peak start and peak end. Setting the threshold value high screens out the detector noise from the peak detection. Too high detection may not be sensitive enough to changes in baseline and may not detect very small or very broad peaks. However, if the value is set too low, then the detection may be too sensitive to small changes in baseline, resulting in

many peaks being observed and detector noise being integrated. In general, use a lower threshold when looking for very small peaks—for example, impurities or very broad peaks (late eluting or chiral peaks). Use a higher threshold value when looking at large sharp peaks, such as in capillary GC or capillary zone electrophoresis (CZE).

14.4.2.5. Signal Drift and Noise. Drift and noise are the rise or fall in the A/D baseline as was shown in Figure 13.2, and they are expressed in microvolts per unit of time. Short-term drift is the rapid shift of the baseline either up or down usually due to gradient profiles or detector equilibration. Longer-term drift is the gradual meandering of the baseline up or down over the entire chromatographic run or across multiple samples and is usually caused by solvent, temperature changes, contamination, and so on.

Positive drift (the upwards drift of the detector signal) can be coped with relatively well by most A/D converters. However, negative drift is a different case. Most data systems measure the input voltage as low as a baseline of −50 or −10 mV, thereby coping with some negative drift. However, because bad instances of negative drift can exceed this, it is necessary to find out to what extent negative drift is accommodated by the A/D unit and whether it has any influence on peak detection. For example, if threshold value is used to detect peaks, then this may be compromised with adverse negative drift with the result that drift is reported as a peak.

Negative drift is an area that few, if any, data systems handle well. The use of the autozero and offset functions can be used to minimize negative drift in an operational situation. A flat baseline with no noise may indicate that the signal has gone below the A/D lower limit and is not really a baseline.

Correcting drift is most important for ongoing performance qualification because it is more likely to degrade as the electronics age than is the linearity. The ASTM standard provides an approach that can be adapted to address these concerns [22].

However, not all the noise comes from the detector and the A/D; therefore chromatographers need to be aware of the sources of noise and how to minimize them [19].

14.4.3. Peak Detection

The key stages in identifying a peak are shown in Figure 14.3. These are: monitoring baselines to detect the possibility and recognition of a peak, the peak apex determination, and the rear inflection when the peak has eluted and the baseline becomes stable again.

Some of the issues that we will discuss about peak detection are as follows:

- How is a peak detected?
- What are the minimum and maximum number of points required for peak detection and definition?

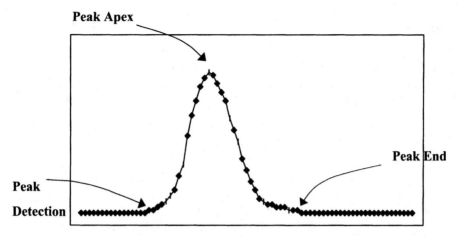

Figure 14.3. Stages in peak identification.

- How are data points bunched to reduce noise or to help define slow eluting peaks toward the end of a run?

The individual data slices in a file are analyzed by the integration software using the parameters defined in the method file. Within the method file, some timed events will determine when the method will start to measure the peaks (integrate inhibit). This allows sections of the chromatogram such as the solvent front to be omitted and allows the chromatographer and data system to concentrate on the peaks of interest.

Let us consider what happens to a single peak of interest and how a typical CDS will define the peak by determining the peak start, the apex, and the peak end. This is the ideal situation.

- *Peak Start:* The CDS algorithm will look for the rise of the detector signal above baseline, assuming that the data capture rate has been correctly set. A peak is detected when the rate of detector signal or slope increase rises above the minimum values set in the method file for the threshold parameter. If the threshold value is set too low, then noise can be detected as peaks; this is especially noticeable when you have a noisy detector signal. Equally so, if you set the threshold value too high, then you will not detect small peaks of interest—hence the need to develop and validate the method to the degree that is adequate for its application.
- *Peak Apex:* The data capture rate must be correctly set for determining the apex of any peak. This is to ensure that the apex is correctly defined when the detector signal is rapidly changing. To ensure this the data capture rate defined must be appropriate to the type of chromatography being undertaken (e.g., LC, capillary GC, or CZE).

• *Peak End:* This is when the peak has eluted and the signal returns to the baseline; however, virtually all peaks in LC and GC are nonsymmetrical. The peak end is where the detector slope is zero within the limits defined by the slope sensitivity of the method. Where there is little noise and the data capture rate is suitable, there is little problem. However, where there is noise in the system or the data capture rate is too high, the placement of the peak end will be incorrect.

Peak Integration. This is the process where peak heights or areas are calculated from the defined peaks. The CDS will place a straight line between peak start and peak end to define the baseline. The signal representing the baseline at every data slice is subtracted from the total microvolt signal, and the remaining values represent the peak. Peak area is the integration of the residual signals over the defined peak. The algorithms used for this will vary from vendor to vendor, are commercially sensitive and are inevitably different, and can produce different results for the same input signals [23]. The peak area units are microvolts per second. Peak height is simply the highest residual value in microvolts which is printed out in the report at the end of the run.

For the most accurate work peak areas are preferred to peak height measurements.

14.4.3.1. Integration of Unresolved Peaks. The two techniques used to measurement of fused peaks are either perpendicular drop or tangent skim (baselines in the latter case usually involve straight lines), as shown in Figure 14.4. They are simply an automation of the manual process of peak measurement. Space does not permit a detailed discussion of the problems, but inaccuracies as well as under- and overmeasurement of peak areas are possible depending on the size ratio of peaks being measured. The reader is referred to the papers by Dyson [19], Papas [23], and Meyer [24].

Comparison of Peak Height and Area Integration. After integration of the peaks the chromatographer has two options to quantify the peak: either area or height. Each option has its advantages and disadvantages, as discussed below.

Peak heights from CDS are usually reported in microvolts, and the distance from the baseline to the peak apex is measured in microvolts. Peak heights can be more accurate than peak areas because there are two main factors in determining the accuracy of the height measurement: (1) the placing of the baseline and (2) the location of the peak apex. Both are measured in microvolts, and the height is simply the difference between the two values. Peak heights are less subject to interference by overlapping peaks, and therefore less chromatographic resolution is required for quantification by height. The main requirement for the use of peak height in an analytical method is Gaussian peak shape.

Peak areas resulting from the integration of the *peak start* through the apex to the *peak end* are measured in microvolts second^{-1}. Area measurement re-

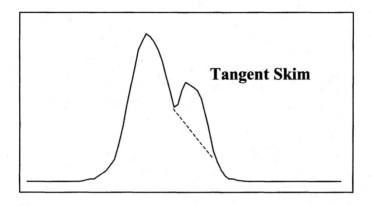

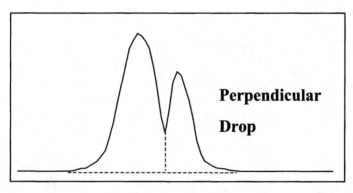

Figure 14.4. Integration of unresolved peaks.

quires accurate peak start and peak end, and this requires the threshold and peak width to be set correctly. Area measurements are more precise than height and are preferred with asymmetrical peaks.

For the majority of analytical work, peak areas are more precise and reflect the peak for quantification and are less suseptible to variations in individual A/D performance.

14.4.3.2. Integration Errors. There are a number of common problems and errors that can occur in integration:

- *Sampling Rate Incorrectly Set.* This will result in insuffient data points to define a peak, and the peak will be poorly identified. The result is that the peak height and the peak area will not be correct. For example, the peak height will probably be underestimated because the peak apex will be missed, and because the peak start will not be seen by the integrator until after the analyte has eluted from the column, peak area will be under-estimated as well.

- *Incorrect Threshold Setting.* As discussed earlier, if the threshold is set too low, then noise will be detected as peaks. In contrast, if the threashold is set too high, then small peaks will not be detected. Therefore, the threshold needs to be set just above the noise level to ensure effective peak detection.
- *Noise Too High.* Where the noise is too high, there are a number of problems that could occur such as incorrect baseline setting and poor peak quantification, especially for small peaks (e.g., related substances) because the peak signal will be lost in the noise.
- *Baseline Drift.* The majority of baselines are fitted by straight lines, so when the baseline is rising, the peak area will be underestimated where the baseline curves.

14.5. CDS WORKFLOW

14.5.1. Sequence of Data System Operation

This section discusses the operation of a chromatography data system from the perspective of the workflow; Figure 14.5 shows the overall sequence of events that a typical data system should perform. This is a generalized approach to the operation of a "typical" data system. While expensive integrators have fewer functions available, minicomputer and PC systems usually have more.

14.5.1.1. Method Files. The start of the data acquisition operation of a chromatography data system is to build a method file. This tells the data system how to acquire data and process and interpret the results. A method file should control the following:

- The data sampling rate of the A/D converter
- When to start and stop the integration of the chromatogram
- Whether peak areas or heights should be used
- Retention time windows and identification of the analytes and internal standard
- Allocation of the method to calculate the analyte amount or concentration

A name, a number, or a mixture of both should identify individual method files within the system. In addition, the system should be able to provide facilities for version control of method files to ensure that control is maintained over the method for the lifetime of its use. Part of the control function must be access control to identify the individuals who can create, modify, or delete analytical methods. If a method has been modified, then copies of the modifications must be stored with the data processed by that method to provide an audit trail for the data and results produced by a version of a method. However, when

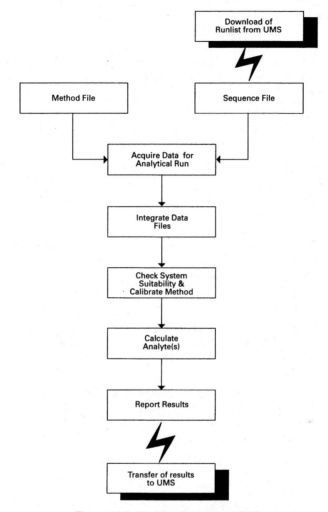

Figure 14.5. Workflow for a typical CDS.

developing methods, flexibility with method files is essential and a default method should be available to acquire data and then feed back to a normal method.

14.5.1.2. Naming Conventions Associated with the Use of a CDS System. When a pharmaceutical laboratory uses a client-server CDS, there will be an urgent need to consider naming conventions for method, sequence, and all data files within the data system. Any CDS must have sufficient capacity for naming all of the files that would be created by the system over a reasonable time period and to aid efficient archiving and unambiguous identification of these files. Therefore for efficient management of data files and methods, naming conventions should be introduced. Any naming convention system must aid users, quality assurance, and regulatory inspectors.

A naming convention should be based on the workflow undertaken by a laboratory. This is to allow efficient archiving of data but also, just as importantly, the efficient retrieval of data. Some ideas might be:

- Organize the directories around the drug projects. This is often how the work is structured and how project teams are organized.
- Major subdivisions of each project should be based around the type of work done—for example, method development, method validation, preformulation, and so on.
- Individual analytical runs can be named so that data file overwrite is prevented.

In any case, it is best if a uniform scheme is established for each laboratory.

14.5.1.3. Sequence File—Sample Queue. A sample queue is often dictated by the method or the lab policy. The sequence file is the run list or order in which the samples, standards, quality control samples, and blanks will be injected into the chromatograph. Each sequence file or each injection must be linked with a method file to process the resulting data. For laboratories with large numbers of samples for a single method, the sequence file will usually be linked with a single method. Smaller laboratories may need the flexibility to link the sequence file with several methods during the course of a single analytical run for maximal use of equipment resources.

Each sample to be analyzed should be identified in the sequence file as one of the following types:

- Unknown or test sample
- Calibration standard
- System suitability sample
- Quality control
- Blank

Depending on the data system involved, at least the first two options are available to a user.

Label. Some of the calibration standards are selected as system suitability test (SST) standards. The SST parameters such as plate number, tailing, and resolution need to be defined for each analytical method and can be calculated by the CDS. Traditionally, the system suitability parameters are calculated before committing the samples for analysis. The key paper by Furman et al. [25] outlines current FDA policy concerning chromatographic analysis for the pharmaceutical industry:

Only after acceptable system precision with replicate injections of a standard solution is obtained, should sample analysis proceed.

This is reasonably explicit: Run the system suitability standards and determine if the system is suitable before committing your samples. However, the 1997 Pharmacopeial Forum [26] lists a proposed change for the *United States Pharmacopeia* General Chapter ⟨621⟩ on chromatography [27] and gives proposed guidance on setting up each injection sequence:

> Replicate injections of the standard preparation required to demonstrate adequate system precision may be made before the injection of samples or may be interspersed among sample injections. System suitability must be demonstrated thoughout the run by injection of an appropriate control at appropriate intervals. The control preparation can be a standard preparation or a solution containing a known amount of analyte and any additional materials useful in the control of the analytical system such as excipients or impurities. Whenever there is a significant change in equipment or in a critical reagent, suitability testing should be performed before the injection of samples. No sample analysis is acceptable unless the requirements of system suitability have been met. Sample analyses obtained while the system fails requirements are unacceptable.

A sequence or run file of the system can be constructed using a series of create, copy, and edit functions to input information such as sample number, sample identity, laboratory number, sample volume, and internal standard amount. There may be additional calculations required, such as dilutions or to account for different sample weights; some data systems also have the facilities for customized calculations linked through the sequence file.

If a LIMS is used within the laboratory environment, a key way of linking the LIMS with the data system will be by downloading of electronic copies of worksheets created by the LIMS (see Figure 14.6). These worksheets can be incorporated into a sequence file by the data system to save repetitive data input. If this is done, it is a prime requirement that the data system offer the

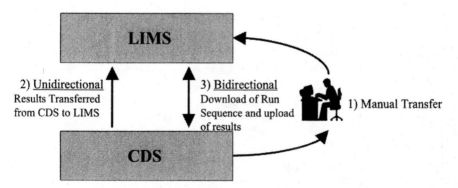

Figure 14.6. Options for interfacing a CDS to a LIMS.

editing ability because the original LIMS file will have been created prior to sample preparation as discussed earlier.

14.5.1.4. Interpretation of Chromatographic Data. After the method file and the sequence file have been set up, the analytical run is started and data are collected. A data file containing the A/D data slices will be obtained for each chromatographic run and sample injected. It is imperative from scientific and regulatory considerations that the data files must not be capable of alteration. Moreover, they must not be overwritten either if the same sample information is assigned to an assay or if the disk becomes full. This is an area for consideration when validating the chromatography data system [18]; you must know what happens to your data files, especially in a regulated environment.

The data system will interpret each data file, identifying the analytes and fitting the peak baselines according to the method parameters. An emerging trend in data systems is the ability of the system to identify whether the peak baselines have been automatically or manually interpreted. This is a useful feature.

Most data systems should be able to provide a "real-time" plot, so that the analyst can review the chromatograms as the analytical run progresses. In addition, the plotting options of a data system should include the following:

- Fitted baselines
- Peak start/stop ticks
- Named components
- Retention times
- Timed events (e.g., integration start/stop)
- Run-time windows and user-defined plotting windows
- Baseline subtract

Each of these options should be enabled or disabled by a user.

An overlay function should be available to enable you to compare results between samples. This will be used to compare chromatograms from the same run sequence as well as chromatograms from different sources. The maximum number of overlays will vary from data system to data system, but a minimum of six to eight is reasonable and practicable. More overlays may be technically possible, but the amount of useful information obtained may be limited. Overlays that can be offset by an amount determined by the user are useful to highlight certain peak information. Ideally, the overlay screen should have hidden lines removed and be able to be printed.

14.5.1.5. Calibration. Calibration is a weak area with most data systems, because chromatographers use many ways to calibrate their methods as evidenced by the multitude of calibration options available. Often these methods are basic and lack statistical rigor, because the understanding of many chromatographers, where calibration is concerned, is poor.

Within a pharmaceutical analysis laboratory, the number of calibration model options that can be successfully used is usually limited to the following:

- Bracketed standards at one concentration or amount for bulk drug or finished product assays
- Multilevel or linear regression for related substances

Within each calibration type, the data system must be flexible and able to cope with variations in numbers of standards used and types of standard bracketing. The incorporation of a blank standard into the calibration curve should always be an option. See also the discussions in Chapters 4 and 12.

Each plot of an analyte in a multilevel or linear regression calibration model must contain an identifier for that calibration line and the analyte to be determined. The calibration curve should show all calibrating standards run in any particular assay. In assays containing more than one analyte, it will be necessary to interpret all the calibration graphs before the calculation of results. Again, many data systems are deficent and offer only one line-fitting method for all analytes in the run, resulting in compromises.

14.5.1.6. *User-Defined Analytical Run Information.* The system should
be capable of collating user-defined parameters (e.g., component, peak area, internal standard ratios, concentrations, etc.) for selected analytes from a sequence of runs. After collation, system-defined and/or user-defined statistical calculations will be carried out on the data generated. The type of calculations required should include mean, standard deviation, analysis of variance, and possibly significance testing (see Chapter 4).

14.5.1.7. *Reports and Collation of Results.* Ideally, the report following an
individual chromatogram should contain both elements that are user-definable and those that are standard; this should enable the laboratory to customize a report. At the end of the analytical run, a user-defined summary report containing information such as sample ID, area or height, baseline, and calculated analyte concentration should be created. This report can either be printed out or transferred to a LIMS for further analysis and interpretation.

14.5.2. Instrument Control

The primary interaction of the CDS with analytical instrumentation is with the output from the detector; however, there are other considerations such as instrument control. These can vary from system to system, and the following options are available:

- Contact closures for the control of chromatographic valves or associated equipment during analysis are usually available for other vendor's equipment.

- When the same vendor makes the data system and the chromatography equipment, control is more sophisticated and more tightly integrated with the data system functions. For instance the whole chromatograph can be operated from the data system method file or instrument file, enabling multiple methods with different conditions to be operated during a single analytical run.
- Communication with the autosampler via BCD or equivalent communication for sample continuity is, in my view, essential, but is usually ignored by many and offered as an option by many vendors.
- Remote monitoring of the chromatography system output including the instrument conditions.
- The ability to list the items of equipment (pump, detector, etc.) used for a particular analysis is a function to help to automate the administrative records associated with an analysis and help meet GMP compliance.
- System initialization and on-line diagnostics to check that the chromatograph runs within instrument specification limits during a run.

14.5.3. Interfacing CDS to Laboratory Information Management Systems

There are three possible ways of linking a CDS to a LIMS as shown in Figure 14.6:

1. Manual data entry
2. Unidirectional or semiautomatic data capture
3. Bidirectional or automatic data capture

Each approach will be reviewed, and the processes involved will be discussed to highlight specific points. The discussion will concern a CDS, but the principles outlined in this section are applicable to other laboratory data systems.

The simplest method of integration is manual; the two systems are essentially separate with no physical connections between them, and they operate independently of one another. The unidirectional link is based on result data being uploaded from the data systems to the LIMS. Selected information from the CDS is transferred to the LIMS; usually this is the individual results and overall system suitability data but usually not the data files associated with the analytical results.

A bidirectional link has two separate actions. The first is the downloading of the sample and run-order information from the LIMS in order to set up the data system sequence file before the analysis starts. After completion of the analysis, the second stage is the uploading of result information as described above.

The computerized system validation requirements of the manual approach is zero for the interface, but there will be an ongoing requirement for transcription

Table 14.4. Process Analysis of CDS Linkage to LIMS

Analytical Process	Manual	Unidirectional	Bidirectional
Work list production	Automatic	Automatic	Automatic
Enter run sequence to data system	Manual	Manual	Automatic
Data collection and calculation	Automatic	Automatic	Automatic
Transfer to LIMS	Manual	Automatic	Automatic
Transcription check	Manual	Automatic	Automatic
Collate results	Automatic	Automatic	Automatic
Prepare report	Automatic	Automatic	Automatic

error checking the data entry that will not be eliminated as the underlying process is error prone as it is under human control.

Electronic interfacing of the data system and the LIMS will eliminate most, if not all, of the data transcription error checking for daily operations; this is the greatest benefit for the laboratory as well as for increased speed of reporting results. However, the interface must be validated, but the effort involved in this will be small compared with the benefits outlined above.

14.5.3.1. Process Analysis of the Workflow. To see the advantages of the three levels of integration, we will look at the work involved in any analysis can be broken down into seven steps or processes. These are listed in Table 14.4. For a manual process, there are three manual stages of the seven, assuming that all the calculations are performed by the data system to the endpoint required by the analyst. The automatic stages proceed with little human involvement, but the manual stages require human input and are slow and error-prone. Thus manual data entry is slow and has several stops and starts.

The unidirectional link has several advantages over the manual link; the worksheet is produced as before, then entered manually as the run sequence into the data system as before. However, the remaining steps are all automatic: data are acquired and reduced by the data system, and the results are transmitted electronically to the LIMS for collation and reporting as before.

Bidirectional integration automates the last manual stage: The work sheet is downloaded into the data system and is incorporated as the sequence file. No manual transcription is required, although there would always be the option to print out the worksheet for use as a means of recording any modifications (e.g., different sample volumes) during sample preparation.

The main requirement for any automation project must be to reduce the human labor content of an analysis. When the labor is reduced, the laboratory is in a position to increase productivity; thus, unless the CDS is interfaced to the LIMS, the latter will not be used properly and the organization will not see any major benefits.

From the process perspective, the advantage of either the unidirectional or bidirectional interfacing is that the transcription checks for data entry are eliminated (see Table 14.4). The other impact of the automation is to increase the quality of the results, because manual stages have been eliminated.

14.6. CONCLUDING REMARKS

Data systems are essential for efficient and effective operation of any pharmaceutical laboratory; this chapter has concentrated on two types: chromatography data system and LIMS. However, the principles outlined in this chapter are applicable to most data systems:

- Understand the operating prinicples of the data system.
- Know the limitations of the data acquisition and processing software.
- Automate as much as possible within the data system and avoid data manipulation outside.
- Transfer results electronically to a LIMS using validated software for quality, reliability, and speed.

REFERENCES

1. *Federal Register* Vol. 58, No. 102, Friday, May 28, 1993, p. 31035 (Notice)/1061; Department of Health and Human Services, Food and Drug Administration [Docket No. 93N-0184], Barr Laboratories, Inc., Refusal to Approve Certain Abbreviated Applications; Opportunity for a Hearing.
2. United States District Court of the District of New Jersey; Civil Action No. 92-1744, *USA v Barr Laboratories, Inc.*, Alfred M. Wolin, USDJ, February 4, 1993.
3. R. D. McDowall and D. C. Mattes, *Anal. Chem.* **1990**, *62*, 1069A.
4. D. C. Mattes and R. D. McDowall, in E. J. Karajalainen (editor), *Scientific Computing and Automation (Europe) 1990, Data Handling in Science and Technology*, Volume 6, Elsevier, Amsterdam, 1990, p. 301.
5. R. D. McDowall, Chemometrics and Intelligent Laboratory Systems, *Lab. Inf. Manage.* **1992**, *17*, 181.
6. *Standard Guide for Laboratory Information Management Systems (LIMS)*, E5178, ASTM, Philadelphia, 1994.
7. R. D. McDowall, *Anal. Chem.* **1993**, *65*, 896A.
8. R. D. McDowall, *Lab. Autom. Inf. Manage.* **1995**, *31*, 57.
9. R. D. McDowall, Chemometrics and Intelligent Laboratory Systems, *Lab. Inf. Manage.* **1992**, *17*, 265.
10. J. H. Golden, *Intell. Instrum. Comput.* **1985**, *3*, 13.
11. R. R. Stein, Chemometrics and Intelligent Laboratory Systems, *Lab. Inf. Manage.* **1991**, *13*, 15.
12. P. D. Lepore, Chemometrics and Intelligent Laboratory Systems, *Lab. Inf. Manage.* **1992**, *17*, 283.

13. S. H. Segalstad, *Lab. Autom. Inf. Manage.* **1995**, *31*, 11.

14. S. H. Segalstad, *Lab. Autom. Inf. Manage.* **1996**, *32*, 23.

15. R. D. McDowall, *Sci. Data Manage.* **1998**, *2*(2), 8.

16. R. D. McDowall, *Sci. Data Manage.* **1998**, *2*(3), 8.

17. D. W. Darkin, Chromatographic Data Acquisition, Chapter 11 in *Laboratory Information Management Systems: Concepts, Implementation and Integration*, R. D. McDowall, (editor), Sigma Press, Wilmslow, G.B. 1988.

18. B. Wikensted, P. Johansson, and R. D. McDowall, *LC-GC Int.* **1999**, *11*, 88.

19. N. Dyson, *Chromatographic Integration Methods*, 2nd edition, *RSC Chromatography Monographs Series* R. M. Smith, (editor), Royal Society of Chemistry, Cambridge, 1998.

20. D. J. Malcolme-Lawes, *Encyclopedia of Analytical Science*, Volume 2, Academic Press, London, 1995, p. 826.

21. C. Burgess, D. G. Jones, and R. D. McDowall, *LC-GC Int.* **1997**, *10*, 791.

22. ASTM E 685-93 (*Annual Book of Standards*, Volume 14.02), American Society for Testing and Materials, West Conshohocken, PA, 1996.

23. A. N. Papas, *Anal. Chem.* **1990**, *62*, 234.

24. V. R. Meyer, *Chimia* **1997**, *51*, 751.

25. W. Furman, T. Layloff, and R. Tezlaff, *JAOAC Int.* **1994**, *77*, 1314–138.

26. *Pharmacopeial Forum* **1997**, *23*, 4513.

27. *United States Pharmacopeia*, 24th edition, United States Pharmacopeia, Inc., ⟨621⟩ 2000.

15

QUALIFICATION OF LABORATORY INSTRUMENTATION, VALIDATION, AND TRANSFER OF ANALYTICAL METHODS

Jonathan B. Crowther, M. Ilias Jimidar, Nico Niemeijer, and Paul Salomons

15.1. INTRODUCTION

Government agencies require accurate information and data recorded both in regulatory filings and in day-to-day operations of pharmaceutical manufacture. From a pharmaceutical laboratory's perspective, analysts need to ensure the accuracy and reliability of the data generated by their test methods. As diagramed in Figure 15.1, there are required and fundamental controls that ensure the overall quality of the analytical test data.

Several of these control processes are discussed in detail in other chapters of this text. In Chapter 12, for instance, development scientists must ensure that they are supplying V-TR^{2}AP (validatable, transferable, rapid, robust, accurate, and precise) test methods to the operating laboratories. These V-TR^{2}AP attributes must be designed into the method at the earliest stages. Chapter 12 also

Analytical Chemistry in a GMP Environment. Edited by J. M. Miller and J. B. Crowther
ISBN 0-471-31431-5 © 2000 John Wiley & Sons, Inc.

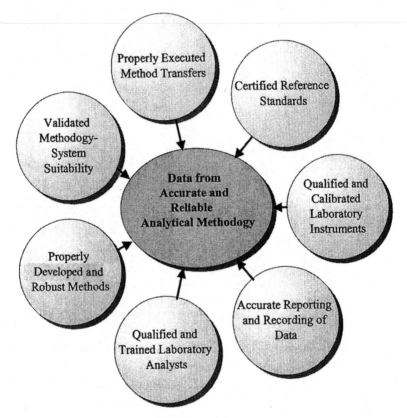

Figure 15.1. Interrelated elements that ensure reliability of data from analytical methodology.

reviews both the importance and requirements for certification of reference standards. Earlier chapter 2 discusses laboratory records while Chapter 14 reviews the attributes of a computerized chromatographic and laboratory data systems.

This chapter will discuss qualification of laboratory instrumentation, focus on validation of analytical methodology, and expand upon the method transfer process discussed earlier in Chapter 12. The interdependent processes represented in Figure 15.2 all correlate to ensure the quality of the reported data.

15.2. INSTRUMENT QUALIFICATION

Validation is often applied to many elements of pharmaceutical manufacture including instrumentation, equipment, facility services, systems, processes, and cleaning. In any good manufacturing practice/good laboratory practice (GMP/

GLP) application, the laboratory instrumentation must be validated for its intended use. In a typical situation, instrument validation is an active partnership between the instrument supplier and the customer. The validation of instrumentation traditionally consist of three parts: installation qualification (IQ), operational qualification (OQ), and performance qualification (PQ). GAMP3 [1] definitions are as follows:

Installation Qualification. "(IQ) is documented verification that all key aspects of software and hardware installation adhere to appropriate codes and approved design intentions and that the recommendations of the manufacturer have been suitably considered. In practice this is ensuring that the system has been installed as specified, and sufficient documentary evidence exist to demonstrate this."

Operational Qualification. "(OQ) is documented verification that the equipment or system operates as intended throughout representative or anticipated operating ranges. In practice this is ensuring that the installed system works as specified and sufficient documentary evidence exists to demonstrate this." OQ is typically performed on individual components of a modular system.

Performance Qualification. "(PQ) is documented verification that the process and/or the total process-related system performs as intended throughout all anticipated operating ranges. In practice this is ensuring that the system in its normal operating environment produces acceptable quality product and sufficient documentary evidence exists to demonstrate this." PQ is "holistic" and is typically performed on a complete system.

A suitable framework for instrument qualification is reviewed in Figure 15.2. As outlined in the GAMP3, the three test components of IQ, OQ, and PQ are linked to the specification developed during the design phase. Because most modern laboratory instrumentation is not custom built [examples of exceptions may be robotic equipment and larger instrumentation such as mass spectrometers and nuclear magnetic resonance (NMR) spectroscopes], the following discussions will focus on IQ, OQ, and PQ components of the qualification process.

15.2.1. Instrumentation Life Cycle

Before discussing the details of instrument validation, it is important to consider the life cycle of analytical instrumentation (Figure 15.3). Initially, justification for new laboratory instrumentation is submitted along with the proposed design and requirements. Upon installation and qualification, the instrument is "operational," requiring periodic maintenance and calibration. In the event

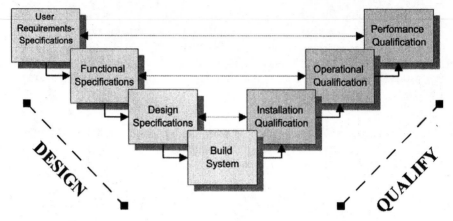

Figure 15.2. Instrument qualification/specification, design, and testing.

that a system change is required (updated software, change of location, new modular component, etc.), a change control is initiated and the system may require partial or full requalification. The system continues in this cycle until it is obsolete or replaced.

15.2.2. Introduction—Qualification Versus Calibration

Upon qualification, the laboratory instrumentation is required to be placed on both a preventive maintenance and periodic calibration program. A current good manufacturing practice (CGMP) auditor will expect instrumentation to be calibrated, properly maintained, and clean. The frequency of calibration, preventive maintenance, and performance verification* depends on the instrument, environment, and function. A typical high-performance liquid chromatography (HPLC) system should be maintained, calibrated, and performance-verified every 6–12 months. Laboratory instruments should be calibrated against a nationally accepted standard [e.g., National Institute for Standards and Technology (NIST)] or a secondary standard that has been calibrated against such a standard.

15.2.3. Prospective Versus Retrospective

Figure 15.4 outlines the general flow of instrument validation.

In general, newer instrumentation will be qualified using a prospective validation approach. Existing laboratory instrumentation is already purchased and

---

* Performance verification is often performed periodically or after instrument maintenance, providing proof that the instrument continues to perform as expected.

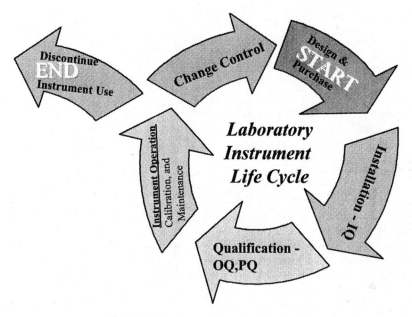

Figure 15.3. Instrumentation life cycle concept.

installed and presumably has a documented history of suitable operation. This preexisting instrumentation may be retrospectively qualified.

Prospective Validation. This type of validation implies that the system is qualified according to the life cycle model as it is being developed and installed and prior to utilization.

Retrospective Validation. This type of validation applies to systems that have been in use for an extended period of time, and there is documented evidence that the system has functioned properly during that period. Typically, the majority of effort in retrospective validation/qualification activity consists of assembling existing documentation. As a general rule, existing instrumentation may be validated with little or no additional experimentation as shown in the retrospective portion of the schematic in Figure 15.4. Data retrieved from documented calibration programs can serve as the operational qualification portion of a retrospective qualification. Retrospective PQ data can also be derived from existing system suitability data and test data from approved methods or procedures. Calibration/Preventive Maintenance (PM) procedures already in place may serve for ongoing system evaluation.

Examples where retrospective qualification may apply include gas chromatographs, HPLC, and spectroscopic instrumentation.

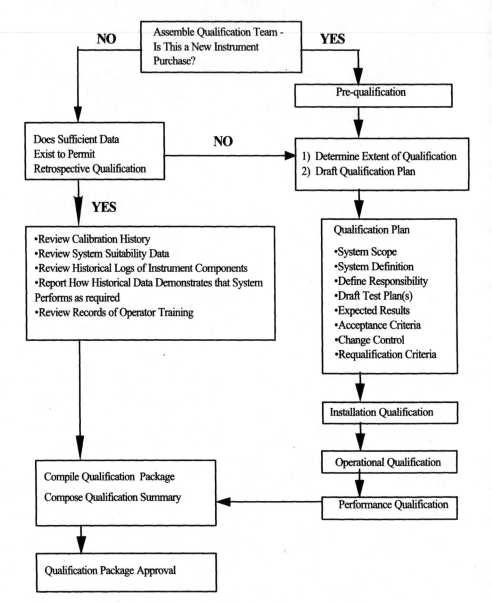

Figure 15.4. Generalized flowchart of instrument validation.

15.3. INSTRUMENT QUALIFICATION PROCESS—ASSEMBLY OF THE QUALIFICATION TEAM

A qualification team may be assembled to monitor project progress and validate the equipment. In preparation for a new system, the team is ideally formed prior to the purchase of the system. The typical qualification team is comprised of the instrument operator (often the team leader), a facilities engineer, and a representative of laboratory management. The extent of instrument qualification along with the size and composition of the qualification team will vary with the complexity of the instrument to be validated. As an example, larger projects may also include members from purchasing, members from engineering, a representative from the instrument vendor, and additional laboratory staff, as required.

The team will define prequalification criteria and acceptance criteria for the instrument. The team will then select vendors upon review of several criteria including:

- Company instrument standardization policy
- Vendor reputation, maintenance programs, and financial status
- Training, start-up support, qualification support
- Initial performance testing (i.e., evaluation or demonstration at the vendor's site)
- Vendor's qualification documentation
- Satisfaction of current users
- ISO 9000 certification

15.4. THE QUALIFICATION PROTOCOL

The qualification team assembles a qualification plan for the system to be validated. This plan should include the following:

- *System Scope.* Define the intended purpose and major functions of the system.
- *System Definition.* Provide a detailed description of the system's instrumentation, hardware/software, utilities, and other components that define the system and its intended functions. Special consideration such as security, special documentation, and so on, should be addressed.
- *Responsibility.* Define the individuals responsible for preparing and drafting the qualification plan and those expected to execute the testing. Define, with a written agreement, the responsibility of the instrument vendor in the qualification process.

- *Qualification Procedures.* An individual qualification procedure will be drafted which will confirm successful instrument installation (IQ). Separate procedures will then test and challenge the intended functions as described in the system definition (OQ and PQ).
- *Expected Results/Acceptance Criteria.* The expected results of each test should be listed in the plan, leaving space for the actual results. The plan should include acceptance criteria for formally accepting the system and suggested action if all acceptance criteria are not met.
- *Change Control/Requalification Criteria.* In accordance with the life cycle approach, the plan should include criteria for requalification of the system upon change in the system. Depending upon the extent of the change, some or full requalification may be required.
- *Maintenance and Calibration.* Maintenance and re-calibration schedules should be defined using the vendor's qualification documentation as a guide. Utilizing an instrument logbook may be appropriate for certain instrumentation applications in addition to the instrument documentation file.

Qualification Plan Approval. The qualification plan should be signed by the qualification team and persons executing the tests. Management will also sign the document. The qualification document, listing acceptance criteria, should be approved prior to system installation.

15.5. IQ PROTOCOL

15.5.1. Installation Qualification

Figure 15.5 shows an example of an installation qualification checklist. This checklist should be completed for each instrument component in addition to the following information when appropriate:

1. Preinstall Phase:
   - Check that all utilities necessary for the instrument are present.
   - A package should be prepared that contains all information necessary to properly install, operate, and maintain the specific instrument; include a system diagram.
   - Review/approve all the existing documentation that was shipped with the system.
   - The facility engineer will ensure that adequate utilities (capacities/quality) and environmental (HVAC) conditions are available.
   - For larger instruments, IQ and OQ of the system may be performed at the vendor's manufacturing facility as an additional QC check, re-

Installation Qualification - EXAMPLE

Component Installation Qualification (Complete for each Component of the System)

System Identity_____

Component Description _____

1. Purchase Order number   _____
2. Identification Number   _____
3. Model Number   _____
4. Serial Number   _____

System Utilities	Required (Y/N)	Specifications	Complete (Y/N)
Gas Supply (including traps, filters, etc.)			
Electrical			
Lighting			
Plumbing/Water			
Steam			
Ventilation/Air Quality			
Heating/Cooling			
IT/other			

Description:

_____
_____

Manufacturer's Specifications:

_____
_____

Instrumentation and Auxiliary Equipment:

_____
_____

Specifications:

_____

Recommended Spare Parts List (Optional):   ☐Attached

_____
_____

Calibration/PM/MaintenanceSOP - ListSOPs

_____

Figure 15.5. An example of an installation qualification checklist.

quested by the team, prior to the system being shipped. Upon in-house installation, IQ, OQ, and PQ checks will be reperformed.

2. Physical Installation Phase: Document exactly how the instrument was installed; verify:
   - Completeness of shipment
   - Environmental specifications—often supervised by a facility engineer
3. Register and verify that new instruments are placed on preventative maintenance and calibration schedules.

15.5.2. Operational Qualification

The individual components as well as the entire system should function according to the manufacturer's specifications and any additional specifications required by the qualification team. This should be accomplished using the established OQ protocol, which tests critical instrument parameters. Proper operation will be verified by performing the test protocols specified in the qualification plan. Test instruments used in OQ must be calibrated. Upon satisfactory completion, acceptance, and reporting of the operational qualification segment, performance qualification may commence. Operational qualification should be performed at both the component and system level.

15.5.3. Performance Qualification

Performance qualification checks the operation of the instrument at the system level. Typically, PQ documents the accuracy, linearity, and precision of the system under typical operating conditions. Experiments typical to system suitability and generation of calibration curves are often used for performance qualification which test the system for its intended function and evaluate performance over the proposed operating range. Ongoing system suitability testing (USP 24) or a test mixture of standards adequately verify the prevailing performance and suitability of the system. Other tests may be included to test the system's robustness.

15.5.4. Ongoing Monitoring

There are certain Standard Operating Procedures (SOPs) that are required to support adequate performance of the instrumentation.

- *Security.* This provides a safeguard to data integrity.
- *Data Archival and Retrieval.* These ensure that data can be reinstalled on the system once archived.
- *System Operation.* This may be a simple list of operator's manuals or a brief listing of operations and parameters.

- *Training.* This is necessary to ensure that only qualified operators use the instrumentation.
- *Maintenance.* Sufficient preventative maintenance and calibration procedures must be instituted so instruments and equipment remain calibrated and suited for their intended purpose.
- *Change Control.* Changes to the system must be documented, using a procedure/form similar to that represented in Figure 15.6. Changes to the system may require partial or complete requalification. Partial requalification would be performed when the instrument is moved, a module is changed, software is upgraded, or system is enhanced. Complete requalification may be necessary for major changes (i.e., change in computer data system operating system, new modular components, etc.).
- *Requalification.* A review of small changes and additions to the system over time may indicate that a requalification is necessary. Requalification can be performed using the previous operation and performance qualification protocols.

15.5.5. Final Qualification Report

The final instrument qualification report may consist of some of the following sections:

- Instrument Qualification Report Approval. Upon completion of the qualification report which includes documentation that the test protocols have been followed and completed, an executive summary is drafted and the qualification report is circulated for approval by the qualification team. Management and QA will also approve the report.
- Executive Summary. The executive summary will clearly state that the instrumentation has been validated. The summary will include historical information on previous qualifications. This summary also includes a list of each of the tested functions of the components as well as the total system and the results. The summary will note any limitation that the qualification testing has placed on the system or any particular function.
- Completed and Approved Test Plan.
- Completed Analytical Instrumentation/Equipment Installation Qualification with Attached Test Data.
- Operational Qualification Test Plans and Results with Attached Test Data.
- Performance Qualification Test Plans and Results with Attached Test Data.

As stated previously, the qualification package may need to be updated periodically to reflect changes, updates, and improvements of the system.

Example - Change Control

Equipment/System Description or Number:

_____

_____

_____

Description of Change and Purpose:

_____

_____

_____

_____

_____

Requalification Test Requirements: ☐   None Required
                                            ☐   See Test Required Below

_____

_____

_____

_____

Results of Test Requirements: ☐   Attached

_____

_____

_____

Certification Documents Updated:

___ Installation Qualification   ___ Qualification
___ Operational Qualification   ___ Calibration
___ Performance Qualification

Figure 15.6. An example of a change control form for laboratory instrumentation.

15.6. INSTRUMENT QUALIFICATION SUMMARY

There are many approaches to analytical instrument qualification. This discussion attempted to highlight the important parameters of IQ, OQ, and PQ through a generalized example. Practical approaches to instrument validation appear throughout recent scientific literature, notably for HPLC system validation [2–4]. While individual approaches may vary depending upon company policy and instrument type, the ultimate goal is the same: to verify, document, and understand the capabilities and limitations of the instrumentation being validated.

15.7. ANALYTICAL METHOD VALIDATION

In its basic form, analytical method validation is a matter of establishing documented evidence that provides a high degree of assurance that the specified method will consistently provide accurate test results that evaluate a product against its defined specification and quality attributes [5]. This section will discuss an approach to method validation that will meet these objectives; these discussions will focus on validation of methods using HPLC for assay and purity determinations; nevertheless, fundamentals of the approach can be applied to most method validation activities.

15.7.1. Introduction to Method Validation

The laboratory analyst must be sure that the methodology that they are using is optimized, capable, and validated. Chapter 12 reviewed the process of developing optimized and capable V-TR^{2}AP methods; this section will discuss validation in similar detail. Properly executed, the validation process can be as significant an effort as method development. Due to the level of resources required, the activities must be properly planned to ensure an efficient and successful validation. The validation stage of method development is a confirmation process; there should be few "surprises" in validation results, because prevalidation evaluation data should suggest that the method will validate successfully. While the requirements of validation have been clearly documented by regulatory authorities [6, 7], the approach to validation is varied and open to interpretation. The approach below will focus on International Conference on Harmonisation (ICH) guidelines.

Prior to beginning validation, controls should exist to ensure that the method has been properly developed and capable of the objectives outlined prior to beginning the method development endeavor. As stated in Chapter 12, validation is considered a GMP activity. Therefore, validation activities must be properly documented and performed on qualified and calibrated instrumentation and equipment. At this stage, there should be documented evidence that the method is robust. A generalized flowchart of the validation process is detailed in Figure 15.7.

15.7.2. Determining the Characteristics of the Validation

Prior to beginning the method validation, a validation team will be assembled to categorize the validation requirements and propose acceptance criteria for each validation parameters. These requirements and criteria will ultimately be documented in a validation protocol. The validation requirements for each method type are clearly reported by the ICH (see Table 15.3 on page 453). Acceptance criteria, on the other hand, must be in-line with the expected product specifications, development experience, and current industry practice [8, 9]. A separate section will further discuss acceptance criteria.

15.7.3. Definitions

It is important to define the terms used in regulatory guidances when discussing method validation. In the definitions listed below, the *italic* portions are quoted directly from the ICH guideline [10].

Accuracy. *"The accuracy of an analytical procedure expresses the closeness of agreement between the value which is accepted either as a conventional true value or an accepted reference value and the value found."* "Accuracy studies measure the effects of the sample matrix on the values obtained and are often carried out in the context of *recovery* studies" [11].

Precision. *"The precision of an analytical procedure expresses the closeness of agreement between a series of measurements from multiple sampling of the same homogeneous sample under prescribed conditions. Precision may be considered at three levels: repeatability, intermediate precision, and reproducibility."*

 Repeatability. *"Repeatability expresses the precision under the same operating conditions over a short interval of time."*

 Intermediate Precision. *"Intermediate precision expresses within-laboratories' variations: different days, different analysts, different equipment, etc."*

 Reproducibility. *"Reproducibility expresses the precision between laboratories."*

Specificity. *"Specificity is the ability to assess unequivocally the analyte in the presence of components which may be expected to be present."*

Detection Limit. *"The detection limit of an individual analytical procedure is the lowest amount of analyte in a sample which can be detected but not necessarily quantitated as an exact value."*

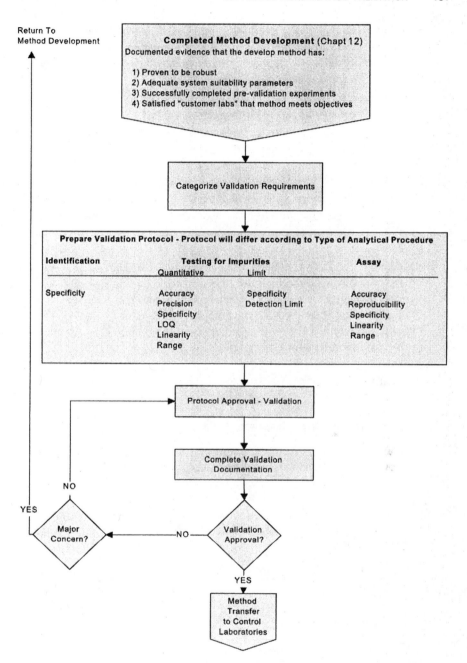

Figure 15.7. Generalized validation flowchart.

Quantitation Limit. *"The quantitation limit of an individual analytical procedure is the lowest amount of analyte in a sample which can be quantitatively determined with suitable precision and accuracy."*

Linearity. *"The linearity of an analytical procedure is its ability (within a given range) to obtain test results which are directly proportional to the concentration (amount) of the analyte in the sample."*

Range. *"The range of an analytical procedure is the interval between the upper and lower concentration (amounts) of the analyte in the sample (including these concentrations) for which it has been shown that the analytical procedure has a suitable level of precision, accuracy, and linearity."*

Robustness. *"The robustness of an analytical procedure is the measure of its capacity to remain unaffected by small, but deliberate, variations in method parameters and provides an indication of its reliability during normal usage."*

Ruggedness. The USP [12] defines ruggedness as "the degree of reproducibility of test results obtained by the analysis of the same samples under a variety of normal test conditions, such as different labs, different analyst, different lots of reagents, ... Ruggedness is a measure of reproducibility of test results under normal, expected operational conditions from laboratory to laboratory and from analyst to analyst."

Sensitivity. The sensitivity of an analytical method is equal to the slope of the calibration line in a linear system.

15.7.4. Method Validation Documentation

The validation documentation typically consists of a protocol, test data, and a final report. One approach to simplifying validation documentation is to focus on a thorough protocol with preapproved acceptance criteria. This protocol may have data tables to enter the test results, requiring only a short executive summary to summarize the results and a reference or attachment of raw data. Often a development lab will use these well-developed and optimized master method validation protocols [13, 14] as templates for subsequent validations. Frequently a copy of the method procedure and, if possible, a method development report is appended to the validation protocol. In general the validation protocol should contain the following:

Validation Protocol

1. Method principle/objective
2. Listing of responsibilities (laboratories involved and their role in the validation)
3. Method categorization according to the ICH or USP
4. List of reagents (including test lots) and standards required

5. Test procedures to evaluate each validation parameter and proposed acceptance criteria
6. Plan or procedure when acceptance criteria are not met
7. Requirements for the final report

Appendices

1. Method Development Report
2. Method Procedure

The validation process cannot proceed until the protocol and all parties involved approve the acceptance criteria.

15.7.4.1. System Suitability. System suitability (SS) is an interfacial validation parameter that ensures that both methodology and instrumentation are performing within expectations prior to the analysis of test samples. System suitability is discussed in some detail in Chapter 12, and it is discussed further here. In the practical sense, SS criteria should be finalized prior to beginning validation experiments.

System Suitability Parameters. The following system suitability parameters should be included in HPLC methods to evaluate and monitor performance:

Resolution. The resolution (R_s) is a measure of how well two peaks are separated. For reliable quantitation, well-separated peaks are essential. The separation of all peaks of interest is checked visually using a synthetic sample solution. The resolution factor (R) between the critical peak pair is calculated according to the formulas described in the *U.S. Pharmacopeia*, 24th edition (USP 24) and the *European Pharmacopoeia*, 3rd edition (EP 3).

Relative Standard Deviation. This serves as a daily evaluation of the repeatability of the system. Often, the relative standard deviation for five replicate injections of a reference standard.

Tailing Factor. The tailing factor is used as an SS test in the case where there is a tendency to tailing of the peak of the active ingredient or one of the related compounds; this is a critical parameter if peak tailing is exacerbated as HPLC columns age. The tailing factor is calculated according to the formula described in USP 24 and EP 3.

LOQ. The system's ability to detect the limit of quantitation (LOQ) should be evaluated with each sample sequence. An injection of the LOQ concentration during SS evaluation may also be used as a rough check of the linearity of the system over the range from LOQ to 100% of the target active concentration.

Additional System Suitability Parameters. Other parameters for system suitability testing can be considered, such as retention factor, number of theoretical plates, and so on.

Setting Limits for System Suitability Tests. The limits of the system suitability test can be based on worst-case conditions of the robustness test [15] or through practical experience obtained during method development. After the evaluation of methodology in other laboratories, as discussed in Chapter 12, the SS limits are often reevaluated. If the SS test limits cannot be routinely met during the prevalidation evaluation, adjustment of these limits should be investigated. Ideally, SS test limits should be properly set to fail just prior to the method providing less acceptable data, yet not too demanding that they fail when acceptable and usable data are obtained. These system suitability limits should be monitored over time to verify that the criteria remain realistic and achievable, while continuing to provide assurance of the suitable performance of the method. Generally, the following should be considered in choosing criteria for system suitability parameters:

Resolution. All peaks of interest should be separated as checked visually. A target value of $R_s > 2.0$ is ideal for a critical peak pair.

Relative Standard Deviation of a Repeated Injection. Maximum 1.0–2.0% for active ingredient and 5.0–15.0% for low levels of related compounds. These values should be adjusted based on the expected performance of the HPLC system.

Tailing Factor. The mean value found in the worst-case condition during robustness testing or the extent of tailing observed during validation experiments that began to produce unacceptable data or difficulties in integration or detection of a closely eluting peak.

LOQ. The LOQ (method threshold LOQ) can be verified at the beginning of the analysis; often a dilution of the standard of the "active" is injected at the LOQ concentration. Data from this sample can be used as an additional check on the recovery of this sample; this suggests that the linearity of the system is acceptable throughout the range. The recovery of the LOQ solution should typically not exceed $50.0\% \leq LOQ \leq 150\%$.

Reference Standard Check. The operating policy of some laboratories require that a duplicate injection of a separately weighed reference solution should be analyzed as a control to serve as a check on the accuracy of the standard weighing. The expected result for the second standard should be $98.0\% \leq$ reference standard $\leq 102.0\%$. Laboratories may also evaluate the change in the system response over time (typically every 10–12 injections) through monitoring variation in the area counts of the reference standard; area counts are anticipated not to change by more than $\pm 2\%$ during a given chromatographic sequence.

15.8. A SYSTEMATIC APPROACH TO VALIDATION EXPERIMENTATION

Depending on the requirements of the validation, there can be a preferred order to efficiently perform the validation experiments. For the specific example of validation of an assay/purity HPLC method, Figure 15.8 suggests a rational approach.

15.8.1. Determination of Method Specificity

Specificity is one of the most important characteristics of a stability-indicating method and should be determined as one of the first validation items. A specific method can accurately measure the analyte of interest even in the presence of potential sample components (placebo ingredients, impurities, degradation products, etc.). When criteria for specificity are not met, this often indicates that the method is not sufficiently developed; furthermore, it is likely that criteria for accuracy, precision, and linearity may also not be fulfilled. A major objective of determining specificity is to ensure "peak purity" of the main compound to be determined; in other words, confirm that no related compound or product ingredient coelutes and interferes with the measurement of the assayed compound. Stressed stability samples are often specified in validation protocols to evaluate peak purity. In addition, the ICH outlines two approaches to further evaluating method specificity for when impurities are and are not available.

15.8.1.1. When Impurities Are Available. Knowledge of degradation and synthetic impurities can be derived from the historical information that has accumulated for the drug substance/product. Ideally a library of impurity and degradation compound reference standards are synthesized and characterized and sufficient quantities are available. These compounds can be spiked into the sample matrix (placebo) to determine if the matrix interferes with the quantitation of the compound(s) of interest.

15.8.1.2. When Reference Compounds Are Not Available. When impurities are not available to check method specificity, one federal guideline [16] defines several conditions under which various drug substances/product types should be stressed to support the suitability of the method. Depending upon matrix and packaging, these include extremes of pH, heat, high oxygen exposure, and light exposure. In the case of drug substances, heat (50°C), light (600 ft-c), acid (0.1 N HCl), base and oxidant (3% H_2O_2) are often used. For drug products, heat, light, and humidity (85%) are used as stress conditions [8]. Analyte peaks are evaluated for peak purity in samples sufficiently stressed to effect 10–15% degradation.

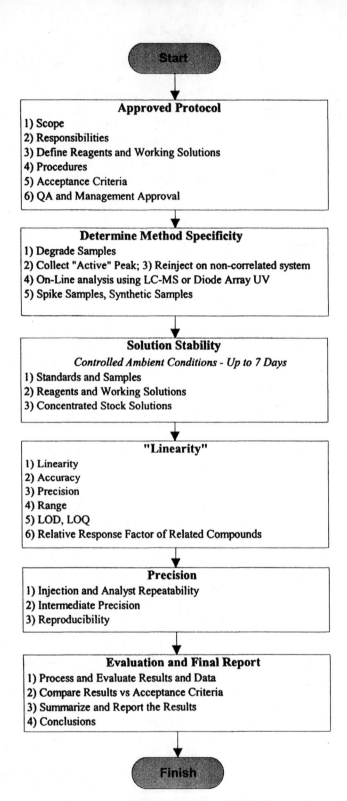

Figure 15.8. Schematic of validation process for an HPLC assay/purity method.

Evaluating Peak Purity. The peak purity in these degraded or spiked samples should be determined by using specific detection techniques, such as diode array ultraviolet (UV) or high-performance liquid chromatography–mass spectroscopy (HPLC-MS). Software to evaluate peak purity is often available on commercial diode array and liquid chromatography–mass spectroscopy (LC-MS) data systems. While both of these detection techniques provide relatively straightforward peak purity evaluation, there are limitations to the appropriateness of this approach. A less direct approach, but perhaps more persuasive, is to isolate the peak of interest and reinject on a chromatographic system that is based upon a different "noncorrelated" separation mechanism; for example, for evaluating a reversed-phase HPLC method, isolate the peak and reinject on an ion-exchange HPLC system. The chromatogram produced from the ion-exchange system is evaluated to observe any secondary peaks that are detected that may have eluted under the peak isolated on the reversed-phase system. More recently, capillary electrophoresis (CE) has been used extensive as a noncorrelated analytical technique to evaluate peaks isolated from reversed-phase methodology.

15.8.2. Demonstration of Linearity and Range; Determination of Relative Response Factor

Linearity is the ability to obtain results that are directly, or by well-defined mathematical transformation, proportional to the concentration of a substance in a sample within a given range [12]. The range is the interval between the upper and the lower levels of the analytical method that have been demonstrated to obtain acceptable accuracy, linearity, and precision. Hence, during linearity experiments the following parameters are typically evaluated:

The relationship between the sample concentrations and the corresponding instrumental signals for the majority of analytical techniques [17] is one of a straight-line (first-order) type. A line that fits best through the coordinates of the measured signals and the corresponding concentrations of the sample represents such a relationship. This line, known as the calibration line, is expressed by an estimated first-order equation:

$$Y = aX + b \tag{15.1}$$

where Y is the measured signal, X is the concentration of the sample, and a and b are the linear regression coefficients of the line, of which a is called the slope of the line and b the intercept.

The calibration lines are usually calculated by ordinary least-squares (OLS) regressions. A precondition for the application of OLS regressions is that the variance of the signal should be independent from the signal itself. This property is also called homoscedasticity. When this is not the case, one is dealing with a heteroscedastic situation. The heteroscedastic property of data can be observed by reviewing a graph that displays residuals (see section 4.8).

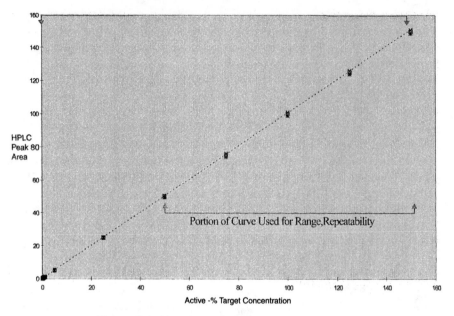

Figure 15.9. Linearity curve (0.05–150%) of the active.

15.8.2.1. Linearity for the Active Component.

The linearity can be demonstrated by analyzing five or more concentrations of the active in the presence of the matrix—for example, 50%, 75%, 100%, 125% and 150% of the normal sample concentration for a stability indicating method. There are also added advantages to evaluating the linearity over the whole range from LOQ to 150%.

Linearity can be established by visual evaluation of a plot of the area as a function of the analyte concentration (Figure 15.9). Furthermore, the correlation coefficient, y intercept, slope, and relative standard deviation (RSD) for all the generated response ratios (= area/concentration) should be calculated. The y intercept should statistically not differ from zero. Low levels of the active (0.05–1.0%) are also examined to determine the LOQ of the active (Figure 15.10).

15.8.2.2. Linearity for the Related Compounds.

The linearity should be demonstrated by analyzing five concentrations in the presence of the matrix: at LOQ, specification level, at an upper level above specification, and at two intermediate concentrations (e.g., 0.1%, 0.25%, 0.5%, 0.75%, and 1.0%).

Linearity should be established by visual evaluation of a plot of the area as a function of the analyte concentration. The correlation coefficient, y intercept, slope, and RSD for all the generated response ratios (= area/concentration) should be calculated. The y intercept should statistically not differ from zero. The slopes of these curves (Figure 15.11) are divided by the slope of the active compound curve (Figure 15.10) to determine the relative response factors

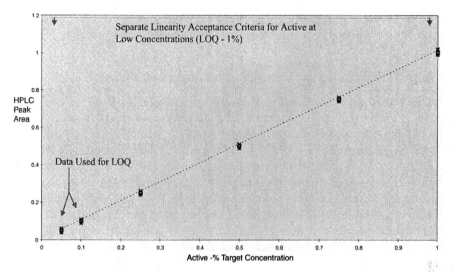

Figure 15.10. Low-level linearity curve of the active for examination of linearity and LOQ/LOD.

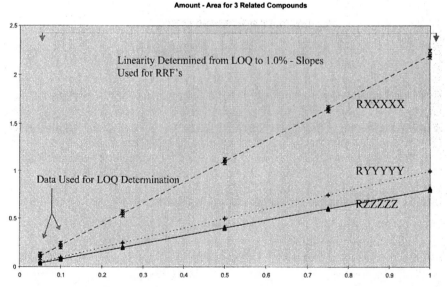

Figure 15.11. Calibration curve for low levels of related substances; used for determining LOQ/LOD and relative response factor (RRF) (slope).

(RRFs); these are recorded in the method procedure if the method does not prescribe the use of external standards for related compounds.

15.8.3. Determination of Detection and Quantitation Limit

In the literature there is some discussion regarding how detection limits are defined, calculated, and interpreted. It is therefore important to state clearly how the detection and quantitation limit should be calculated in the validation protocol [17]. Both the USP and the ICH guideline similarly define LOD and LOQ in ways that are widely accepted in the industry. In the most straightforward case, any compound detected with a response at about three times the noise response level is construed to be at its LOD. For LOQ, the value is very commonly taken as 10-fold the noise response level. Therefore, LOD and LOQ can be simply determined from the known amount (concentration) of an analyte that produces such responses when the noise level can be easily measured. In fact, some chromatography data systems can be programmed to report baseline noise. An alternative method of determination is described by the ICH guidelines as follows:

$$LOD = \frac{3.3 \times \text{Std. error}_{y \text{ intercept}}}{\text{Slope}} \tag{15.2}$$

$$LOQ = \frac{10 \times \text{Std. error}_{y \text{ intercept}}}{\text{Slope}} \tag{15.3}$$

This method can be conveniently applied to regression data obtained in linearity studies. However, parameters estimated by this approach must often be verified experimentally. The target limit of quantitation and detection may be stricter for drug substance than for drug product. In pharmaceutical analysis of the active drug substance, the target value for LOQ is typically set at 0.05%.

15.8.4. Demonstration of Accuracy of the Method

This is defined as the closeness of agreement between a test result and the accepted reference value (combination of random and systematic errors). The accuracy is usually examined by determination of the trueness of a test result, which is the closeness of agreement between the average value of a large number of test results and the true result or an accepted reference value. The measure of the trueness is expressed by the bias, which is the difference between the expectation of the test results and an accepted reference value.

The accuracy of a method can be determined by performing recovery experiments, standard addition calibration procedures, testing reference materials, and so on. It is also possible to compare the test results of a new method with that of an existing fully validated reference method [18] through "cross-validation" experiments.

Accuracy is often determined by recovery studies in which the analytes are spiked into a solution containing the matrix. The matrix (placebo in formulations) should be found not to interfere with the assay of the compound(s) of interest. For stability-indicating HPLC methods it is necessary to determine the accuracy of the active ingredient and for all related compounds. It is possible to determine the accuracy of each related compound separately, but it is more efficient to validate these related compounds in a combined spiked mixture of all the related compounds at their appropriate levels. The analyst should be certain that the impurity standards used to spike the solutions are pure and do not contain significant impurities that would affect the results.

15.8.5. Determination of Method Precision

The precision of an analytical procedure expresses the closeness of agreement among a series of measurements obtained from multiple sampling of the same homogeneous sample under the prescribed conditions. Precision may be considered at three levels:

Injection Repeatability. The precision as measured by multiple injections ($n = 10$) [8] of 100% level of reference standard and indicates the performance of the HPLC instrument using the chromatographic conditions on one particular day and in one lab. The specification as the relative standard deviation, RSD(%), set here will determine the lowest variation limit of the analytical results. Injection repeatability indicates the performance of the HPLC instrument using the chromatographic conditions on one particular day and in one lab.

Analysis Repeatability. The analysis repeatability expresses the precision under the same operating conditions over a short interval of time. It typically consists of multiple preparation and measurements of a homogeneous sample by the same analyst on the same day. The analysis repeatability can be determined by assaying 15 individual sample preparations covering the specified range for the procedure (5 concentration levels/ 3 replicates each).

Intermediate Precision. The intermediate precision expresses the effects of random events on the precision of the analytical procedure within the same lab. The procedure requires repeating the analysis of one technician by a qualified second technician, on a different instrument, using a different lot of HPLC column, on a different day. On the second day of analysis, all reagents and mobile phases are prepared freshly.

15.8.6. Target Acceptance Criteria

Table 15.1 contains an example of validation target acceptance criteria for a stability-indicating HPLC method for the assay of the active drug substance

Table 15.1. Typical Acceptance Criteria for HPLC Method Validations

Parameter	Limit: Active Ingredient	Limit: Related Compounds
Linearity		
Correlation coefficient	>0.99	>0.98
y intercept (relative to the active or related compound)	±2%	±15.0%
RSD response ratios	≤3.0%;	≤10.0%
Visual	Linear	Linear
Linearity over the whole range		
Correlation coefficient	>0.99	
y intercept (relative to the active)	±10%	
RSD response ratios	≤5.0%	
Visual	Linear	
Accuracy		
Active ingredient		
Recovery of each over the whole range	96.0–104.0%	
Mean recovery per concentration	98.0–102.0%	
Related compounds (mean recovery)		
$LOQ \leq x < 2 \times LOQ$	50.0–150.0%	50.0–150.0%
$2 \times LOQ \leq x < 10 \times LOQ$	70.0–130.0%	70.0–130.0%
$10 \times LOQ \leq x < 20 \times LOQ$	80.0–120.0%	80.0–120.0%
$\geq 20 \times LOQ$	90.0–110.0%	90.0–110.0%
Precision		
Active ingredient	RSD:	
Injection repeatability	1.0%	
Analysis repeatability	2.0%	
Intermediate precision	3.0%	
Related compounds (analysis repeatability):	RSD:	RSD:
$LOQ \leq x < 2 \times LOQ$	25.0%	25.0%
$2 \times LOQ \leq x < 10 \times LOQ$	15.0%	15.0%
$10 \times LOQ \leq x < 20 \times LOQ$	10.0%	10.0%
$\geq 20 \times LOQ$	5.0%	5.0%
Range	Acceptance criteria are listed under linearity, accuracy and precision	Acceptance criteria are listed under linearity, accuracy and precision

Table 15.1 (Continued)

Parameter	Limit: Active Ingredient	Limit: Related Compounds
Recovery intermediate precision		
Active ingredient		
For both technicians: mean recovery	95.0–105.0%	
Difference of the mean recovery between the two determinations	≤3.0%	
Related compounds		
%Difference of the mean recovery between the two determinations		
$LOQ \leq x < 2 \times LOQ$		50.0–150.0%
$2 \times LOQ \leq x < 10 \times LOQ$		70.0–130.0%
$10 \times LOQ \leq x < 20 \times LOQ$		80.0–120.0%
$\geq 20 \times LOQ$		90.0–110.0%
Precision active		
Individual	2.0%	
Pooled (both analysts)	3.0%	
Precision related compounds		RSD
$LOQ \leq x < 2 \times LOQ$		30.0%
$2 \times LOQ \leq x < 10 \times LOQ$		20.0%
$10 \times LOQ \leq x < 20 \times LOQ$		15.0%
$\geq 20 \times LOQ$		10.0%
LOD	≤0.03%	≤0.03%
LOQ	≤0.05%	≤0.05%
Accuracy	50.0–150.0%	50.0–150.0%
Precision	25.0%	25.0%
Stability of analytical solutions		
Difference in concentration	≤2.0%	
Related compounds		
%Change of the mean concentration		
$LOQ \leq x < 2 \times LOQ$		50.0–150.0%
$2 \times LOQ \leq x < 10 \times LOQ$		70.0–130.0%
$10 \times LOQ \leq x < 20 \times LOQ$		80.0–120.0%
$\geq 20 \times LOQ$		90.0–110.0%

and its related compounds. These criteria or others cited in the literature [8, 9] can be used as a general guideline when considering acceptance criteria for a validation protocol.

15.8.6.1. Establishment of Acceptance Criteria—Plan If Criteria Are Not Met. When one or more items during method validation fail to meet the acceptance criteria, lab management should decide the following:

- Whether the results can still be accepted with justification
- Whether a retest should be performed on the same sample preparation(s)
- Whether the test should be repeated (reanalysis)
- Whether the concern is significant, and the method needs adaptation (through additional method development) after which the test or validation is repeated.
- Whether as a result of the failure to meet acceptance criteria, a limitation could be put to the method—for example, range of method limited to 80– 120%.

All deviations to the validation procedure should be documented and authorized by lab management and QA department. A list of deviations, if any, is included in the final validation report.

15.8.6.2. Method Robustness. Robustness can be considered a component of method validation and is mentioned here briefly in addition to the detailed discussion in Chapter 12. The need for the development of fast, reliable, and robust analytical procedures in pharmaceutical analysis is an absolute requirement.

"Robustness" should be differentiated from "ruggedness," a term that is only utilized by the USP [12] and not by the ICH [10]. According to the USP, ruggedness is the degree of reproducibility of test results obtained under a variety of normal test conditions and expressed as the percentage RSD. The conditions can be: different laboratories, analysts, instruments, reagents, experimental periods or a combination of those conditions. As mentioned before, the ICH does not use this term. In fact, the ICH considers ruggedness as a part of the method precision.

In robustness testing, an experimental design approach is often preferred. The aim of an experimental design is to get as much as possible relevant information in the shortest possible time from a limited number of experiments. Different kind of designs can be employed in robustness testing—for example, factorial designs and Placket–Burmann designs. The choice of a design depends on the purpose of the test and the number of factors. The procedure for experimental design used in robustness testing is beyond the scope of this chapter. Briefly, however, an example of eight factors that can be evaluated using a design include:

1. Column type
2. Temperature
3. pH
4. Flowrate
5. Concentration (ionic strength of buffer)
6. % Organic
7. Slope of the gradient
8. injection volume

In a Placket–Burmann approach, eight factors are evaluated using only 12 analyses where parameters are randomly varied (Table 15.2) [19]. In addition, nominal method conditions are performed at the start and at the end of the experimental design runs. This way, a possible drift in the chromatographic system can be detected.

In general, the purpose of robustness testing is to indicate factors that can significantly influence the outcome of the studied responses. This gives an idea of the potential problems that might occur when the method is repeated in different laboratories upon transfer. Therefore, one can anticipate these problems by controlling the significant factors adequately—for example, by including a "precautionary statement" (ICH [1]) in the method description. Additionally, it is possible to determine system suitability limits for the system suitability parameters, a procedure that is recommended by the ICH.

15.8.7. Final Method—Minor Method Refinement

Once the extensive validation experiments are complete minor changes to the method description may be required. Typically, these may include adding validation data (RRFs for the related substances, LOQs, etc.) but may also include slight changes to the system suitability requirements due to data from multiple laboratories. There should not be fundamental changes that would change the principles of the methodology or necessitate revalidation unless a portion of the validation failed, suggesting minor method adjustment and reperforming required validation experiments.

15.8.8. Validation Summary

Due to the current accuracy and precision in analytical instrumentation, reagents, and capabilities of modern data processing systems, even poor methods may validate to be acceptable. Validation does not necessarily certify a method as "good," robust, or suitable for a control environment; these must be established into the method during development. It, however, is a necessary and important step in both proving and documenting the capabilities of the method.

According to the validation life cycle [20], test methods may require additional validation or revalidation when regulatory agencies issue new require-

Table 15.2. An Example of a Plackett–Burman Screening Design for 11 Variables and 12 Experimental Runs[a,b]

Experiment	A	B	Dum1	C	D	E	Dum2	F	G	H	Dum3
Start	0	0	0	0	0	0	0	0	0	0	0
1	1	1	−1	1	1	1	−1	−1	−1	1	−1
2	−1	1	1	−1	1	1	1	−1	−1	−1	1
3	1	−1	1	1	−1	1	1	1	−1	−1	−1
4	−1	1	−1	1	1	−1	1	1	1	−1	−1
5	−1	−1	1	−1	1	1	−1	1	1	1	−1
6	−1	−1	−1	1	−1	1	1	−1	1	1	1
7	1	−1	−1	−1	1	−1	1	1	−1	1	1
8	1	1	−1	−1	−1	1	−1	1	1	−1	1
9	1	1	1	−1	−1	−1	1	−1	1	1	−1
10	−1	1	1	1	−1	−1	−1	1	−1	1	1
11	1	−1	1	1	1	−1	−1	−1	1	−1	1
12	−1	−1	−1	−1	−1	−1	−1	−1	−1	−1	−1
End	0	0	0	0	0	0	0	0	0	0	0

[a]An example of eight factors that can be evaluated using this design include: 1. Column type; 2. Temperature; 3. pH; 4. Flowrate; 5. Concentration of the buffer; 6. % Methanol; 7. Slope of the gradient; 8. Injection volume.

[b]The nominal conditions are performed at the start and at the end of the experimental design runs. This way a possible drifting of the "chromatographic" system can be detected. The experiments of the design are performed at random and can usually be selected as such by the software program.

Table 15.3. Validation Requirements as Specified by the ICH

Type of Analytical Procedure Validation Parameter	Identification	Testing for Impurities		Assay Dissolution, Potency, Content U
		Quantitative	Limit	
Accuracy	–	+	–	+
Precision				
Repeatability	–	+	–	–
Intermediate Reproducibility[b]	–	+[a]	–	+
Specificity[c]	+	+	+	+
Detection limit	–	*	+	–
Quantitation limit	–	+	–	–
Linearity	–	+	–	+
Range	–	+	–	+
Robustness[d]				

– indicates that this characteristic is not evaluated; + signifies this characteristic is normally evaluated; * indicates this parameter may be required, depending on the nature of the specific test.

[a] In cases where reproducibility has been performed, intermediate precision is not needed.

[b] Reproducibility expresses precision between laboratories.

[c] Lack of specificity in one procedure (i.e., dissolution) could be compensated by other supporting analytical procedures.

[d] Robustness is not listed in the table, but it should be considered at an appropriate stage in the development of an analytical procedure.

ments or when changes are made to the methodology. Method changes and additional validation activities may be required when there are any of the following:

- Instrument changes
- Product changes
- Method modifications
- Analysts changes
- Outdated technology

15.8.9. Method Transfer

As stated in Chapter 11, the transfer of analytical methods may be a component of the overall process of technology transfer involving transfer of several methods simultaneously, or limited to transfer of an existing method to qualify an additional testing site. In simplest terms, the analytical method transfer is conducted to ensure the following:

- A clear understanding of the analytical methodology among the participating laboratories

- Training of the receiving lab when required
- Demonstration of the receiving laboratory's ability to perform the method
- Sufficient documentation of the completed and successful transfer

Analytical method transfer should be performed using a validated procedure; this transfer data can be useful in determining "intermediate precision" of the method. The transferring laboratory should ensure that the recipient laboratory(s) is CGMP compliant; a record of successful audit by QA personnel is essential especially if the laboratory is a contractor.

At this stage of development of a V-TR^{2}AP method, the method is expected to meet the requirements of the proposed testing laboratory. The method should be efficient and practical; reagents, standards, and necessary equipment should be readily available to the testing laboratory. In addition, the validation documentation should be completed and approved along with a background method development report.

15.8.9.1. Introduction—Requirements for Formal Transfer and Definitions

General—Definitions. The method transfer process involves several laboratories; successful transfer requires efficient communication of responsibilities between all parties involved. The responsibilities of the various laboratories involved can be summarized as follows:

> *Transferring Laboratory.* The laboratory designated to coordinate the transfer of the method to the receiving laboratory. The transferring laboratory must be qualified to perform the method, but is not necessarily the laboratory who developed and/or validated the method.
>
> *Receiving Laboratory.* The laboratory to which the methods are to be transferred.
>
> *Development Laboratory.* The laboratory who developed and validated the method.

The roles and responsibilities of each laboratory must be defined and agreed upon. Figure 15.12 is a suggested and very general timeline of the transfer process. The smooth flow of protocols and supplies is critical to timely transfer.

15.8.10. Transfer Documentation

The method documentation package is typically extensive and should contain the following:

- *A Detailed Method.* The method procedure should be unambiguous with

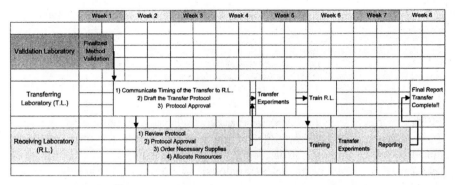

Figure 15.12. Generalized procedure/timeline for method transfer.

example chromatograms of the standard, a typical sample, and a system suitability sample with example calculations.

- *A Method Development Report.* The details of a thorough method development report were reported in Chapter 11. The report should review the method development and provide some justification for the choice of key operational parameters and choice of related compounds for purity methods.
- *The Transfer Protocol.* This protocol details the requirements, timing, responsibilities, and the acceptance criteria of the transfer.
- *Validation Summary.* The transfer documentation should include a summary of the validation activities and the results.

The transfer package should be presented to the receiving lab in an understandable format, and a review of this documentation is considered "training" or part of technology transfer.

15.8.11. Method Transfer Protocol

A qualified individual from the transferring laboratory typically writes the method transfer protocol. Management and QA representatives of both the transferring and receiving laboratory must approve the transfer plan. As a guide, the transfer plan may contain the following:

- Clear responsibilities of each party.
- Purpose. List all methods to be transferred for the given product and indicate rationale for product methods not included in the transfer.
- Scope.
- Materials (chemicals, instrumentation, standards, etc.) required in the transfer. Standards should be supplied with certificates of analysis.

- Protocols for testing which outline the experiments to be performed.
- Acceptance criteria for the tests.
- Method transfer documentation.
- Copies of the methods.
- Requirements and form of the final report.

The batches tested should be selected to challenge the method transfer (i.e., have sufficient level of impurities/degradants for HPLC impurity methods, samples spiked with impurities, etc.). The batches should test the range of products that are likely to be encountered by the receiving laboratory. The protocol should also include tests to document the receiving laboratory's ability to perform the LOQ determination.

The protocol may require training if the technique is new to the receiving laboratory (CZE, LC/MS, etc.); such training should be documented.

The acceptance criteria should be based on the data obtained for intermediate precision during validation and knowledge of the robustness of the method.

15.8.12. Method Transfer Experimental

Transfer experiments are not initiated until the protocol is reviewed and approved by all parties involved. The experiments must be carefully planned to ensure that proper resources, all testing reagents, standards, and laboratory equipment are available. Because transfer experiments are considered a CGMP activity, all instrumentation utilized must be calibrated and qualified; the experiments and data obtained must be properly documented and reviewed.

15.8.13. Transfer Summary and Approval

Once the method transfer experimentation is complete, the data are compiled and analyzed and the final report is drafted. The transfer report should indicate whether the transfer was successful, and all data should be recorded and reviewed. The report should indicate the file or location of the raw data. Any deviations to the protocol should be appended to the report.

15.9. CHAPTER SUMMARY

This chapter reviewed the validation of instrumentation, validation of analytical methods, and successful transfer of the methods. Successful completion of each component is both a regulatory requirement and good laboratory practice. To ensure that the data are both accurate and reliable, qualified and trained laboratory analysts must perform methods on qualified equipment, using suitable standards.

REFERENCES

1. Good Automated Manufacturing Practice, GAMP Forum, March 1998, Version 3.0.

2. W. Maxwell and J. Sweeney, Applying the Validation Timeline to HPLC System Validation, *LC-GC* **1994**, *12*, 678–682.

3. W. B. Furman, T. P. Layloff, and R. R. Tetzlaff, Validation of Computerized HPLC Systems, paper presented at the Workshop on Antibiotics and Drugs in Feeds, 106th AOAC Annual Meeting and Exposition, Cincinnati, OH, August 30, 1992.

4. P. Bedson and D. Rudd, The Development and Application of Guidance on EQ of Analytical Instruments: High Performance Liquid Chromatography, *Accred. Qual. Assur.* **1999**, *4*, 50–62.

5. Guideline on General Principles of Validation, U.S. Department of Health and Human Services, Food and Drug Administration, May 1987.

6. International Conference on Harmonization, Guideline on Validation of Analytical Procedures: Methodology, *Federal Register*, **1996**, *61*, 59–62.

7. *Reviewers Guidance, Validation of Chromatographic Methods*, Center for Drug Evaluation and Research, Food and Drug Administration, 1994.

8. J. M. Green, A Practical Guide to Analytical Method Validation, *Anal. Chem.* **1996**, *68*, 305A–309A.

9. D. R. Jenke, Chromatographic Method Validation: A Review of Current Practices and Procedures. II. Guidelines for Primary Validation Parameters, *J. Liq. Chrom. Rel. Technol.* **1996**, *19*, 737–757.

10. International Conference on Harmonization, Guideline on Validation of Analytical Procedures: Definitions and Terminology, Availability. *Federal Register* **1995**, *60* (40), 11260–11262.

11. Gerald C. Hokanson, A Life Cycle Approach to the Validation of Analytical Methods During Pharmaceutical Product Development, Part I: The Initial Method Development Process, *Pharm. Technol.* **1994**, *18*(9), 118–130.

12. Validation of Compendial Methods, *The United States Pharmacopeia*, 24th edition, USP 24, Section ⟨1225⟩, 2000.

13. C. DeSain, Master Method Validation Protocols, *BioPharm* **1992**, *6*, 30–34.

14. J. D. Johnson and G. E. Van Buskirk, Analytical Method Validation, *J. Validation Technol.* **1995**, *2*, 88–105.

15. Y. Vander Heyden, M. Jimidar, E. Hund, N. Niemeijer, R. Peeters, J. Smeyers-Verbeke, D. L. Massart, and J. Hoogmartens, Determination of System Suitability Limits with a Robustness Test, *J. Chromatogr. A* **1999**, *845*, 145–154.

16. *Guideline for Submitting Samples and Analytical Data for Methods Validation*, U.S. Food and Drug Administration, 1987.

17. D. L. Massart, B. G. M. Vandeginste, S. N. Deming, Y. Michotte, and L. Kaufman, *Chemometrics: A Textbook, Data Handling in Science and Technology*, Volume 2, Elsevier Science Publishers, Amsterdam, 1988.

18. M. Jimidar, C. Hartmann, N. Cousement, and D. L. Massart, *J. Chromatogr. A*, **1995**, *706*, 479.

19. M. Jimidar, N. Niemeijer, R. Peeter, and J. Hoogmartens, Robusteness Testing of a

Liquid Chromatography Method for Determination of Vorazole and Its Related Compounds in Oral Tablets, *J. Pharm. Biomed. Anal.* **1998**, *18*, 479–485.

20. Gerald C. Hokanson, A Life Cycle Approach to the Validation of Analytical Methods During Pharmaceutical Product Development, Part II: Changes and the Need for Additional Validation, *Pharm. Technol.* **1994**, *18*(10), 92–100.

APPENDIX **I**

LIST OF SYMBOLS AND ACRONYMS

A	Peak area or
	Absorbance (spectroscopy)
A_s	Surface area of stationary phase in column
AA	Atomic absorption (spectroscopy); also AAS
A/D	Analog-to-digital
ADME	Absorption/distribution/metabolism/excretion of a drug formulation
AIA	Analytical Instrument Association
API	Active pharmaceutical ingredient
ATR	Attenuated total reflectance
AUC	Area under curve
BP	Bonded phase (stationary phase) or
	British Pharmacopeia
BPC	Bonded-phase chromatography
C	Concentration of analyte
c	Speed of light
CDS	Chromatography data system
CE	Capillary electrophoresis
CEC	Capillary electrochromatography
CGMP	Current good manufacturing practice (GMP is used in this book to represent both CGMP and GMP)
CI	Chemical ionization (mass spectrometry) or Confidence interval
CL	Confidence Limit
CTA	Clinical trial approval
C.V.	Curriculum vitae
CZE	Capillary zone electrophoresis
d	Distance between maxima of two adjacent peaks (resolution definition)

459

d_c	Column inside diameter
d_f	Thickness of liquid phase (film)
d_p	Particle diameter
D	Minimum detectability of a detector or
	Diffusion coefficient in general
D_G	Diffusion coefficient in the gas phase
D_L	Diffusion coefficient in liquid stationary phase
D_M	Diffusion coefficient in the mobile phase
D_S	Diffusion coefficient in the stationary phase
DAD	Diode array detector
DP	Drug product
DSC	Differential scanning calorimetry
DQ	Design qualification
E	Peak height or
	Cell voltage or
	Energy
E^0	Standard cell potential
ECD	Electron capture detector or
	Electrochemical detector
EI	Electron impact ionization (mass spectrometry)
EOF	Electroosmotic flow
EP	*European Pharmacopeia*
EPC	Electronic pressure control
F	Mobile-phase flow rate, measured at column outlet
$\mathscr{F}$	Faraday constant
f	Relative detector response factor
F_c	Mobile-phase flow rate, corrected
FDA	Food and Drug Administration
FID	Flame ionization detector
FS	Fused silica (column material)
FTIR	Fourier transform infrared
GAMP	Good automated manufacturing practice
GC	Gas chromatography
GLC	Gas–liquid chromatography
GLP	Good laboratory practice
GLPC	Gas–liquid partition chromatography
GMP	Good manufacturing practice
GSC	Gas–solid chromatography
H	Plate height (HETP) or
	Column dispersivity
$\mathscr{H}$	Enthalpy

h	Planck's constant or
	Reduced plate height
HETP	Height equivalent to one theoretical plate
HPLC	High-performance liquid chromatography
HPTLC	High-performance thin-layer chromatography
I	Ionic strength or
	Electrical current
IC	Ion chromatography
ICH	International Conference on Harmonisation
ICP	Inductively coupled plasma (spectroscopy)
ID	Inside diameter
IEC	Ion-exchange chromatography
IM	Intramuscular
IND	Investigational New Drug
IPC	Ion pair chromatography
IQ	Installation qualification
IR	Infrared
IRF	International Registration File
IS	Internal standard
IUPAC	International Union of Pure and Applied Chemistry
IV	Intravenous
IVIV	In-vivo/In-vitro
j	Mobile-phase compression (compressibility) correction factor (GC)
JP	*Japanese Pharmacopeia*
k	Retention factor (capacity factor)
K_a	Acid equilibrium constant
K_b	Base equilibrium constant
K_c	Distribution constant (chromatography) in which the concentration in the stationary phase is expressed as mass of substance per volume of the phase
K_{eq}	Equilibrium constant, general
K_w	Equilibrium constant, water
KF	Karl Fischer (titration)
L	Column length
LC	Liquid chromatography
LIMS	Laboratory information management system
LLE	Liquid liquid extraction
LOD	Limit of detection or
	Loss on drying
LOQ	Limit of quantitation

M	Mass flow rate of analyte
MA	Moisture analysis
MDQ	Minimum detectable quantity
MEKC	Micellular electrokinetic chromatography
MP	Mobile phase
MS	Mass spectroscopy
MSD	Mass selective detector (mass spectrometer)
MSDS	Material Safety Data Sheet
MW	Molecular weight (mass)
N	Noise of a detector or
	Plate number (number of theoretical plates)
n	Number of electrons in reaction or number of measurements
NDA	New drug application
NF	*National Formulary*
NIR	Near-infrared (spectroscopy)
NIST	National Institute of Standards and Technology (formerly NBS)
NME	New molecular entity
NMR	Nuclear magnetic resonance (spectroscopy)
NP	Normal phase (LC)
NPD	Nitrogen-Phosphorus detector
OD	Outside diameter
ODS	Octadecylsilane
OLS	Ordinary least squares
OOS	Out of specification
OOT	Out of trend
OQ	Operational qualification
OT	Open tubular (column)
p	Pressure in general; partial pressure
p^0	Equilibrium vapor pressure of pure liquid
PAI	Pre-approval inspection
PC	Paper chromatography or
	Personal computer
PD	Pharmacodynamic
PDA	Photodiode Array
pH	$-\log a_{H^+}$, a measure of acidity
PK	Pharmacokinetic
pK_a	$-\log K_a$
pK_b	$-\log K_b$
PLB	Pilot laboratory batch
PLOT	Porous-layer open-tubular (column)
PM	Preventative maintenance
ppb	Parts per billion

ppm	Parts per million
ppt	Parts per thousand
PQ	Performance qualification
psi	Pounds per square inch (pressure)
PTGC	Programmed temperature gas chromatography
QA	Quality assurance
QC	Quality control
r_c	Column inside radius
R	Retardation factor in column chromatography; fraction of a sample component in mobile phase
$\mathscr{R}$	Gas constant
R_s	Peak resolution
RH	Relative humidity
RI	Refractive index
ROI	Residue on Ignition
RRF	Relative response factors (for detectors)
RP	Reversed phase (LC)
RSD	Relative standard deviation
S	Detector sensitivity
SCOT	Support coated open tubular (column)
SEC	Size exclusion chromatography
SFC	Supercritical fluid chromatography
S/N	Signal-to-noise ratio
SOP	Standard operating procedure
SP	Stationary phase
SPE	Solid-phase extraction
SPME	Solid-phase microextraction
t	Time in general
t_M	Mobile-phase hold-up time; it is also equal to the retention time of an unretained compound
t_R	Peak elution time
t_R'	Adjusted retention time
t_R^0	Corrected retention time
T	Temperature in general (always in kelvin) or Transmittance (spectroscopy)
T'	Significant temperature (in PTGC)
T_c	Column temperature
TCD	Thermal conductivity detector
TF	Tailing factor
TGA	Thermogravimetric Analysis
TK	Toxicokinetic

TLC	Thin-layer chromatography
TOC	Total organic carbon (analysis)
TS	Test solutions

u	Mobile-phase velocity or
	Average linear carrier gas velocity
USP	*United States Pharmacopeia*
USP/NF	*United States Pharmacopeia/National Formulary*
UV	Ultraviolet
UV-VIS	Ultraviolet and visible spectroscopy

V	Volume in general
v	Velocity of solute (chromatography)
V_G	Interparticle volume of column in GC
V_L	Liquid-phase volume
V_M	Mobile-phase hold-up volume; it is also equal to the retention volume of an unretained compound
V_M^0	Corrected gas hold-up volume
V_N	Net retention volume
V_R	Total retention volume
V_R'	Adjusted retention volume
V_R^0	Corrected retention volume
V_s	Volume of stationary phase in column
VTR^{2}AP	Methods that are validatable, transferable, robust, reliable, accurate, and precise
VWD	Variable wavelength detector

W	Weight (mass) of analyte
w_b	Peak width at base
w_h	Peak width at half-height
w_i	Peak width at inflection point
WCOT	Wall-coated open tubular (column)
WLS	Weighted least squares

X_a	Mole fraction of solute A

GREEK SYMBOLS

α	Separation factor (relative retardation)
β	Phase ratio or
	Beta ray (electron)
γ	Activity coefficient or
	Tortuosity factor (rate equation for packed columns)

λ	Wavelength or
	Packing factor (rate equation for packed columns)
υ	Frequency or
	Reduced mobile-phase velocity
π	Pi electron type
σ	Standard deviation of a Gaussian peak or
	Sigma bonding electron type
σ^2	Variance of a Gaussian peak
τ	Time constant (detector)
ω	Obstruction factor (rate equation for packed columns)

APPENDIX II

GLOSSARY OF TERMS USED IN ICH DOCUMENTS*

METHOD VALIDATION

ICH Definitions for Validation are listed in Chapter 15.

STABILITY TESTING

Accelerated testing: Studies designed to increase the rate of chemical degradation or physical change of an active drug substance or drug product by using exaggerated storage conditions as part of the formal, definitive, storage program. These data may be used in conjunction with long-term stability studies to assess longer-term chemical effects at nonaccelerated conditions and to evaluate the impact of short-term excursions outside the label storage conditions. Results from accelerated testing studies are not always predictive of physical changes.

Drug substance: The unformulated drug substance that may be subsequently formulated with excipients to produce the drug product (also called active pharmaceutical ingredient (API), active substance, or medicinal substance).

Bracketing: The design of stability schedule so that at any time point only the samples on the extremes (container size or dosage strength) are tested. The design assumes that the stability of the intermediate condition samples is represented by that of the extremes. Where a range of dosage strengths is to be tested, bracketing designs may be particularly applicable if the strengths are very closely related in composition (e.g., common granulation). Where a range of sizes of immediate containers are to be evaluated, bracketing may be applicable if the material of the container and the type of closure are the same for all sizes.

---

*Taken from reference 14 in Chapter 3 and from ICH Guidelines.

Climatic zones: The concept of dividing the world into four zones based on defining the prevalent annual climatic conditions (temperature, humidity).

Dosage form preparation: A pharmaceutical product type (e.g., tablet, capsule, solution, cream, etc.) that contains a drug ingredient generally, but not necessarily, in association with excipients.

Drug product: (finished product) The dosage form in the final immediate packaging intended for marketing.

Excipient: Anything other than the drug substance in the dosage form.

Expiry/expiration date: The date placed on the container/labels of a drug product designating the time during which a batch of the product is expected to remain within the approved shelf-life specification if stored under defined conditions and after which it must not be used.

Formal (systematic) studies: Formal studies are those undertaken to a pre-approved stability protocol that embraces the principles of the ICH guidelines.

Long-term (real-time) testing: Stability testing of the physical, chemical, biological, and microbiological characteristics of a drug product and a drug substance, covering the expected duration of the labeled shelf-life and retest period.

Mass balance; material balance: The process of adding together the assay value and the levels of degradants to see how closely these add up to 100% of the initial value, with due consideration for the margin of analytical precision. This is a useful scientific guide for evaluating data, but it is not achievable in all circumstances. The focus may instead be on assuring the specificity of the assay, the completeness of the investigation of routes of degradation, and the use, if necessary, of identified degradants as indicators of the extent of degradation via particular mechanisms.

Matrixing: The statistical design of a stability schedule so that only a fraction of the total number of samples are tested at any specified sampling point. At a subsequent sampling point, different sets of samples of the total number would be tested. The design assumes that the stability of the samples tested represents the stability of all samples. The differences in the samples for the same drug product should be identified as, for example, covering different batches, different strengths, and so on. In all cases, batches are tested initially and at the end of the long-term testing. Matrixing can cover reduced testing when more than one variable is being evaluated. Thus, the design of the matrix will be dictated by the factors needing to be covered and evaluated. This potential complexity precludes inclusion of specific details and examples, and it may be desirable to discuss design in advance with the regulatory authority where possible.

Mean kinetic temperature: When establishing the mean value of the temperature, the formula of J. D. Haynes* (reference 17 in Chapter 3) can be used to

* J. D. Haynes, *J. Pharm, Sci*, **1971**, *60*, 927–929.

calculate the mean kinetic temperature. It is higher than the arithmetic mean and takes into account the Arrhenius equation from which it is derived.

New molecular entity; new active substance: A substance which has not previously been registered as a new drug substance with the national or regional authority concerned.

Pilot plant scale: The manufacture of either a drug substance or the drug product by a procedure fully representative of and simulating that to be applied on a full manufacturing scale. For solid oral dosage forms, this is generally taken to be one-tenth that of full production or 100,000 units, whichever is greater.

Primary stability data: Data on the drug substance or drug product stored in the proposed packaging/container closure under storage conditions that support the proposed retest or expiration date.

Retest date: The date when samples of the drug substance should be re-examined to ensure that material is still suitable for use.

Retest period: The period of time during which the drug substance can be considered to remain within the specification and therefore acceptable for use in the manufacture of a given drug product, provided that it has been stored under the specified conditions. After this period, the batch should be retested for compliance with specification and then used immediately.

Shelf life/expiration dating period: The time interval that a drug product is expected to remain within the approved shelf life specification provided that it is stored under the conditions defined on the label and in the proposed container/closure.

Specification—release: The combination of physical, chemical, biological and microbiological test requirements that determine a drug product suitable for release at the time of its manufacture.

Specification—check/shelf life: The combination of physical, chemical, biological, and microbiological test requirements that a drug substance must meet up to its retest date or that a drug product must meet throughout its shelf life.

Storage condition tolerances: The acceptable variation in temperature and humidity of storage facilities. Air-handling equipment must be capable of controlling temperature to a range of $\pm 2°C$ and relative humidity to $\pm 5\%$. The actual temperatures and humidities should be monitored during stability storage. Short-term spikes occurring when facility doors are opened are unavoidable and hence acceptable. The effect of excursion due to equipment failure should be addressed by the applicant and reported if judged to impact stability results. Excursions that exceed $\pm 2°C$ and/or $\pm 5\%$ RH for more than 24 hr should be described in the study report, and the impact should be assessed.

Stress testing of drug substance: These studies are designed to elucidate intrinsic stability characteristics. Such testing is part of the development strategy and is normally carried out under more severe conditions than is normally used for accelerated tests. Stress testing is conducted to provide data on forced decomposition products and decomposition mechanisms for the drug substance. The severe conditions that may be encountered during marketing can be covered by stress testing of definitive batches of drug substance. These studies should establish the inherent stability characteristics of the molecule, such as the degradation pathways, and lead to identification of degradation products and hence support the suitability of the proposed analytical procedures. The detailed nature of the studies will depend on the individual drug substance and type of drug product. This testing is likely to be carried out on a single batch of material and to include the effect of temperatures in 10°C increments above the accelerated test conditions and with elevated humidity. Oxidation and photolysis on the drug substance, plus its susceptibility to hydolysis across a wide range of pH values when in solution or suspension, are also measured. Results from these studies form an integral part of the information provided to regulatory authorities. Light testing should be an integral part of stress testing. Some degradation pathways can be complex and, when forced, some decomposition products may be observed which are unlikely to be formed under accelerated or long-term testing. This information may be useful in developing and validating suitable analytical methods, but it may not always be necessary to examine specifically all degradation products.

Stress testing of drug product: Light testing should be an integral part of stress testing also for the drug product.

Supporting stability data: Data other than primary stability data.

Enantiomers: Compounds with the same molecular formula as the drug substance, but which differ in the spatial arrangement of atoms within the molecule and/or nonsuperimposable mirror images.

Extraneous substance: An impurity arising from any source extraneous to the manufacturing process.

Herbal products: Medicinal products containing exclusively plant material and/or vegetable drug preparations as active ingredients.

Intermediate: A material produced during the steps of the synthesis of a new drug substance which must undergo further molecular change before it becomes a new drug substance.

Ligand: An agent with a strong affinity to a metal ion.

Polymorphism: The occurrence of different crystalline forms of the same drug substance.

Qualification: The process of acquiring and evaluating data that establish the biological safety of an individual impurity or a given impurity profile at the level(s) specified.

Reagent: A substance other than a starting material or solvent which is used in the manufacture of a new drug substance.

Safety information: The body of information that establishes the biological safety of an individual impurity or a given impurity profile at he level(s) specified.

Solvent: An inorganic or an organic liquid used as a vehicle for the preparation of solutions or suspensions in the synthesis of a new drug substance.

Starting material: A material used in the synthesis of a new drug substance which is incorporated as an element into the structure of an intermediate and/or of the new drug substance. Starting materials are normally commercially available and of defined chemical and physical properties and structure.

Validated limit of quantitation: For impurities at a level of 0.1% the validated limit of quantitation should be less than or equal to 0.05%. Impurities limited at higher levels may have higher limits of quantitation.

Genotoxic carcinogens: Carcinogens that produce cancer by affecting genes or chromosomes.

Neurotoxicity: The ability of a substance to cause adverse effects on the nervous system.

Reversible toxicity: The occurrence of harmful effects that are caused by a substance and which disappear after exposure to the substance ends.

Teratogenicity: The occurrence of structural malformations in a developing fetus when a substance is administered during pregnancy.

Immediate (primary) pack: That constituent of the packaging that is in direct contact with the drug substance or drug product and that includes any appropriate label.

Marketing pack: The combination of immediate pack and other secondary packaging such as a carton.

Forced degradation testing: Those studies undertaken to degrade the product deliberately. These studies (often performed during development) are used to evaluate the overall photosensitivity of the material for method development and/or degradation pathway elucidation.

Confirmatory studies: Those studies undertaken to establish photostability characteristics under standardized conditions.

Degradation product: A molecule resulting from a chemical change in the drug molecule brought about over time and/or by the action of for example water, light, temperature, or pH or by reaction with an excipient and/or the immediate container/closure system (also called a decomposition product).

Degradation profile: A description of the degradation products observed in the drug product or drug substance.

Development studies: Studies conducted to scale-up, optimize, and validate the manufacturing process for a drug substance or product.

Identified impurity: An impurity for which a structural characterization has been achieved.

Impurity: Any component of the drug substance or product that is not the chemical entity defined as the drug substance or an excipient in the drug product.

Impurity profile: A description of the identified and unidentified impurities present in a drug substance or product.

New drug substance: The designated therapeutic moiety which has not been previously registered in a region or a member state. It may be a complex, ester, or salt of a previously approved drug (also called a new molecular entity or new chemical entity).

Potential impurity or degradation product: An impurity which, from theoretical considerations may arise during or after manufacture or storage of the drug product. It may or may not actually appear in the drug substance or drug product.

Qualification: The process of acquiring and evaluating data which establish the biological safety of an individual impurity or a given impurity profile at the level(s) specified.

Reaction product: Product arising from the reaction of drug substance with an excipient in the drug product or immediate container/closure system.

Safety information: The body of information that establishes the biological safety of an individual impurity or a given impurity profile at the level(s) specified.

Specified degradation product: Identified or unidentified degradation product that is selected for inclusion in the new drug product specification and is individually listed and limited in order to ensure the safety and quality of the new drug product.

Toxic impurity: An impurity having significant undesirable biological activity.

Unidentified degradation product: An impurity that is defined solely by qualitative analytical properties such as chromatographic retention time.

Unspecified degradation product: A degradation product that is not recurring from batch to batch.

APPENDIX **III**

UNIVERSAL TESTS, DOSAGE-FORM-SPECIFIC TESTS, AND ACCEPTANCE CRITERIA

Universal Tests and Acceptance Criteria

Area of Application	Test/Criteria
New drug substances	1. Description: qualitative statement about state and color.
	2. Identification: specific and discriminate between compounds of closely related structure which are likely to be present; include tests for ions if a salt; enantiomers.
	3. Assay: specific and stability indicating; may use same method for assay and impurities (if nonspecific assay method, include suitable impurity test).
	4. Impurities: include organic, inorganic, and residual solvents.
	5. Physicochemical properties (pH, melting point, refractive index, etc.); particle size and where applicable size distribution.
	6. For solid-state forms, include polymorphs or solvates where bioavailability, stability or dissolution could be affected.
	7. Chiral impurities should be treated according to the principles in ICH impurity guidelines (enantioselective assay and specific identity tests).
	8. Water content is important if drug substance is hygroscopic, degraded by moisture or a stoichiometric hydrate.
	9. Microbial limits may include total aerobic counts and absence of objectionable organisms. Compendial methods, sterility, and endotoxin testing should be used as appropriate.
New drug products	1. Description: qualitative statement of unit of use (e.g., size, shape, and color).

Area of Application	Test/Criteria
New drug products (*Continued*)	2. Identification: establish identity of a new drug substance in a new drug product by a specific technique. 3. Assay: specific and stability-indicating; may use same method for assay and impurities (if nonspecific assay method, include suitable impurity test). 4. Impurities: include organic, inorganic, and residual solvents, particularly degradation products; state limits for individual identified, unidentified, and total impurities; reduce or eliminate testing when drug substance does not degrade in a given formulation.

Dosage-Form-Specific[a] Tests and Acceptance Criteria

Dosage Form	Tests and Criteria

Solid oral products	1. Dissolution/disintegration
	  • Disintegration my be appropriate for rapidly dissolving products
	  • Single-point measurements normally suitable for immediate release products
	  • Use multiple time points for extended release products
	  • Two-stage testing may be appropriate for delayed release products
	2. Hardness/friability
	  • Normally an in-process control not included in specifications
	3. Uniformity of dosage units
	  • Includes uniformity of content and mass
	  • Use compendial methods
	  • May be performed in-process
	  • Included in regulatory specifications
	4. Water content
	  • Method specific for water preferred
	5. Microbial limits
	  • Test drug product unless components tested and manufacturing process carries no significant risk of contamination

Oral liquid products	1. Uniformity of dosage units
	  • May apply to single and multiple dose packages
	  • Weight variation or fill may suffice
	  • Uniformity of mass testing generally acceptable for dry powders for reconstitution
	2. pH
	  • Provide test and specification
	3. Microbial limits
	  • As with solid orals

Dosage Form	Tests and Criteria
Oral liquid products (*Continued*)	4. Preservative effectiveness • Base specifications on level needed to maintain product quality • Support with preservative efficacy testing • In-process testing may suffice 4. Antioxidant preservative effectiveness • Release testing normally performed • May be necessary on stability • In-process testing may suffice • Include in specifications 5. Extractables Testing generally not required if development and stability data show no significant extractables 6. Alcohol Content • May be assayed or calculated where declared quantitatively on the label 7. Dissolution • May be appropriate for oral suspensions and dry powder for resuspension • Use compendial procedures if possible 8. Particle size distribution • May be required for oral suspensions • Describe particle size distribution • May replace dissolution test • Different release and shelf life limits may be required • Investigate potential for particle growth 9. Redispersibility • May be appropriate for suspensions which settle on storage • Define amount of shaking to resuspend 10. Rheological properties • Specify acceptable viscosity range where appropriate • May conduct skip lot testing or eliminate test based on developmental data 11. Specific gravity • Establish test method and specification as appropriate 12. Reconstitution time • Should be provided for dry powders which require reconstitution
Parenteral products	1. Uniformity of dosage units • As with oral liquids 2. pH • As with oral liquids

Dosage Form	Tests and Criteria
Parenteral products (*Continued*)	3. Sterility • Should have specification • Compendial procedures preferred • Parametric release may be proposed when justified 4. Endotoxins/pyrogens • Perform endotoxins testing when required • Pyrogen testing may be proposed as alternative to endotoxin when justified 5. Particulate matter • Should have appropriate specification, which may include limits for visible particulates and/or clarity of solution 6. Antimicrobial preservative effectiveness • As with oral liquids 7. Antioxidant preservative effectiveness • As with oral liquids 8. Extractables • As with oral liquids 9. Functionality testing • For items such as prefilled syringes, autoinjector cartridges, etc. • May include pressure, seal integrity, tip/cap removal force, etc., as appropriate 10. Osmolarity • Control when tonicity of product is declared on labeling • Development data may justify in-process, or skip-lot testing or direct calculation 11. Particle size distribution • May be appropriate for injectable suspensions • Development data should be used to select between dissolution and particle size distribution • Investigate the potential for particle growth upon storage 12. Redispersability • As with oral liquids 13. Reconstitution time • As with oral liquids

a Concepts described above should also be considered for inhalation dosage forms (aerosols, solutions, sprays), topical formulations (creams, ointments, gels), and transdermal systems.

APPENDIX IV

USP CHROMATOGRAPHIC PHASES

Stationary Phases for GC

	USP Designation	Commerical Equivalent
G1	Dimethylpolysiloxane oil	OV-101
G2	Dimethylpolysiloxane gum	OV-1
G3	50% Phenyl–50% methylpolysiloxane	OV-17
G4	Diethylene glycol succinate polyester	DEGS
G5	3-Cyanopropylpolysiloxane	OV-105
G6	Trifluoropropylmethylpolysiloxane	OV-202, OV-210
G7	50% 3-Cyanopropyl–50% phenylmethylsilicone	OV-225. DB-225
G8	90% 3-Cyanopropyl–10% phenylmethylsilicone	OV-2330, DB-23
G9	Methylvinylpolysiloxane	SE-52
G10	Polyamide	Polyamide
G11	Bis(2-ethylhexyl)sebacate polyester	—
G12	Phenyldiethanolamine succinate polyester	—
G13	Sorbitol	—
G14	Polyethylene glycol (MW 950–1050)	Carbowax 1000
G15	Polyethylene glycol (MW 3700)	Carbowax 4000
G16	Polyethylene glycol compound (MW 15000)	Carbowax 20M
G17	75% Phenyl–25% methylpolysiloxane	OV-25
G18	Polyalkylene glycol	—
G19	25% Phenyl–25% cyanopropylmethylsilicone	—
G20	Polyethylene glycol (MW 380–420)	Carbowax 400
G21	Neopentyl glycol succinate	—
G22	Bis(2-ethylhexyl) phthalate	DOP
G23	Polyethylene glycol adipate	EGA
G24	Diisodecyl phthalate	—
G25	Polyethylene glycol compound TPA	Carbowax 20M-TPA
G26	25% 2-Cyanoethyl–75% methylpolysiloxane	—
G27	5% Phenyl–95% methylpolysiloxane	—
G28	25% Phenyl–75% methylpolysiloxane	—
G29	3,3'-Thiodipropionitrile	—
G30	Tetraethylene glycol dimethyl ether	—
G31	Nonylphenoxypoly(ethyleneoxy)ethanol	Nonoxynol 30

	USP Designation	Commerical Equivalent
G32	20% Phenylmethyl–80% dimethylpolysiloxane	—
G33	20% Carborane–80% methylsilicone	Silar-5C
G34	Diethylene glycol succinate polyester stabilized with phosphoric acid	—
G35	A high-molecular-weight compound of a polyethylene glycol and a diepoxide that is esterified with nitroterephthalic acid	OV-351, FFAP
G36	1% Vinyl–50% phenylmethylpolysiloxane	Restek Rt$_x$-5
G37	Polyimide	—
G38	Phase G1 containing a small percentage of a tailing inhibitor	SP 2100/0.1% Carbowax 1500
G39	Polyethylene glycol (average MW ~1500)	Carbowax 1500
G40	Ethyleneglycol adipate	EGA
G41	Phenylmethyldimethylsilicone	OV-3
G42	35% Phenyl–65% dimethylvinylsiloxane	—
G43	6% Cyanopropylphenyl–94% dimethylpolysiloxane	OV-1701 Restek Rt$_x$/Mt$_x$ 1301
G44	2% Low-molecular-weight petrolatum hydrocarbon grease and 1% solution of potassium hydroxide	—
G45	Divinyl benzene–ethylene glycol–dimethylacrylate	—
G46	14% Cyanopropylphenol–86% methylpolysiloxane	—

Solid Supports for GC

	USP Designation	Commercial Equivalent
S1A	Silaceous earth; flux-calcined; may be acid-washed and silanized	Chromosorb W-HP
S1AB	Silaceous earth; like S1A but both acid- and base-washed	Chromosorb W-AW-BW
S1C	Crushed firebrick; calcined and acid-washed; may be silanized	Chromosorb P-AW Chromosorb P-AW-DMDCS
S1NS	Silaceous earth; untreated	Chromsorb P-NAW
S2	Styrene–divinylbenzene copolymer	Chromsorb 101
S3	Copolymer of ethylvinylbenzene and divinylbenzene having a nominal surface area of 400–600 m^2/g and an average diameter of 0.0075 μm	Porapak Q
S4	Styrene–divinylbenzene copolymer; aromatic –O and –N groups	Porapak R HayeSep R
S5	40- to 60-mesh, high-molecular-weight tetrafluoroethylene polymer	Chromosorb T

	USP Designation	Commercial Equivalent
S6	Styrene–divinylbenzene copolymer; normal surface area of 250–350 m^2/g; average pore diameter of 0.0091 μm	Porapak P, Chromosorb 102
S7	Graphitized carbon; nominal surface area of 12 m^2/g	Carbopack C
S8	Copolymer of 4-vinylpyridine and styrene–divinylbenzene	Porapak S
S9	A porous polymer based on 2,6-diphenyl-*p*-phenylene oxide	Tenax TA
S10	A highly polar cross-linked copolymer of acrylonitrile and divinylbenzene	HayeSep C
S11	Graphitized carbon; nominal surface area of 100 m^2/g; small amounts of petrolatum and polyethylene glycol compound	SP1500 on Carbopack B
S12	Graphitized carbon; nominal surface area of 100 m^2/g	Carbopack B

HPLC Stationary Phases

USP Designation	Equivalent[a]
L1—Octadecyl silane chemically bonded to porous silica or ceramic microparticles, 3–10 μm in diameter.	Luna 5 μ C18 (2) Pinnacle ODS (Restek) Chromegabond WR-C18 (ES)
L2—Octadecyl silane chemically bonded to silica gel of a controlled surface porosity that has been bonded to a solid spherical core, 30–50 μm in diameter.	—
L3—Porous silica particles, 5–10 μm in diameter.	Luna 5 μ silica (2) Chromegaspher (ES)
L4—Silica gel of controlled surface porosity bonded to a solid spherical core, 30–50 μm in diameter.	—
L5—Alumina of controlled surface porosity bonded to a solid spherical core, 30–50 μm in diameter.	—
L6—Strong cation-exchange packing; sulfonated fluorocarbon polymer coated on a solid spherical core, 30–50 μm in diameter.	—
L7—Octylsilane chemically bonded to totally porous silica particles, 3–10 μm in diameter.	Luna 5 μ C8 (2) Pinnacle octyl (Restek)
L8—An essentially monomolecular layer of aminopropylsilane chemically bonded to totally porous silica gel support, 10 μm in diameter.	PhenoSphere 10 μ NH_2 Chromegabond amino (ES)

USP Designation	Equivalent[a]
L9—10-μm irregular, totally porous silica gel having a chemically bonded, strongly acidic cation-exchange coating.	Maxsil 10 μ SCX Chromegabond P-SCX (ES)
L10—Nitrile groups chemically bonded to porous silica particles, 3–10 μm in diameter.	Luna 5 μ CN 100 Å Chromegabond CN (ES)
L11—Phenyl groups chemically bonded to porous silica particles, 5–10 μm in diameter.	Luna 5 μ phenyl-hexyl Pinnacle phenyl (Restek)
L12—Strong anion-exchange packing made by chemically bonding a quaternary amine to a solid silica spherical core, 30–50 μm in diameter.	—
L13—Trimethylsilane chemically bonded to porous silica particles, 3–10 μm in diameter.	Develosil TMS-UG 130 Å Pinnacle methyl (Restek)
L14—Silica gel 10 μm in diameter having a chemically bonded, strongly basic quaternary ammonium anion-exchange coating.	Spherex 10 μ SAX 100 Å Chromegabond SAX (ES)
L15—Hexylsilane chemically bonded to totally porous silica particles, 3–10 μm in diameter.	PhenoSphere C6 Chromegabond C6 (ES)
L16—Dimethylsilane chemically bonded to porous silica particles, 5–10 μm in diameter.	Maxsil 5 μ RP2 60 Å Chromegabond C2 (ES)
L17—Strong cation-exchange resin consisting of sulfonated cross-linked styrene–divinylbenzene copolymer in the hydrogen form, 7–11 μm in diameter.	Rezex RHM monosaccharide
L18—Amino and cyano groups chemically bonded to porous silica particles, 3–10 μm in diameter.	Partisil 5 μ, 10 μ PAC Chromegabond AKN; (ES)
L19—Strong cation-exchange resin consisting of sulfonated cross-linked styrene–divinylbenzene copolymer in the calcium form, about 9 μm in diameter.	Rezex RCM Rezex RCU
L20—Dihydroxypropane groups chemically bonded to porous silica particles, 5–10 μm in diameter.	Spherex 5 μ Diol Chromegabond diol (ES)
L21—A rigid, spherical styrene–divinylbenzene copolymer, 5–10 μm in diameter.	Hamilton 5 μ, 10 μ PRP-1
L22—A cation-exchange resin made of porous polystyrene gel with sulfonic acid groups, about 10 μm in size.	Rezex ROA
L23—An anion-exchange resin made of porous polymethacrylate or polyacrylate gel with quaternary ammonium groups about 10 μm in size.	Shodex IEC QA-825
L24—Semirigid hydrophilic gel consisting of vinyl polymers with numerous hydroxyl groups on the matrix surface, 32–63 μm in diameter.	Fractogel TSK-HW-40F(Merck)

USP Designation	Equivalent[a]
L25—Packing having the capacity to separate compounds with a molecular weight range from 100 to 5000 (as determined by polyethylene oxide) applied to neutral, anionic, and cationic water-soluble polymers. A polymethacrylate resin base, cross-linked with polyhydroxylated ether (surface contained some residual carboxyl functional groups), was found suitable.	Shodex OHpak SB-802.5
L26—Butyl silane chemically bonded to totally porous silica particles, 5–10 μm in diameter.	Prime sphere 5 μ C4 Chromegabond C4 (ES)
L27—Porous silica particles, 30–50 μm in diameter.	—
L28—A multifunctional support, which consists of a high-purity, 100 Å, spherical silica substrate that has been bonded with anionic (amine) functionality in addition to a conventional reversed-phase C8 functionality.	Protec C8 (ES)
L29—Gamma alumina, reverse-phase, low carbon percentage by weight, alumina-based polybutadiene spherical particles, 5 μm in diameter with a pore volume of 80 Å.	Gammabond ARP-1 (ES)
L30—Ethyl silane chemically bonded to totally porous silica particles, 3–10 μm in diameter.	Maxsil 5 μ RP2 60 Å Chromegabond C2-E (ES)
L31—A strong anion-exchange resin-quaternary amine bonded on latex particles attached to a core of 8.5-μm macroporous particles having a pore size of 2000 Å and consisting of ethylvinylbenzene cross-linked with 55% divinylbenzene.	—
L32—A chiral ligand-exchange packing-L-proline copper complex covalently bonded to irregularly shaped silica particles, 5–10 μm in diameter.	Nucleosil chiral-1
L33—Packing having the capacity to separate proteins by molecular size over a range of 4000 to 40,000 daltons. It is spherical, silica-based, and processed to provide pH stability.	BioSep-SEC S2000, S3000
L34—Strong cation-exchange resin consisting of sulfonated cross-linked styrene–divinylbenzene copolymer in the lead form about 9 μm in diameter.	Rezex RPM monosaccharide
L35—A zirconium-stabilized spherical silica packing with a hydrophilic (diol-type) molecular monolayer bonded phase having a pore size of 150 Å.	BioSep-SEC-S2000

USP Designation	Equivalent[a]
L36—A 3,5-dinitrobenzoyl derivative of L-phenlyglycine covalently bonded to 5 μ aminopropyl silica.	—
L37—Packing having the capacity to separate proteins by molecular size over a range of 2000 to 40,000 daltons. It is a polymethacrylate gel.	PolySep-GFC-P 3000
L38—A methacrylate-based size-exclusion packing for water-soluble samples.	PolySep-GFC-P 1000
L39—A hydrophilic polyhydroxymethacrylate gel of totally porous spherical resin.	PolySep-GFC-P
L40—Cellulose, tri-3,5-dimethlyphenylcarbamate-coated porous silica particles, 5 μm to 20 μm in diameter.	—

[a] Available from Phenomenex unless otherwise noted. ES stands for ES Industries.

INDEX